Lecture Notes in Networks and Systems

Volume 101

The series "Lecture Notes in Networks and Systems" publishes the latest developments in Networks and Systems—quickly, informally and with high quality. Original research reported in proceedings and post-proceedings represents the core of LNNS.

Volumes published in LNNS embrace all aspects and subfields of, as well as new challenges in, Networks and Systems.

The series contains proceedings and edited volumes in systems and networks, spanning the areas of Cyber-Physical Systems, Autonomous Systems, Sensor Networks, Control Systems, Energy Systems, Automotive Systems, Biological Systems, Vehicular Networking and Connected Vehicles, Aerospace Systems, Automation, Manufacturing, Smart Grids, Nonlinear Systems, Power Systems, Robotics, Social Systems, Economic Systems and other. Of particular value to both the contributors and the readership are the short publication timeframe and the world-wide distribution and exposure which enable both a wide and rapid dissemination of research output.

The series covers the theory, applications, and perspectives on the state of the art and future developments relevant to systems and networks, decision making, control, complex processes and related areas, as embedded in the fields of interdisciplinary and applied sciences, engineering, computer science, physics, economics, social, and life sciences, as well as the paradigms and methodologies behind them.

**** Indexing: The books of this series are submitted to ISI Proceedings, SCOPUS, Google Scholar and Springerlink ****

More information about this series at http://www.springer.com/series/15179

Annamária R. Várkonyi-Kóczy
Editor

Engineering for Sustainable Future

Selected papers of the 18th International Conference on Global Research and Education Inter-Academia – 2019

Springer

Editor
Annamária R. Várkonyi-Kóczy
Obuda University
Budapest, Hungary

ISSN 2367-3370 ISSN 2367-3389 (electronic)
Lecture Notes in Networks and Systems
ISBN 978-3-030-36840-1 ISBN 978-3-030-36841-8 (eBook)
https://doi.org/10.1007/978-3-030-36841-8

This Springer imprint is published by the registered company Springer Nature Switzerland AG
The registered company address is: Gewerbestrasse 11, 6330 Cham, Switzerland

Preface

Inter-Academia Community links Shizuoka University and 13 universities in Central and Eastern Europe. The collaborating institutions are involved in the active exchange of research, students, and doctoral courses to improve academic progress and achievement in engineering. The IA Community is a general term that describes the community comprised of the universities, faculty members, students, and staff work with IA.

The international conference on Global Research and Education also called Inter-Academia is keystone of the IA program. The Inter-Academia conference has taken place every year since 2002. IA activities are typically carried out by the faculty of Shizuoka University and its sister universities, but IA welcomes a wide range of participants of other research institutions and industries from all over the world to promote the exchange of knowledge by providing a forum for exchanging novel research results and opinions. The IA Community welcomes the participation of new members in this conference. The growing interests, the enthusiasm of the participants have proved that the Conference has become an internationally recognized scientific event providing a good platform for the annual meeting of Inter-Academia Community and the wider scientific and research community mostly from Europe and Asia.

This year IA conference was organized in Budapest and Balatonfüred, Hungary, on September 4–7, 2019 by Óbuda University. The Conference contained 57 reviewed papers from which 35 has been selected to incorporate in this volume after a new, two-circle review procedure. Present book does not intend to be an overall survey on the engineering and educational fields of interest of Inter-Academia Community but tries to find topics which represent new, hot, and challenging problems.

The book begins with 7 chapters investigating selected problems of *Machine Intelligence and Computer Science*. In the second part of the volume *Modeling and Simulation* is treated by also 7 papers. The third part is devoted to *Measurement, Monitoring, and Identification* and contains 5 chapters. The fourth topic attached is *Electronics and Nanoelectronics* with 6, while the fifth is *Bio- and Environmental*

Engineering with 5 problems. Finally, the book includes 5 papers belonging to the field of *Chemical Processes and Material Science*.

The editor is grateful to the authors for their excellent work. On behalf of the organizing committee, we also would like to express our appreciation to the reviewers for offering their time in reviewing the papers. Their effort made possible to match the prestige of the books published by Springer Verlag. A particular acknowledgment goes to Dr. Thomas Ditzinger (Editorial Director), Prof. Janusz Kacprzyk (Series Editor), and to Ms. Varsha Prabakaran (Project Manager), to Ms. Monika Jetzin (Director of Trivent Conference Office), and Ms. Olga Hoffmann (Event manager of Trivent Conference Office) for the assistance and excellent collaboration during the development of this volume.

We hope that the reader will share our excitement and find the volume "*Engineering for Sustainable Future*" both inspiring and useful.

Budapest, Hungary
September 2019

Annamária R. Várkonyi-Kóczy

Contents

Bio- and Environmental Engineering

The Effects of Strip Cropping Systems on Physico-Chemical Properties of Soil in the Moldavian Plain

Maria-Mihaela Ciobăniță[1(✉)], Gerard Jităreanu[1], Petronela Andriescu[2], Paul-Marian Gherasim[2], Manuela Filip[1], and Costică Ailincăi[1]

[1] Pedotechnics Department, University of Agricultural Sciences and Veterinary Medicine of Iasi, Iași, Romania
maria.ciobanita@yahoo.com
[2] Faculty of Hydrotechnical Engineering, Geodesy and Environmental Engineering, "Gheorghe Asachi" Technical University of Iasi, Iași, Romania

Abstract. The research was carried out on the sloping land at the Agricultural Research Station of Podu-Iloaiei (41°18′52″ N latitude, 27°25′45″ E longitude) and the Experimental Farm of the Agricultural University of Iasi (47°12′62″ N latitude, 27°51′52″ E longitude) on a cambic chernozem (The World Reference Base for Soil Resources (WRB, 2006) and The Romanian Soil Taxonomy System (SRTS-2012). The experiments carried out the Experimental Farm of the Agricultural University of Iasi and at the Podu-Iloaiei Agricultural Research Station, during 2001–2018, had the following objectives: study of water runoff and soil losses by erosion, in different crops; annual rate of erosion processes under the influence of anti-erosion protection of different crops; influence of water runoff and soil erosion on organic matter and mineral element losses from soil. Measurements made after 36 years after the placement of perennial grass strips show that the slope of the platforms has decreased by 21.0–35.0% compared to the initial slope of the terrain, by the agro-terracing process, and the slope of the taluses with perennial grasses, increased by 214–397%. On 16% slope lands, the mean annual soil losses by erosion were comprised between 4.762 and 9.326 t ha^{-1} $year^{-1}$ in row crops (soybean and sunflower) and between 1.564 and 2.453 t ha^{-1} $year^{-1}$ in wheat and rape crops. The crop structure, which determined the diminution in mean soil losses by erosion until 2.800 t ha^{-1} $year^{-1}$ included 20% winter wheat, 20% of peas, 20% maize and 40% perennial grasses and legumes.

Keywords: Water erosion · Cropping system · Organic carbon

1 Introduction

The soil is subject to various degradation processes such as erosion, decline in organic carbon, compaction, acidification and other threats. One of the most important objectives of sustainable development is to protect soil against erosion caused by water and wind. The rate of soil loss through water erosion (t ha^{-1} $year^{-1}$) and the agricultural

A. R. Várkonyi-Kóczy (Ed.): INTER-ACADEMIA 2019, LNNS 101, pp. 3–18, 2020.
https://doi.org/10.1007/978-3-030-36841-8_1

area affected by soil erosion, hectare (ha) are two sub-indices showing the state of soil degradation in the countries of the European Union. According to Joint Research Centre (JRS) studies, around 11.4% of the European Union (EU) territory is affected by erosion, from moderate to high, i.e. more than 5 t ha^{-1} $year^{-1}$. In the EU, more than 150 million hectares of soil are affected by erosion and 45% of the European soils have a low content of organic matter [1]. In the EU 28, due to severe erosion of the surface of 12 million hectares where about 0.43% of crop productivity is lost, the annual cost was estimated at around 1.25 billion euros [2]. The Directive 2006/42/EC proposes the identification of zones with erosion-degraded soils and organic matter in decline, for meeting the requirements of the United Nations Convention to Combat Desertification (UNCCD) in Northern Mediterranean and Central and Eastern European Country Parties.

The main problems requiring agro-environment measures in Romania are the degradation degree of fields by erosion (6.3 million ha), deterioration of soil structure and compaction (on 44% of the total farming area) [3]. In Romania, soil erosion is the most expensive degradation process, which affects almost 63% of the total area and 56% of the arable area from Romania. Average soil losses through erosion, recorded in Romania (2.84 t ha^{-1} $year^{-1}$), are about, three times larger, than the EU-28 average tolerable losses, considered by most researchers of 0.3 to 1 t ha^{-1} $year^{-1}$.

The investigations concerning the potential erosion, conditioned by geomorphologic, soil and climatic factors, have shown that in the NE region of Romania, the mean soil losses by erosion were of 18.5 t ha^{-1} $year^{-1}$, values which corresponded to a moderate erosion risk. The investigations concerning the effective erosion, based on direct determinations and complex analyses, have shown that in the NE region, the effective erosion had a mean value of 4.5 t ha^{-1} $year^{-1}$ [4, 5]. The Organization for Economic Co-operation and Development (OECD) considers that tolerable soil losses by erosion are up to 6 t ha/year. Bazzoffi (2018) considers that, in order to ensure agricultural activity and environmental requirements, the value the tolerable soil losses should not exceed 3 t ha^{-1} $year^{-1}$ [6].

The purpose of the study was to characterize the slope sections and the effect of crop rotation and erosion based on measured soil parameters. The parameters analyzed were soil texture, bulk density, soil penetration resistance, aggregate stability (physical characteristics) as well as soil pH, soil organic matter (SOM), basic cations (Ca, Mg, K), N and P (chemical characteristics). The characterization of sloping terrain by sections of one-third of the slope, was found in several studies that evaluated the effects of crop rotation on soil erosion and agricultural production [4, 7]. The depletion of synthetic organic chemical (SOC), due to soil degradation by erosion, determines nutrient loss, soil structure degradation, biodiversity loss, and the decline in essential biotic and abiotic processes required for productivity [8].

2 Materials and Methods

2.1 Study Area

The research was carried out on the sloping land at the Agricultural Research Station of Podu-Iloaiei (41°18′52″ N latitude, 27°25′45″ E longitude) and the Experimental Farm of the Agricultural University of Iasi (47°12′62″ N latitude, 27°51′52″ E longitude) on a cambic chernozem The World Reference Base for Soil Resources (WRB, 2006) and The Romanian Soil Taxonomy System (SRTS-2012).

The experiments carried out at the Podu-Iloaiei Agricultural Research Station and the Experimental Farm of the Agricultural University of Iasi, during 2001–2018, had the following objectives: study of water runoff and soil losses by erosion, in different crops; annual rate of erosion processes under the influence of anti-erosion protection of different crops; influence of water runoff and soil erosion on organic matter and mineral element losses from soil.

Experiments were conducted in the hydrographic basin of Podu-Iloaiei, with catchment area of 159 ha, mean height of 119.4 m and mean slope of 12%. The area of the hydrographic basin has been anti-erosion arranged since 1983. Mixed cropping systems were used, made of grass strip, crop strips and agro-terraces. The width of cultivated strips was of 200–250 m on 5–10% slopes, 100–150 m on 10–15% slopes and 50–100 m on 15–18% slopes.

Investigations conducted at the University of Agricultural Sciences and Veterinary Medicine of Iaşi (in the Ezareni catchment area), followed the influence of different crops on water runoff and nutrient losses, due to soil erosion. These experiments were carried out on a 12% slope field, on a Cambic Chernozem with clayey loam texture (394 g clay, 325 g loam and 281 g sand) and a mean nutrient supply.

The climate is temperate continental with large thermal amplitude and uneven and commonly torrential rainfall prevalent during the vegetative season. The climatic conditions in the Moldavian Plain were characterized by a mean multiannual temperature of 9.6 °C and a mean rainfall amount, on 60 years, of 558.4 mm, of which 160.6 mm during September–December, and 397.8 mm during January–August.

2.2 The Methods Used to Determine Soil Erosion

At the beginning of the sampling, the study of the researched area was carried out and the Global Positioning System (GPS) was used to identify the geographic location and the coordinate system in the researched areas. Data obtained on soil properties were recorded on thematic maps drawn up at the scale of 1: 2000. The determination of runoff and soil losses by erosion was carried out by means of loss control plots with a collecting area of 100 m^2 (25 × 4 m) and by means of a hydrological section equipped with spillway and limn graph and devices for sampling water and soil loss by erosion.

The data was obtained in plots at the scale of the Podu Iloaiei catchment with the surface of 159 ha and an average slope of 14%, permitted the soil erosion estimation on the whole area of the catchment, where the experiments are placed, too. The catchment is provided in downstream section with triangular waste weir, limnigraph and devices for water and soil samplings. Total nitrogen, nitrate, phosphorus and potassium content were determined in soil and water samples, lost by erosion in different crops, thus establishing the losses of nutritive elements.

The experimental design consisted of two plots equipped with anti-erosion works, with a slope of 12% (Ezăreni) and 17% (Podu-Iloaiei) each with a width of 270 m and a length of about 700 m. Various methods are used to determine soil erosion, such as parcels for erosion control, monitoring and measurement of total water and soil leakage in the catchment area, the Revised Universal Soil Loss Equation (RUSLE 2015) model and the use of radioactive isotopes. These methods, when used individually, have certain disadvantages because they do not always provide reliable results [10]. Combining several methods to determine soil erosion helps to obtain data more real needed to control soil erosion.

Determination of erosion by plots for erosion control, for different natural rainfall events and measurement of total water and soil leakage across the river basin are reliable methods for establishing the crop structure and anti-erosion work.

2.3 Soil Physical and Chemical Properties

Bulk density was determined by the cylinder method [11], aggregate stability was determined using the Yoder type wet-sieving device [12]. Bulk density was calculated depending on the inner diameter of the core sampler, sampling depth and the oven dried weight of the soil samples. Core samples of depths 0–5 and 15–20 cm were taken by steel core sampler of a 100 cm^3 volume, 5 cm in diameter and 5 cm in height, with three replications, for bulk density and hydraulic conductivity measurements. Soil samples were collected during June of 2017 and 2018. Soil macro-aggregates were measured using wet-sieving method [12]. For aggregate stability analysis, soil samples were taken from the same depth of the soil pits. These soil samples were air-dried at laboratory temperature, sieved over a series of sieves into seven size fractions (10, 5, 3, 2, 1, 0.5 and 0.25 mm). These size fractions (dry sieve) were used for determination of water-stable aggregates (WSA). Macro-aggregates sizes were follows: >5, 5–3, 3–2, 2–1, 1–0.5, 0.5–0.25.

Determination of soil penetration resistance was performed in the field using the Eijkelkamp penetrologger. Composite soil samples from experimental site were collected from 0–20 cm depth before sowing or harvest. The samples were analyzed for organic matter by methods of soil analysis established from Nelson and Sommers [13] and mineral nitrogen by semi-Kjeldahl digestion method [14]).

The content of mobile phosphorus from soil was determined by Egner-Riechm Domingo method, in solution of ammonium acetate-lactate (AL) and potassium was measured in the same solution of acetate-lactate (AL) at flame photometer.

The positions of the locations were measured by Global Positioning System (GPS) for will enable the same positions to be studied again in future.

Analysis of Variance (ANOVA) methods were applied in the data analysis, Statistical package for the social sciences (SPSS 1999). The level of variation in organic material (OM), pH, Ca, AP, K, Na, Mg, and TN, in the three positions that are upper, mean and lower slope, was performed with variance analysis.

3 Results and Discussion

3.1 The Intensity of Erosion Processes

In Romania, soil erosion classes and eroded arable areas have been established by the Institute for Research on Pedology and Agrochemistry (IRPA). In Romania, about 2/3 of the country's surface and over 36% of the arable land (Table 1) is situated on the slopes, where the soil is in different degrees of erosion.

Table 1. The arable surface (ha) affected by erosion in Romania, IRPA, 2015

Clasa	Soil erosion t ha^{-1} $year^{-1}$	Surface, ha	Surface, %
Slight eroded soil	2–8	944763	28.0
Moderately eroded	9–16	1013884	30.1
Strongly eroded	17–23	749420	22.2
Very strongly eroded	24–30	454150	13.5
Excessively eroded	>=31	210729	6.2
Total		3372946	100

The measurements carried out in 2018 on strip grass and crop strips in the Podu-Iloaiei and Ezareni basins aimed at assessing changes of the slope of the agro-terraced land (Fig. 1).

On the land with a 16.55% overall slope, from at the Podu-Iloaiei watershed A, through the agro-terraced process produced due to erosion and soil tillage, the slope of cultivated stripes decreased, depending on the slope of the land, to 9.98–10.57% and slope of the taluses increased to 44.59–63.83% (Table 2, Fig. 1).

Average slope on the 9 agro-terraces platforms decreased from 16.55, which represents the average slope of the land, to 10.36% (37.4%) and the mean slope of the taluses increased to 51, 88% (213.5%).

In the Podu Iloaiei catchment area A, arable land has a slope of 15.8% on the upper third, 19% on the middle and 14% on the lower third of the slope (Fig. 2). On the concave slope the water erosion with the highest intensity takes place in the upper and middle third of the slopes. In the sub-basin Podu Iloaiei, Iasi, area B, on a land with a general slope of 13.94% through the agro-terraced process, after 36 years since the establishment of the strips, the slope of the cultivated strips decreased to 10.70–11.38% and the slope of the taluses increased to 62.03–81.89% (Table 3).

Average slope on the 5 agro-terraces platforms decreased from 13.94, which represents the average slope of the land, to 11.04% (–20.8%) and the mean slope of the taluses increased to 69.42% (+397.9%) (Table 3).

Measurements made after 36 years after the placement of perennial grass strips show that the slope of the platforms has decreased by 11.0–35.0% compared to the initial slope of the terrain, by the agro-terracing process, and the slope of the taluses with perennial grasses, increased by 214–397%.

At the Research Center for the Prevention of Soil erosion in Perieni, after a 12-year agro-terracing period, it was found that the slope of the agro-terrace platforms

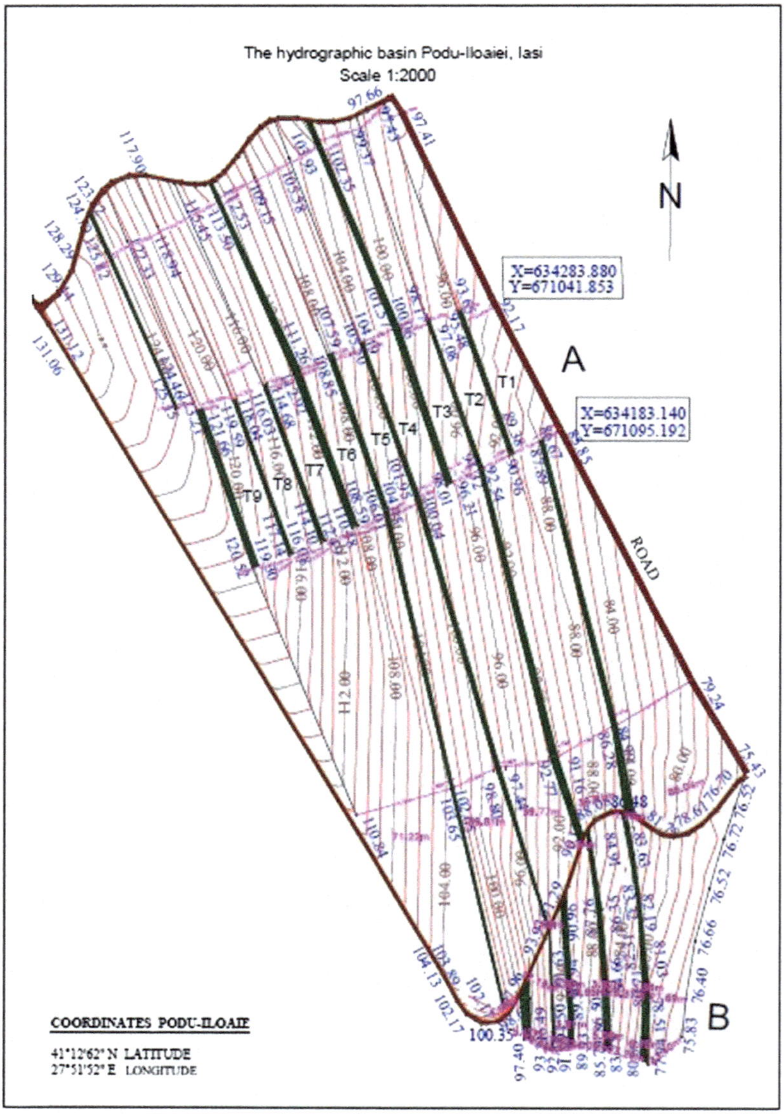

Fig. 1. The hydrographic basin map Podu Iloaiei, with sub-basins A and B.

decreased by 35–60%, compared to the initial slope of the land, and the slopes of the taluses increased by 41–62% [15].

The results on runoff and soil loss by erosion in different crops, which were determined by the help of plots for runoff control have shown that, during 1998–2017, of the total of 571.2 mm rainfall, 374.4 mm (65.5%) caused runoff between 10.2 mm in perennial grasses on the second growing season and 64.2 mm in sunflower (Table 4).

Table 2. The slope of the land (%) at three transects at the Podu-Iloaiei watershed A

Terrace (Te)	Transect 1	Transect 2	Transect 3	Average
Te-1	6.62	6.56	7.82	7.00
Te-2	8.13	11.23	12.64	10.67
Te-3	9.11	8.46	10.14	9.24
Te-4	1.62	9.24	8.12	9.33
Te-5	1.32	11.24	8.74	10.77
Te-6	1.50	12.31	11.14	12.65
Te-7	1.21	16.50	11.10	12.60
Te-8	9.21	9.84	13.20	10.75
Te-9	9.12	9.26	12.23	10.20
Average	9.98	10.52	10.57	10.36
LSD 5% = 1.1	LSD 0.1% = 1.7		LSD 0.1 = 2.1%	
Taluse (Tl)	Transect 1	Transect 2	Transect 3	Average
Tl-1	55.62	48.21	78.95	60.93
Tl-2	45.52	42.31	59.48	49.10
Tl-3	39.21	52.12	41.11	44.15
Tl-4	46.23	53.16	64.32	54.57
TL-5	41.32	45.21	56.24	47.59
Tl-6	31.24	37.22	43.25	37.24
Tl-7	52.3	54.26	89.68	65.41
TL-8	57.32	56.84	98.79	70.98
TL-9	32.54	35.62	42.64	36.93
Average	44.59	47.22	63.83	51.88
Overall slope of the land	15.97	16.73	16.95	16.55
LSD 5% = 1.4	LSD 0.1% = 2.3		LSD 0.1 = 3.4%	

The crop rotation and the degree of coverage of the land by plants play an essential role in influencing of erosion processes. To replace the lost phosphorus due to water erosion on agricultural land in the EU, for the amount of eroded soil established after Revised Universal Soil Loss Equation, (RUSLE 2015) and for a price of phosphate diammonium 440 € per ton, it would take 3 to 17-million-euro per year [1].

Annual soil losses by erosion registered at the same period were between 0.291 t/ha in perennial grasses on the second growing season and 9.326 t ha^{-1} $year^{-1}$ in sunflower. In the last 20 years' erosion was within the "allowable limits" of 2–3 t ha^{-1} $year^{-1}$ in perennial grasses on the second growing season, pea, wheat and winter rape.

On 16% slope lands, the mean annual soil losses by erosion were comprised between 4.762 and 9.326 t ha^{-1} $year^{-1}$ in row crops (soybean and sunflower) and between 1.564 and 2.453 t ha^{-1} $year^{-1}$ in wheat and rape crops.

Erosion affects soil fertility by removing together with eroded soil, significant organic carbon and mineral element amounts which in maize and sunflower crops reach

a) b)

Fig. 2. Agro-terraces on the field with a 17% slope, upstream (a), and middle slope (b).

Table 3. The slope of the land (%) at three transects at the Podu-Iloaiei watershed B

Terrace (Te)	Transect 1	Transect 2	Transect 3	Average
Te-1	8.62	7.62	8.04	8.09
Te-2	11.20	12.23	12.47	11.97
Te-3	11.74	12.53	12.64	12.30
Te-4	11.77	11.56	11.87	11.73
Te-5	10.15	11.24	11.86	11.08
Average	10.70	11.04	11.38	11.04
LSD 5% = 1.2	LSD 0.1% = 1.9		LSD 0.1 = 2.6%	
Taluse (Tl)	Transect 1	Transect 2	Transect 3	Average
Tl-1	48.20	47.34	60.80	52.11
Tl-2	75.60	74.22	87.61	79.14
Tl-3	44.60	52.37	89.35	62.11
Tl-4	76.60	68.21	96.34	80.38
TL-5	76.70	68.00	75.34	73.35
Average	64.34	62.03	81.89	69.42
Overall slope of the land	13.67	13.78	14.38	13.94
LSD 5% = 1.6	LSD 0.1% = 3.1		LSD 0.1 = 4.74%	

21.2–21.9 kg ha^{-1} of nitrogen, 1.2–1.3 kg ha^{-1} of phosphorus and 2.1–2.3 kg ha^{-1} of potassium, representing on the average between 12–14% of chemical fertilizers necessary for these crops (Table 5).

Table 4. Mean annual runoff and soil losses due to erosion, recorded in different crops

Crop	Rainfall causing runoff, mm	Water runoff, mm	Erosion, t ha^{-1} $year^{-1}$	Runoff coefficient	Mean turbidity, $g\ l^{-1}$
Field	374.4	87.6	18.245	0.23	20.83
Sunflower	374.4	64.2	9.326	0.17	14.53
Maize	374.4	60.3	8.940	0.16	14.83
Bean	374.4	52.6	5.462	0.14	10.38
Soybean	374.4	34.2	4.762	0.09	13.92
Pea	374.4	28.7	2.135	0.08	7.44
Wheat	337.2	15.6	1.564	0.05	10.03
Rape	337.2	16.8	2.453	0.05	14.60
I st year grasses	374.4	31.8	1.862	0.08	5.86
II nd year grasses	326.4	10.2	0.291	0.03	2.85

Table 5. Mean annual losses of nutritive elements in different crops

Crop	Nitrogen from water runoff, kg ha^{-1}	Nitrogen from eroded soil, kg ha^{-1}	Total N, kg ha^{-1}	Phosphorus kg ha^{-1}	Potassium kg ha^{-1}	Total NPK, kg ha^{-1}
Field	7.814	26.090	33.904	2.098	4.379	40.381
Sunflower	8.539	13.336	21.875	1.343	2.332	25.550
Maize	8.502	12.695	21.197	1.225	2.155	24.577
Bean	6.943	7.975	14.918	0.601	1.092	16.616
Soybean	4.651	6.953	11.604	0.543	1.038	13.185
Pea	3.588	3.181	6.769	0.233	0.427	7.429
Wheat	2.059	2.283	4.343	0.175	0.391	4.909
Rape	2.268	3.557	5.825	0.275	0.537	6.637
I st year grasses	3.593	2.756	6.349	0.214	0.464	7.027
II nd year grasses	1.153	0.428	1.580	0.033	0.072	1.85

Adoption of a suitable crop structure comprising 16.7% winter wheat, 16.7% pea, 16.6% maize and 50% perennial grasses, has reduced soil losses by erosion to 2.447 t ha^{-1} $year^{-1}$ (Table 4).

For Romania, on the land area of 1146700 ha, with a severe erosion of more than 11 t ha^{-1} $year^{-1}$ (10.5% of the agricultural area), the loss of crop productivity was calculated by Panagos et al. [2], at 74.058 million Euros.

Estimating the impact of erosion on society is very difficult because erosion causes quantitative losses, quantified by the value of agricultural products, washed nutrients, product losses and planting materials, costs for repairing damage caused by erosion, onsite and offsite (sedimentation, floods, landslides and eutrophication of water) etc.

From the investigations carried out on erosion, based on direct determinations, we found that the erosion in the Moldavian Plain, in peas-wheat-maize rotation, had a mean value of 4.213 t ha^{-1} $year^{-1}$ (Table 6).

Table 6. Mean annual water and soil runoff by erosion registered in different crops rotation

Crops rotation	Water runoff, mm	Erosion, t ha^{-1} $year^{-1}$	Total NPK, kg ha^{-1}	SOC kg ha^{-1}	Row plants, %
1. B-W-M-Sf-W	41.7	5.371	13.336	83.4	60
2. W-M	38.0	5.252	12.770	83.0	50
3. P-W-M-Sf+Pg	37.2	4.529	11.392	74.8	40
4. P-W-M	34.9	4.213	10.770	67.7	33
5. P-W-M-Sf+2Pg	33.9	3.888	9.955	63.0	33
6. P-W-M-Pg	30.5	3.330	8.770	53.8	25
7. P-W-M +2Pg	27.9	2.800	7.570	43.8	20
8. R-W-M +2Pg	25.5	2.863	7.382	42.2	20
9. P-W-M +3Pg	26.2	2.447	6.770	37.2	16.6

1. Beans-wheat-maize-sunflower-wheat (B-W-M-Sf-W);
2. Wheat-maize rotation (W-M);
3. Peas-wheat-maize-sunflower + a field with perennial grasses (P-W-M-Sf + Pg);
4. Peas-wheat-maize (P-W-M);
5. Peas-wheat-maize-sunflower + two fields with perennial grasses(P-W-M-Sf + 2Pg);
6. Peas-wheat-maize + a field with perennial grasses (P-W-M-Pg);
7. Peas-wheat-maize + two fields with perennial grasses (P-W-M + 2Pg);
8. Rape-wheat-maize + two fields with perennial grasses (R-W-M + 2Pg);
9. Peas-wheat-maize + three fields with perennial grasses (P-W-M + 3Pg).

From the investigations carried out on erosion, based on direct determinations, we found that the erosion in the Moldavian Plain, in peas-wheat-maize rotation, had a mean value of 4.213 t ha^{-1} $year^{-1}$ (Table 6).

From the results obtained, it was found out that the highest losses of nutritive elements were registered in two-year rotation (wheat-maize) (83.0 kg/ha organic carbon and 12.77 kg ha^{-1}, Nitrogen, Phosphorous, and Potassium, [NPK]. These amounts decreased very much at the same time with the increase in rotation of crops protecting soil against erosion, such as pea, wheat, alfalfa and perennial grasses (Table 6). These results are necessary for establishing the crop structure and dimensioning the anti-erosion works, which determine the decrease in soil erosion below the tolerable limit, of 1–3 t ha^{-1} $year^{-1}$of eroded soil.

The crop structure, which determined the diminution in mean soil losses by erosion until 2.800 t/ha included 20% winter wheat, 20% of peas, 20% maize and 40%

perennial grasses and legumes (Alfalfa + Lolium perene). On the land with a slope of 16%, lowering the percentage of row plants from 60% to 20% has reduced the amount of eroded soil by 48.9% (2.800 t ha^{-1} $year^{-1}$) (Table 6).

3.2 The Effects of Cropping Systems on Physico-Chemical Properties of Soil

Soil erosion caused a reduction in the percentage of aggregate by 31.0% to slightly eroded soil (upper slope) and 39% in the strongly eroded (middle slope) soil (Table 7). The percentage of water stable aggregates was comprised between 64.1% in non-eroded soil, at the bottom of slope land and 33.3% at the highly eroded soil.

The bulk density of the soil was 1.30 g cm^{-3} (100%) at the lower slope of the terrace platform and 1.38 g cm^{-3} (106%) on the middle slope (Table 8). The highest values of bulk density (T-7 - maize, 1.45 g cm^{-3}) were recorded on the number seven terrace where the erosion is stronger and the lowest values at the lower slope (T-1 wheat, 1.30 g cm^{-3}).

Table 7. The influence of crops and soil erosion on the degree of water stability of aggregates, aggregates >0.25 mm (%)

Terrace (T)	Depth, cm	Upper slope	Middle slope	Lower slope	Mean
	0–10	63.4	55.9	50.3	56.5
T-I- Wheat	10–20	66.5	58.2	58.2	61.0
	20–30	62.4	54.3	55.1	57.3
Mean		64.1	56.1	54.5	58.3
	0–10	56.3	54.6	51.9	54.3
T-III- Maize	10–20	58.4	57.2	57.2	57.6
	20–30	57.3	52.9	55.8	55.3
Mean		57.3	54.9	55.0	55.7
	0–10	44.3	39.5	32.7	38.8
T-V- Pea	10–20	45.9	40.7	35.4	40.7
	20–30	46.2	39.6	33.7	39.8
Mean		45.5	39.9	33.9	39.8
	0–10	43.5	39.5	32.7	38.6
T-VII- Wheat	10–20	46.2	40.1	35.2	40.5
	20–30	42.5	38.7	31.9	37.7
Mean		44.1	39.4	33.3	38.9
	0–10	49.3	44.3	49.3	47.6
T-IX- Maize	10–20	51.2	46.3	50.1	49.2
	20–30	48.6	47.2	49.7	48.5
Mean		49.7	45.9	49.7	48.4
	LSD 5% = 4.3%; LSD 1% = 5.5%; LSD 0.1% = 7.5%				

Depending on the agro-technical requirements, the values of the soil penetration resistance were classified as very low (below 1.1 megapascal pressure unit (MPa)), low (1.1–2.5 megapascal pressure unit (MPa)), medium (2.5–5.0 megapascal pressure unit

Table 8. The bulk density (g cm^{-3}) on the sloping land at the Podu-Iloaiei, Iasi county

Terrace (T)	Depth, cm	Upper slope	Middle slope	Lower slope	Mean
	0–10	1.30	1.29	1.28	1.29
T-I- Wheat	10–20	1.31	1.31	1.29	1.30
	20–30	1.31	1.32	1.31	1.31
Mean		1.31	1.31	1.29	1.30
	0–10	1.34	1.36	1.37	1.36
T-IV- Maize	10–20	1.38	1.37	1.39	1.38
	20–30	1.41	1.40	1.42	1.41
Mean		1.38	1.38	1.39	1.38
	0–10	1.35	1.37	1.38	1.37
T-VII- Pea	10–20	1.39	1.39	1.42	1.40
	20–30	1.42	1.44	1.45	1.44
Mean		1.39	1.40	1.42	1.40
	LSD 5% = 0.03; LSD 1% = 0.05; LSD 0.1% = 0.07 g cm^{-3}				

(MPa)), high (5.1 to 10.0 megapascal pressure unit (megapascal pressure unit (MPa)) very high (10.1–15.0 megapascal pressure unit MPa)) and extremely high (over 15 megapascal pressure unit (MPa). From this point of view, for all the tested variants the penetration resistance was very low and low. The highest values of the soil penetration resistance (T-7 1.9 megapascal pressure unit (MPa)) were recorded on the number seven terrace where the erosion is stronger and the lowest values at the lower slope (T-2 1.24 megapascal pressure unit (MPa)) (Table 9).

Investigations, carried out under different climatic conditions, have shown that the use of crop rotations with legumes and perennial grasses contributed to the improvement of physical, chemical and biological characteristics of soil [2, 4, 7, 8, 16–18].

The results on the evolution of synthenic organic material (SOC) and mineral elements content from soil, after 36 years of using the rotation grain legumes-wheat-maize, have shown that a good supply in mineral elements was found only at the bottom of the slope. On 2/3 of slope length, erosion has caused the diminution in SOC content, from 1.73% to 1.02%, according to soil erosion (Table 10). At the same time with the increase in erosion process, the content of mobile potassium and phosphorus from soil diminished from 49 to 15 parts per million, (ppm P) and, respectively, from 296 to 142 parts per million, (ppm K). These data have shown that on 16–18% slope lands, the use of three-year rotation, with crop mineral fertilization was not enough to maintain soil fertility under favorable limits.

In the critical erosion zone on valley side, once with the erosion of the horizon from soil surface, soil degradation processes intensified, even on chernozems, which were

Table 9. The soil penetration resistance (MPa) on the sloping land at the Podu-Iloaiei, Iasi county

Slope third	Depth, cm	Terrace I	Terrace II	Terrace III	Average
	0–10	1.10	0.80	0.80	0.90
	10–20	1.40	1.00	1.10	1.17
Lower	20–30	1.20	1.20	1.30	1.23
	30–40	1.30	1.50	1.60	1.47
	40–50	1.40	1.70	1.60	1.57
Mean		1.28	1.24	1.28	1.27
Slope third	Depth, cm	Terrace IV	Terrace V	Terrace VI	Average
	0–10	0.90	0.80	1.20	0.97
		1.20	1.30	1.20	1.23
Middle	20–30	1.40	1.40	1.50	1.43
	30–40	1.80	1.60	1.70	1.70
	40–50	1.70	1.70	1.80	1.73
Mean		1.40	1.36	1.48	1.41
Slope third	Depth, cm	Terrace VII	Terrace VIII	Terrace IX	Average
	0–10	0.90	0.90	1.20	1.00
	10–20	1.40	1.30	1.20	1.30
Upper	20–30	1.70	1.90	1.90	1.83
	30–40	1.80	1.70	1.70	1.73
Mean	40–50	1.90	1.80	1.60	1.77
	LSD 5% = 0.01; LSD 1% = 0.03; LSD 0.1% = 0.05 MPa				

more resistant to erosion (Table 10, Terrace VII). The average values of the nutrient content at three transverse profiles on the cambic chernozem on the slope lands at the Podu-Iloaiei and Ezareni watersheds, Iasi County, are presented in Table 11.

Cambic chernozem in the Experimental Farm of the Agricultural University of Iasi has a clay- loam texture, a pH value of 6.8, a humus content of 3.17% (1.84% organic carbon) and a medium nutrient level.

The mass of total carbon in the Cambic Chernozem from the Moldavian Plain has registered significant increases in case of organo - mineral fertilization and in 4-year crop rotation + reserve field, cultivated with perennial grasses and legumes. On soils from the Moldavian Plateau, most of them situated on slope the proper use of different organic resources, determine the improvement in the content of organic carbon from soil. Cambic chernozem in the Podu-Iloaiei watershed has a medium organic carbon content (1.56–1.89%), is well supplied with mobile potassium (246–312 parts per million, (ppm)) and moderate with mobile phosphorus (36–52 parts per million, (ppm)) and nitrogen (0.116–0.126%) (Table 11).

Table 10. Change of main soil chemical characteristics on a 16% slope, as influenced by soil erosion and rotation

Horizon/Terrace	Depth, cm	pH, (H_2O)	SOC %	N total %	P_{-AL} mbile ppm	K_{-AL} mbile ppm
T 1 - Weakly eroded cambic chernozem situated at the lower third of slope						
Ap	0–20	6.8	1.73	0.158	49	296
Am	20–35	6.9	1.40	0.129	35	264
A/B	35–52	7.0	0.75	0.114	4	98
Bv	52–71	7.2				
T V - Highly eroded cambic chernozem situated at the upper third of slope						
Am	0–15	7.2	1.26	0.138	26	158
A/B	15–29	7.4	1.19	0.129	5	124
Bv	29–40	7.6	0.46			
C	40–62					
T VII - Highly eroded cambic chernozem situated at the upper third of slope, in the critical erosion zone						
A/C	0–16	7.4	1.02	0.124	15	142
C1	16–30	7.6	0.67	0.112	6	131
C2	30–59	8.2				
LSD 5%		0.2	0.01	0.002	3.2	7.1

Table 11. Change of the main soil chemical properties on a 14% slope, as influenced by soil erosion

Nutrients	Trans. 1	Trans. 2	Trans. 3	Mean	LSD 0.5%
Ezareni, Iasi					
SOC, %	1.72	1.56	1.89	1.72	0.01
N t %	0.119	0.116	0.126	0.120	0.001
P-AL, mg kg^{-1}	46	36	52	44	3.9
K-AL, mg kg^{-1}	267	246	312	275	8.6
Podu-Iloaiei, Iasi					
SOC, %	1.68	1.85	1.98	1.84	0.01
N t %	0.13	0.17	0.18	0.16	0.001
P-AL, mg kg^{-1}	39	51	64	51	4.1

4 Conclusions

The mean annual losses of soil by erosion, recorded during 1998–2017, were of 0.291 t ha^{-1} $year^{-1}$ in perennial grasses in the second growth year, 1.564 t ha^{-1} $year^{-1}$ in wheat, 8.940 t ha^{-1} $year^{-1}$ in maize and 9.326 t ha^{-1} $year^{-1}$ in sunflower. From the results obtained on erosion in different crop rotations, we have found that in 16% slope fields from the Podu-Iloaiei, Iasi watershed, soil losses by erosion diminished below the allowable limit of 2.0 t ha^{-1} $year^{-1}$ only in case of 3 or 4 year-crop rotations with two or three reserve fields, cultivated with legumes and perennial grasses, which protect soil.

The crop structure, which determined, during 1998–2017, the diminution in mean soil losses by erosion until 2.800 t ha^{-1} $year^{-1}$ included 20% straw cereals (wheat), 20% annual legumes (pea), 20% row crops (maize) and 40% perennial grasses and legumes (Alfalfa + Lolium perene). On land with a slope of 16%, lowering the percentage of row plants from 60% to 16.6% has reduced the amount of eroded soil by 54.4% (2.924 t ha^{-1} $year^{-1}$).

The 36-year use of 4-year crop rotations + reserve field, cultivated with perennial grasses and legumes, has determined the increase in the mass of total carbon from soil by 13.4% (2.2 C g kg^{-1}), in comparison with maize continuous cropping.

Acknowledgements. The authors would like to acknowledge the CNFIS-FDI-2019-0267 project for funding The 18th International Conferences on Global Research and Education, special thanks also go to the researchers of the Agricultural Research Station of Podu Iloaiei, Iasi County, for their support in carrying out our investigations.

References

1. Montanarella, L.: Moving ahead from assessments to actions: could we win the struggle with soil degradation in Europe? In: 5th International Conference on Land Degradation, Valenzaro, Bari, Italy, 18–22 September 2008, pp. 5–9. Ideaprint- Bari, Italy (2008). ISBN 2-85352-399-2
2. Panos, P., Gabriele, S., Pasquale, B., Emanuele, L., Luca, M., Francesco, B.: Cost of agricultural productivity loss due to soil erosion in the European Union: From direct cost evaluation approaches to the use of macroeconomic models. Land Degrad. Dev. **29**, 471–484 (2018)
3. Jităreanu, G., Ailincăi, C., Bucur, D.: Soil fertility management in North-East Romania. J. Food Agric. Environ. **5**(3&4), 349–353 (2007)
4. Ailincăi, C., Jităreanu, G., Bucur, D., Mercus, A.: Protecting the soil from erosion by cropping systems and fertilization in Moldavian Plateau. Int. J. Food Agric. Environ. - JFAE **9**(1), 570–574 (2011)
5. Daniel, B., Gerard, J., Costica, A.: Effects of long-term soil and crop management on the yield and on the fertility of eroded soil. Int. J. Food Agric. Environ. - JFAE **9**(2), 207–209 (2011)
6. Paolo, B.: Soil erosion tolerance and water runoff control: minimum environmental standards. Reg. Environ. Chang. **9**, 169–179 (2009)
7. Szabó, B., Szalai, Z., Centeri, C., Dobó, Z., Jakab, G.: Effects of land use types with different intensities on soil erosion, nutrient loss and other soil parameters. In: 13th Alps-Adria Scientific Workshop, Villach, Ossiacher See, Austria 2014, vol. 63(Suppl.), pp. 221–224 (2014)
8. Adhikari, K., Hartemink, A.: Linking soils to ecosystem services-a global review. Geoderma **262**(2016), 101–111 (2016)
9. Lal, R.: Sustainable intensification for adaptation and mitigation of climate change and advancement of food security in Africa. In: Lal, R., Singh, B.R., Mwaseba, D.L., Kraybill, D., Hansen, D.O., Eik, L.O. (eds.) Sustainable Intensification to Advance Food Security and Enhance Climate Resilience in Africa. Springer, New York (2015)

10. Parsons, A.J.: How reliable are our methods for estimating soil erosion by water? Sci. Total Environ. **676**(2019), 215–221 (2019)
11. Blake, G.R., Hartge, K.H.: Methods of Soil Analysis. Part 1. Physical and Mineralogical Methods, vol. 2, pp. 366–375. Agronomy (1986)
12. Kemper, W.D., Rosenau, R.C.: Aggregate Stability and Size Distribution. Methods of Soil Analysis. Physical and Mineralogical Methods, 2nd ed., pp. 425–442. Agronomy (1986)
13. Nelson, D.W., Sommers, L.E.: Total carbon, organic carbon, and organic matter. Agronomy monograph 9. In: Page, A.L., Miller, R.H., Keeney, D.R. (eds.) Methods of Soil Analysis, Part 2, Chemical and Microbial Properties. p. 539. Agronomy Society of America, Madison, Wisconsin (1982)
14. Bremner, J.M., Mulvaney, C.S.: Nitrogen-total. Agronomy monograph 9. In: Methods of Soil Analysis, Part 2, Chemical and Microbial Properties, p. 595. Agronomy Society of America, Madison, Wisconsin (1982)
15. Popa, A., Stoian, G., Popa, G., Oatu, O.: Combaterea eroziunii solului pe terenurile arabile. Editura Ceres, Bucureşti (1984)
16. Orgiazzi, A., Ballabio, C., Panagos, P., Jones, A., Fernández-Ugalde, O.: LUCAS Soil, the largest expandable soil dataset for Europe: a review. Eur. J. Soil Sci. **69**(1), 140–153 (2018)
17. Horn, R., Fleige, H.: Risk assessment of subsoil compaction for arable soils in Northwest Germany at farm scale. Soil Tillage Res. **102**, 201–208 (2009)
18. Ionita, I., Margineanu, R.: Application of 137-Cs for measuring soil erosion/deposition rates in Romania. Acta Geol. Hisp. **35**, 311–319 (2000)

Systematic Review of Deep Learning and Machine Learning Models in Biofuels Research

Sina Ardabili[1], Amir Mosavi[2,3](✉), and Annamária R. Várkonyi-Kóczy[4]

[1] Institute of Advanced Studies Koszeg, Koszeg, Hungary
[2] Kalman Kando Faculty of Electrical Engineering, Obuda University, Budapest, Hungary
amir.mosavi@kvk.uni-obuda.hu
[3] School of the Built Environment, Oxford Brookes University, Oxford OX3 0BP, UK
[4] Department of Mathematics and Informatics, J. Selye University, Komarno, Slovakia

Abstract. The importance of energy systems and their role in economics and politics is not hidden for anyone. This issue is not only important for the advanced industrialized countries, which are major energy consumers but is also essential for oil-rich countries. In addition to the nature of these fuels, which contains polluting substances, the issue of their ending up has aggravated the growing concern. Biofuels can be used in different fields for energy production like electricity production, power production, or for transportation. Various scenarios have been written about the estimated biofuels from different sources in the future energy system. The availability of biofuels for the electricity market, heating, and liquid fuels is critical. Accordingly, the need for handling, modeling, decision making, and forecasting for biofuels can be of utmost importance. Recently, machine learning (ML) and deep learning (DL) techniques have been accessible in modeling, optimizing, and handling biodiesel production, consumption, and environmental impacts. The main aim of this study is to review and evaluate ML and DL techniques and their applications in handling biofuels production, consumption, and environmental impacts, both for modeling and optimization purposes. Hybrid and ensemble ML methods, as well as DL methods, have found to provide higher performance and accuracy.

Keywords: Biofuels · Deep learning · Big data · Machine learning models

Nomenclatures

ANN	Artificial neural network
ELM	Extreme learning machine
ML	Machine learning
SVM	Support vector machine
WNN	Wavelet neural networks
DL	Deep learning
ARIMA	Autoregressive integrated moving average

A. R. Várkonyi-Kóczy (Ed.): INTER-ACADEMIA 2019, LNNS 101, pp. 19–32, 2020.
https://doi.org/10.1007/978-3-030-36841-8_2

FFNN	Feed-forward neural networks
MLP	Multi layered perceptron
DT	Decision tree
RSM	Response surface methodology
BPNN	Back propagation neural network
CM	Centroid mean
ANFIS	Adaptive neuro fuzzy inference system
ANP	Analytic network process
RF	Random forest
NRTL	Non-random two-liquid
RNN	Recurrent neural network
PLS	Partial least squares
DA	Discriminant analysis
PCA	Principal component analysis
LDA	Linear discriminant analysis
SVR	Support vector regression
LS	Least-squares
SB	Sparse Bayesian
MCDM	Multi criteria decision making
GP	Genetic programming
MLR	Multi linear regression
SWARA	Step-wise Weight Assessment Ratio Analysis
MOORA	Multi Objective Optimization by Ratio Analysis

1 Introduction

The global energy systems are highly dependent on fossil fuels [1, 2]. The importance of energy systems and their role in economics and politics is not hidden for anyone [3, 4]. This issue is not only important for the advanced industrialized countries, which are major energy consumers but is also essential for oil-rich countries [5]. Because countries have to understand the fact that fossil fuel resources are limited resources. In addition to the nature of these fuels, which contains polluting substances, the issue of their ending up has aggravated the growing concern. Therefore owing to depleting non-renewable energy resources, pollution, and environmental damage, the world is turning towards renewable energy resources [6]. Fossil fuels remain as one of the major energy resources worldwide [7]. Heavy dependence on fossil fuels has caused an energy crisis. Using fossil fuel for economic activities leads to GHG emissions from almost all regions of the world [8]. Renewable resources like biofuels make an attractive contribution towards meeting the growing demand for energy supply [9–11]. Owing to environmental concerns and the rise and fluctuations in the fossil fuel resources, worldwide interests have moved towards biodiesel, a clean and renewable alternative for fossil fuels [12, 13]. Biofuels can be used in different fields for energy production like electricity production, power production, or transportation [14]. The economy of biofuels and related refineries will be shaped by policies that have shaped the economy of hydrocarbon and its refineries over the last century [15–18]. Due to the environmental benefits of biofuels, their contribution to the automotive fuel market is increasing sharply. Various scenarios have been written about the estimated biofuels from different sources in the future energy system. The availability of biofuels for the electricity market, heating, and liquid fuels is very important. Therefore the need for handling, modeling, decision making, and forecasting for biofuels can be one of the main challenges for scientists [19–22]. Figure 1 shows the research trend in literature considering biofuels. Note that, since 2015 the research in this realm has stopped been progressing.

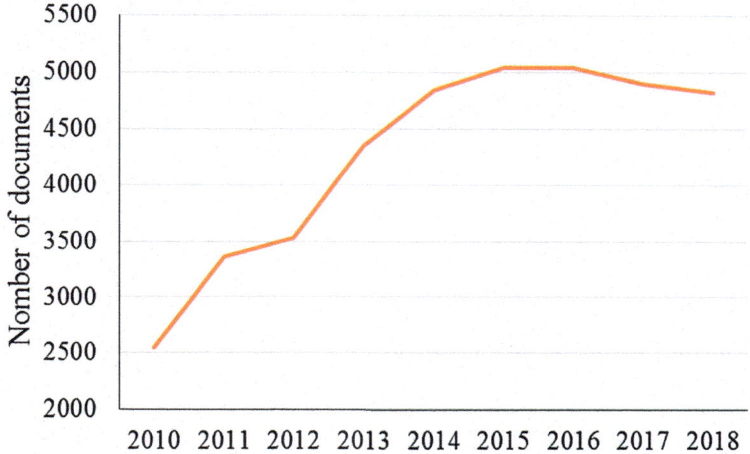

Fig. 1. The research trend in literature considering biofuels research (source: web of science)

Recently, machine learning and deep learning techniques have been accessible in modeling, optimizing and handling the biodiesel production, consumption, and its environmental impacts by considering the effect of parameters on biofuel production yield because the production of the desired product needs an effective use of experimental model [23]. These methods provide an independent modeling approach to the nature of the process or its mathematical models and are able to model the process with high accuracy [9, 11, 24, 25].

The primary purpose of this study is to present a review in a specific field to find the strengths and weaknesses of the field and to provide a complete background. The main aim of this study is to evaluate the ML and DL techniques developed for handling biofuels production, consumption, and environmental impacts, both for modeling and optimization purposes. The study initially explains and defines different biofuels. Then provides a general survey about the characteristics and the basis of the developed studies. In the next stage, explains the state of art of the DL and ML techniques employed in the field. Finally, concludes the results and achievements and proposes the strengths and weakness of different DL and ML techniques.

2 ML and DL Methods in Biofuels Research

The application of ML and DL methods in various scientific and engineering domains have been previously investigated [26–40]. Generally, the ML methods are reported to be further advancing to through ensemble and hybrid techniques [40–77]. On the other hand, the DL methods are still considered as a new phenomenon and are slowly progressing.

In this section, the most popular ML and DL methods in biofuels research are identified and reviewed. During the past decade, the application of these intelligent algorithms has been dramatically increases in biofuels research. Figure 2 represents the increasing demand and popularity of using DL and ML in handling biofuels. It is apparent that since 2010, the use of DL and ML has been increasing until the year 2017. Since then, it starts to decline. The reason can be found in the overall decrease in the number of literature in biofuels research. We made three classifications of the methods, i.e., neural networks-based methods, single ML methods, and a separate group for deep learning, ensembles, and hybrid models.

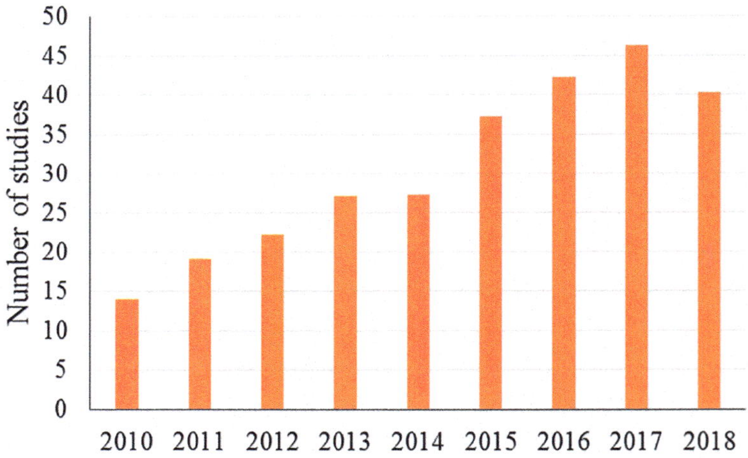

Fig. 2. Demand and popularity of using DL and ML in biofuels research (source: web of science)

2.1 Neural Networks-Based ML Models for Biofuels Research

This section includes the application of artificial neural networks (ANNs), Multilayer perceptron (MLP), Extreme learning machines (ELM), feedforward neural networks (FFNNs) and Backpropagation ANNs in biofuels research (Table 1).

Table 1. Top studies developed by ANN-based methods in biofuel

References	Contribution	Method	Application domains
[59]	To optimize the prediction of liquid-liquid equilibria which is employed in the simulation of the biofuel process by the use of a novel non-random two-liquid-ANN method	NRTL-ANN	– NRTL – Biofuels
[9]	To develop different types of MLP networks for the estimation of enzyme function	MLP	– Enzyme function – Machine learning
[60]	To develop the ANN method for the prediction of un-measurable variables during hydrogen and methane production through the anaerobic digestion process	RNN	– Biofuels – RNN
[61]	To develop a comprehensive survey about the use of ANN in the optimization and estimation of variables in the biofuels production process	ANN	– ANN – Biofuel production
[25]	To employ ANN methods for the prediction of the cetane number of biofuel samples in the presence of furanic additives	ANN	– Machine learning – Biofuels

Reynel-Ávila et al. [59] developed an innovative hybrid non-random two-liquid-ANN method in order to increase the estimation performance of the liquid-liquid equilibria, which is used to simulate the biofuel process. Non-random two-liquid method is considered as a thermodynamic method to be used in a multi-component system. Therefore, hybridization of this method with the ANN method can improve the system accuracy for the regression and fitting proposes. Evaluation of the proposed method has been performed using RMSD factor for measuring the agreement between target and estimated values. This method as a flexible method, could successfully cope with the estimation task as well as increasing the accuracy of estimation.

Concu et al. [9] developed a study in order to employ different machine learning techniques for the estimation of protein function through a conversion process as a type of enzyme for considering in bioethanol production. The developed machine learning techniques included the single method containing different architectures of MLP methodology. Results have been evaluated using accuracy, sensitivity, and specificity. Methods have a different number of neurons in the hidden layer. The accuracy of the proposed MLP method was acceptable, as well as its higher sustainability. Camberos et al. [60] developed a recurrent neural network method in order to estimate un-measurable variables during hydrogen and methane production through the anaerobic digestion process. The reason was the ability of the recurrent ANNs method in predicting the behavior of unknown and sophisticated systems. The method was a single method which benefited the external disturbances as well as the parameter uncertainties. The results have been evaluated using mean square error. Based on results, the proposed RNN method could successfully provide a high performance in confrontation with the complex system. Also, the method provided a high sustainability by a high stability in the presence of the external distributions.

Sewsynker-Sukai [61] did a comprehensive survey about the application of ANN, as one of the most popular and applied machine learning methods, in the field of biofuels for optimization and estimation purposes. This study also presents a brief explanation of the comparison of the performance of ANN with another method and discussing the architectures of the developed ANN methods. Comparisons were performed using the coefficient of determination as to the performance factor. Based on results, developing ANN methods in this field provides a high production performance as well as reducing the time and the cost consuming. Reduction of the time and cost in the biofuels production and consuming processes also increases the sustainability and reliability of the system. Therefore, ANN can be a useful tool for handling biofuels and for managing the production and consuming processes for policymakers in the future researches. Kessler et al. [25] presented a study to estimate the cetane number of biofuel samples in the presence of furanic additives. Results have been evaluated using RMSE values. ANN as a predictive method could be successfully applied for the prediction of cetane number with a low error.

Table 2. The comparison results of ANN-based methods for biofuels handling

Method	Application	Accuracy	Reliability	Sustainability	References
Hybrid NRTL-ANN	Estimation	++	++	++	[59]
MLP	Estimation	++	+	+	[9]
RNN	Estimation	+++	+++	+++	[60]
ANN	Estimation	++	+	+	[61]
ANN	Optimization	++	++	+	[61]
ANN	Estimation	++	+	+	[25]

Different applications of ANN tools in different fields of biofuels have been already discussed. However, there is a need for metrics and different criteria for the evaluation of the performance of each method. Table 2 present a brief comparison of the accuracy, reliability, and sustainability of methods developed for handling biofuels using different types of ANN methods. These factors have been prepared and presented based on different aspects which have been concluded by the reviewed studies.

2.2 Further Single ML Methods for Biofuels Research

This section includes support vector machines (SVM), decision trees (DTs) regression tree (RTs), Bayesian, k-means, and k-nearest neighbors (Table 3).

Table 3. Top studies developed by SVM based methods in biofuel

References	Contribution	Method	Application domains
[62]	To develop PLS-DA, SVM and PCA-LDA methods for the classification of biofuels	PLS-DA, SVM, and PCA-LDA	– Classification – SVM
[63]	To develop prediction models for the estimation of biofuels pellet quality using LSSVM and PLSR methods as non-destructive methods	LSSVM and PLSR	– LSSVM – PLSR
[64]	To develop a fuzzy method for the prediction of a cetane number of biodiesel fuel samples	Fuzzy	– Fuzzy – Cetane number
[65]	To develop a hybrid modeling method for the estimation of the engine performance fuelled by biofuel	SBELM	– Hybrid SBELM – Engine performance
[66]	To employ hybrid machine learning techniques for the estimation of biofuel production yield and optimization production process	ELM-RSM and SVM-RSM	– Biofuel production – Hybrid machine learning

Mancini et al. [62] developed three methods including partial least squares discriminant analysis, SVM, and principal component analysis linear discriminant analysis for the classification of biofuels. Based on results, all the methods could successfully cop with the classification task but SVM has the best classification performance. Feng et al. [63] developed non-destructive prediction methods for the estimation of the quality of the biofuel pellet using partial least-squares regression and a least-squares support vector machine as non-destructive diagnosis methods to be compound with successive projections algorithm. The performance of the methods have been compared using the determination coefficient and root mean square error values. Based on results, the best method was identified to be SPA-LSSVM method as a hybrid diagnosis method. This method employs the advantages of both LSSVM and SPA methods, consequently.

Faizollahzadeh et al. [64] developed a Sugeno based fuzzy method for the prediction of biodiesel fuel cetane number in the presence of Carbon number, Double bond, Saponification number, and Iodine value. The performance of the developed model has been calculated using the determination coefficient, and root mean square error. The developed model has high accuracy in both training and testing steps, but one of the most critical factors for this method was its lower processing time and its user-friendly application. These factors increase the method of sustainability factor to be employed in future researches. Wong et al. [65] developed a novel hybrid sparse Bayesian-based extreme learning machine technique for the estimation of the engine performance fuelled by biofuel as well as the calibration of the ECU. The proposed method has been also compared with the performance of ELM, Bayesian ELM and back propagation neural network in terms of mean absolute percentage error and standard deviation. The proposed hybrid method has an acceptable accuracy in both training and testing steps compared with that for the ELM, BPNN, and BELM methods. The proposed method also has a higher performance in the estimation of engine emissions.

Faizollahzadeh et al. [66] developed an innovative hybrid ELM-RSM and EVM RSM methods for the prediction of biofuel production yield and optimization of the production process for accessing a higher production yield. The developed methods have been compared with SVM, ANN and ANFIS methods in term of performance factors for the prediction phase. Based on results, hybrid ELM-RSM methods could provide higher performance by increasing the production yield compared with that of the other methods. This study also indicates the importance and strength of the hybrid method over single methods. In fact, this method benefits the highest prediction capability of ELM method in parallel with the optimization capability of the RSM. Therefore this study highlights the highest performance of hybrid techniques in comparison with single ones. Table 4 presents the comparison results of SVM based methods for biofuels handling.

Table 4. The comparison results of SVM based methods for biofuels handling

Method	Application	Accuracy	Reliability	Sustainability	References
SVM	Classification	++	++	++	[62]
PCA-LDA	Classification	++	+	+	[62]
PLS-DA	Classification	+	+	+	[62]
SPA-LSSVM	Classification	+++	+++	++	[63]
Fuzzy	Estimation	+++	+++	++	[64]
SBELM	Estimation	+++	+++	+++	[65]
ELM	Estimation	+	+	+	[65]
BELM	Estimation	++	+	+	[65]
ELM-RSM	Optimization	+++	+++	+++	[66]
SVM-RSM	Optimization	+++	++	++	[66]

2.3 Deep Learning, Machine Learning, Ensembles, and Hybrid Models for Biofuels Research

In this section, the more sophisticated ML methods in addition to DL are presented. Here may include neuro-fuzzy models, various DL models, and ensemble MLs (Table 5).

Table 5. Top studies developed by machine and deep learning-based methods in biofuel

References	Contribution	Method	Research domain
[67]	To develop a novel Multi-criteria decision making for improving the energy management system and increasing the energy efficiency	MCDM	– Decision making – Energy management
[68]	To develop different methods including hybrid and single methods for the prediction of short term energy parameters	ANFIS-CM, Genetic programming, M5Tree, RF and MLR	– Hybrid machine learning – Energy systems
[22]	To develop a comprehensive survey about the application of machine learning and deep learning methods in energy systems	Machine learning and deep learning methods	– Hybrid machine learning – Single machine learning – Energy systems – Deep learning

Erdogan et al. [67] developed a novel multi-criteria decision-making system for choosing the best biodiesel fuel for a compression ignition engine in terms of engine performance and combustion characteristics. Based on results, the hybrid Step-wise Weight Assessment Ratio Analysis- Multi-Objective Optimization by Ratio Analysis method and hybrid Analytic network process- Multi-Objective Optimization by Ratio Analysis provided the best performance for choosing the best fuel sample.

Deo et al. [68] developed different hybrid and single machine learning techniques for the prediction of sub-tropical photo-synthetically active radiation. The developed methods included ANFIS integrated with centroid mean, random forest, genetic programming, M5Tree, and multiple linear regression. Methods have been compared in terms of mean absolute error and root mean square error. Results indicated that the hybrid ANFIS-CM followed by GP methods could provide the lowest error as well as the highest sustainability.

Mosavi et al. [22] developed a comprehensive survey about the application of machine learning methods, including single and hybrid methods in the energy systems. The study has been developed in order to present a comprehensive state of the art of machine learning and to discuss their advantage and disadvantages, in detail. Methods have been compared in terms of root mean square error, determination coefficient, correlation coefficient, and mean absolute percentage error. Based on results, hybrid machine learning have the best performance for prediction and optimization, which can help policymakers for developing accurate energy management systems (Table 6).

Table 6. The comparison results of DL and ML-based methods for biofuels handling

Method	Application	Accuracy	Reliability	Sustainability	References
SWARA-MOORA	Decision making	+++	+++	+++	[67]
ANP-MOORA	Decision making	+++	+++	+++	[67]
ANFIS-CM	Classification	+++	+++	+++	[68]
GP	Classification	+++	++	++	[68]
Hybrid ML	Estimation	+++	+++	+++	[22]
Single ML	Estimation	+++	++	++	[22]
Hybrid ML	Optimization	+++	+++	+++	[22]
Single ML	Optimization	+++	+++	++	[22]

3 Conclusion

This paper studies the applications and progress of ML and DL methods biofuels research. This study presents an in-depth survey and analysis of the 'hybrid model' and ensemble models that integrate two or more techniques. Survey shows that the single ML methods except for ANNs, have not been popular. However, the ensemble and hybrid models have emerged and continue to advance for higher accuracy and better performance. DL techniques also will bring a tremendous amount of intelligence for better prediction models. In general, modeling, forecasting, and decision making about the future of biofuels help for developing sustainable energy resources, which are low-cost resources with low environmental impacts. ML and DL techniques have been successfully employed in all fields of sciences and have improved the process. The various combinations of the hybrid and ensemble methods are found to be the most effective in handling biofuels.

Acknowledgments. This publication has been supported by the Project: "Support of research and development activities of the J. Selye University in the field of Digital Slovakia and creative industry" of the Research & Innovation Operational Programme (ITMS code: NFP313010T504) co-funded by the European Regional Development Fund.

References

1. Landis, D.A., et al.: Biomass and biofuel crop effects on biodiversity and ecosystem services in the North Central US. Biomass Bioenergy **114**, 18–29 (2018)
2. Shaosen, S., et al.: Experimental and artificial intelligence for determination of stable criteria in cyclic voltammetric process of medicinal herbs for biofuel cells. Int. J. Energy Res. (2019)
3. Wong, K.I., Wong, P.K.: Optimal calibration of variable biofuel blend dual-injection engines using sparse Bayesian extreme learning machine and metaheuristic optimization. Energy Convers. Manag. **148**, 1170–1178 (2017)
4. Wong, K.I., Wong, P.K.: *Adaptive air-fuel ratio control of dual-injection engines under biofuel blends using extreme learning machine*. Energy Convers. Manag. **165**, 66–75 (2018)
5. Zhang, F., et al.: Integrating GIS with optimization method for a biofuel feedstock supply chain. Biomass Bioenergy **98**, 194–205 (2017)

6. Afsharzade, N., et al.: Renewable energy development in rural areas of Iran. Renew. Sustain. Energy Rev. **65**, 743–755 (2016)
7. Fardad, K.: Producing Biogas from Medicinal Plants, in Biosystem Engineering. University of Mohaghegh Ardabili, Ardabil, Iran (2017)
8. Jebli, M.B., Youssef, S.B.: The role of renewable energy and agriculture in reducing CO_2 emissions: evidence for North Africa countries (2015)
9. Concu, R., et al.: PTML model of enzyme subclasses for mining the proteome of biofuel producing microorganisms. J. Proteome Res. (2019)
10. De Bortoli, A.L., Pereira, F.N.: Obtaining a reduced kinetic mechanism for Methyl Butanoate. J. Math. Chem. **57**(3), 812–833 (2019)
11. del Rio-Chanona, E.A., et al.: Deep learning-based surrogate modeling and optimization for microalgal biofuel production and photobioreactor design. AIChE J. **65**(3), 915–923 (2019)
12. Demirbas, A.: Global biofuel strategies. Energy Educ. Sci. Technol. **17**(1/2), 27 (2006)
13. Xue, J., Grift, T.E., Hansen, A.C.: Effect of biodiesel on engine performances and emissions. Renew. Sustain. Energy Rev. **15**(2), 1098–1116 (2011)
14. Seyyed aram, A., Najafi, B.: The effect of biodiesel of butanol alcohol and waste oil on performance and emission of diesel engine, in biosystem engineering. Thesis of M.Sc., University of Mohaghegh Ardabili (2016)
15. Nicoletti, J., Ning, C., You, F.: Incorporating agricultural waste-to-energy pathways into biomass product and process network through data-driven nonlinear adaptive robust optimization. Energy **180**, 556–571 (2019)
16. Opgenorth, P., et al.: Lessons from two Design-Build-Test-Learn cycles of dodecanol production in Escherichia coli aided by machine learning. ACS Synth. Biol. **8**(6), 1337–1351 (2019)
17. Rezk, H., et al.: Improving the environmental impact of palm kernel shell through maximizing its production of hydrogen and syngas using advanced artificial intelligence. Sci. Total Environ. **658**, 1150–1160 (2019)
18. Salmeron, J.L., Ruiz-Celma, A.: Elliot and symmetric elliot extreme learning machines for Gaussian noisy industrial thermal modelling. Energies **12**(1), 90 (2019)
19. Ghaderi, M., et al.: An analysis of noise pollution emitted by moving MF285 Tractor using different mixtures of biodiesel, bioethanol and diesel through artificial intelligence. J. Low Freq. Noise Vib. Act. Control **38**(2), 270–281 (2019)
20. Habyarimana, E., et al.: Towards predictive modeling of sorghum biomass yields using fraction of absorbed photosynthetically active radiation derived from sentinel-2 satellite imagery and supervised machine learning techniques. Agronomy **9**(4), 203 (2019)
21. Kumar, S., et al.: Thermozymes: adaptive strategies and tools for their biotechnological applications. Bioresour. Technol. **278**, 372–382 (2019)
22. Mosavi, A., et al.: State of the art of machine learning models in energy systems, a systematic review. Energies **12**(7), 1301 (2019)
23. Muthukumaran, C., et al.: Process optimization and kinetic modeling of biodiesel production using non-edible Madhuca indica oil. Fuel **195**, 217–225 (2017)
24. Anderson, R., et al.: An integrated modeling framework for crop and biofuel systems using the DSSAT and GREET models. Environ. Model Softw. **108**, 40–50 (2018)
25. Kessler, T., et al.: Artificial neural network based predictions of cetane number for furanic biofuel additives. Fuel **206**, 171–179 (2017)
26. Choubin, B., et al.: An ensemble prediction of flood susceptibility using multivariate discriminant analysis, classification and regression trees, and support vector machines. Sci. Total Environ. **651**, 2087–2096 (2019)
27. Dehghani, M., et al.: Prediction of hydropower generation using Grey wolf optimization adaptive neuro-fuzzy inference system. Energies **12**(2), 289 (2019)

28. Dineva, A., et al.: Review of soft computing models in design and control of rotating electrical machines. Energies **12**(6), 1049 (2019)
29. Dineva, A., et al.: Multi-Label Classification for Fault Diagnosis of Rotating Electrical Machines (2019)
30. Farzaneh-Gord, M., et al.: Numerical simulation of pressure pulsation effects of a snubber in a CNG station for increasing measurement accuracy. Eng. Appl. Comput. Fluid Mech. **13**(1), 642–663 (2019)
31. Ghalandari, M., et al.: Investigation of submerged structures' flexibility on sloshing frequency using a boundary element method and finite element analysis. Eng. Appl. Comput. Fluid Mech. **13**(1), 519–528 (2019)
32. Ghalandari, M., et al.: Flutter speed estimation using presented differential quadrature method formulation. Eng. Appl. Comput. Fluid Mech. **13**(1), 804–810 (2019)
33. Karballaeezadeh, N., et al.: Prediction of remaining service life of pavement using an optimized support vector machine (case study of Semnan-Firuzkuh road). Eng. Appl. Comput. Fluid Mech. **13**(1), 188–198 (2019)
34. Menad, N.A., et al.: Modeling temperature dependency of oil - water relative permeability in thermal enhanced oil recovery processes using group method of data handling and gene expression programming. Eng. Appl. Comput. Fluid Mech. **13**(1), 724–743 (2019)
35. Mohammadzadeh, S., et al.: Prediction of compression index of fine-grained soils using a gene expression programming model. Infrastructures **4**(2), 26 (2019)
36. Mosavi, A., Edalatifar, M.: A hybrid neuro-fuzzy algorithm for prediction of reference evapotranspiration. Lecture Notes in Networks and Systems, pp. 235–243. Springer (2019)
37. Mosavi, A., Lopez, A., Varkonyi-Koczy, A.R.: Industrial applications of big data: state of the art survey. In: Luca, D., Sirghi, L., Costin, C. (eds.) pp. 225–232. Springer (2018)
38. Mosavi, A., Ozturk, P., Chau, K.W.: Flood prediction using machine learning models: literature review. Water (Switzerland) **10**(11), 1536 (2018)
39. Mosavi, A., Rabczuk, T.: Learning and intelligent optimization for material design innovation. In: Kvasov, D.E., et al. (eds.) pp. 358–363. Springer (2017)
40. Mosavi, A., Rabczuk, T., Varkonyi-Koczy, A.R.: Reviewing the novel machine learning tools for materials design. In: Luca, D., Sirghi, L., Costin, C. (eds.) pp. 50–58. Springer (2018)
41. Aram, F., et al.: Design and validation of a computational program for analysing mental maps: aram mental map analyzer. Sustainability (Switzerland) **11**(14), 3790 (2019)
42. Asadi, E., et al.: Groundwater Quality Assessment for Drinking and Agricultural Purposes in Tabriz Aquifer, Iran (2019)
43. Asghar, M.Z., Subhan, F., Imran, M., Kundi, F.M., Shamshirband, S., Mosavi, A., Csiba, P., Várkonyi-Kóczy, A.R.: Performance Evaluation of Supervised Machine Learning Techniques for Efficient Detection of Emotions from Online Content. Pre-prints 2019, 2019080019. (https://doi.org/10.20944/preprints201908.0019.v1)
44. Bemani, A., Baghban, A., Shamshirband, S., Mosavi, A., Csiba, P., Várkonyi-Kóczy, A.R.: Applying ANN, ANFIS, and LSSVM Models for Estimation of Acid Solvent Solubility in Supercritical CO_2. Preprints (2019), 2019060055. (https://doi.org/10.20944/preprints201906.0055.v2)
45. Choubin, B., et al.: Snow avalanche hazard prediction using machine learning methods. J. Hydrol. **577**, 123929 (2019)
46. Mosavi, A., et al.: Prediction of multi-inputs bubble column reactor using a novel hybrid model of computational fluid dynamics and machine learning. Eng. Appl. Comput. Fluid Mech. **13**(1), 482–492 (2019)
47. Mosavi, A., Varkonyi-Koczy, A.R.: Integration of machine learning and optimization for robot learning. In: Jablonski, R., Szewczyk, R. (eds.) pp. 349–355. Springer (2017)

48. Nosratabadi, S., et al.: Sustainable business models: a review. Sustainability (Switzerland) **11**(6), 1663 (2019)
49. Qasem, S.N., et al.: Estimating daily dew point temperature using machine learning algorithms. Water (Switzerland) **11**(3), 582 (2019)
50. Rezakazemi, M., Mosavi, A., Shirazian, S.: ANFIS pattern for molecular membranes separation optimization. J. Mol. Liq. **274**, 470–476 (2019)
51. Riahi-Madvar, H., et al.: Comparative analysis of soft computing techniques RBF, MLP, and ANFIS with MLR and MNLR for predicting grade-control scour hole geometry. Eng. Appl. Comput. Fluid Mech. **13**(1), 529–550 (2019)
52. Shabani, S., Samadianfard, S., Taghi Sattari, M., Shamshirband, S., Mosavi, A., Kmet, T., Várkonyi-Kóczy, A.R.: Modeling Daily Pan Evaporation in Humid Climates Using Gaussian Process Regression. Preprints (2019), 2019070351. https://doi.org/10.20944/preprints201907.0351.v1
53. Shamshirband, S., Hadipoor, M., Baghban, A., Mosavi, A., Bukor J., Varkonyi-Koczy, A. R.: Developing an ANFIS-PSO Model to predict mercury emissions in Combustion Flue Gases. Preprints (2019), 2019070165. https://doi.org/10.20944/preprints201907.0165.v1
54. Shamshirband, S., et al.: Ensemble models with uncertainty analysis for multi-day ahead forecasting of chlorophyll a concentration in coastal waters. Eng. Appl. Comput. Fluid Mech. **13**(1), 91–101 (2019)
55. Shamshirband, S., Mosavi, A., Rabczuk, T.: Particle swarm optimization model to predict scour depth around bridge pier. arXiv:1906.08863 (2019)
56. Taherei Ghazvinei, P., et al.: Sugarcane growth prediction based on meteorological parameters using extreme learning machine and artificial neural network. Eng. Appl. Comput. Fluid Mech. **12**(1), 738–749 (2018)
57. Torabi, M., et al.: A Hybrid clustering and classification technique for forecasting short-term energy consumption. Environ. Prog. Sustain. Energy **38**(1), 66–76 (2019)
58. Torabi, M., et al.: A hybrid machine learning approach for daily prediction of solar radiation. Lecture Notes in Networks and Systems, pp. 266–274. Springer (2019)
59. Reynel-Ávila, H.E., Bonilla-Petriciolet, A., Tapia-Picazo, J.C.: An artificial neural network-based NRTL model for simulating liquid-liquid equilibria of systems present in biofuels production. Fluid Phase Equilib. **483**, 153–164 (2019)
60. Camberos, S.A., et al.: Neuronal modeling of a two stages anaerobic digestion process for biofuels production. IFAC-PapersOnLine **51**(13), 408–413 (2018)
61. Sewsynker-Sukai, Y., Faloye, F., Kana, E.B.G.: Artificial neural networks: an efficient tool for modelling and optimization of biofuel production (a mini review). Biotechnol. Biotechnol. Equip. **31**(2), 221–235 (2017)
62. Mancini, M., Taavitsainen, V.M., Toscano, G.: Comparison of three different classification methods performance for the determination of biofuel quality by means of NIR spectroscopy. J. Chemom. (2019)
63. Feng, X., et al.: Rapid and non-destructive measurement of biofuel pellet quality indices based on two-dimensional near infrared spectroscopic imaging. Fuel **228**, 197–205 (2018)
64. Faizollahzadeh Ardabili, S., Najafi, B., Shamshirband, S.: Fuzzy logic method for the prediction of cetane number using carbon number, double bounds, iodic, and saponification values of biodiesel fuels. Environ. Prog. Sustain. Energy (2019)
65. Wong, K.I., et al.: Sparse Bayesian extreme learning machine and its application to biofuel engine performance prediction. Neurocomputing **149**(Part A), 397–404 (2015)
66. Faizollahzadeh Ardabili, S., et al.: Using SVM-RSM and ELM-RSM approaches for optimizing the production process of methyl and ethyl esters. Energies **11**(11), 2889 (2018)

67. Erdoğan, S., et al.: The best fuel selection with hybrid multiple-criteria decision making approaches in a CI engine fueled with their blends and pure biodiesels produced from different sources. Renew. Energy, 653–668 (2019)
68. Deo, R.C., et al.: Adaptive Neuro-Fuzzy Inference System integrated with solar zenith angle for forecasting sub-tropical Photosynthetically Active Radiation. Food Energy Secur. **8**(1), e00151 (2019)
69. Ardabili, S., Mosavi, A., Mahmoudi, A., Gundoshmian, T.M., Nosratabadi, S., Var-konyi-Koczy, A.: Modelling temperature variation of mushroom growing hall using artificial neural networks. Preprints (2019)
70. Gundoshmian, T.M., Ardabili, S., Mosavi, A., Varkonyi-Koczy, A.: Prediction of combine harvester performance using hybrid machine learning modeling and response surface methodology. Preprints (2019)
71. Ardabili, S., Mosavi, A., Varkonyi-Koczy, A.: Systematic review of deep learning and machine learning models in biofuels research. Preprints (2019)
72. Ardabili, S., Mosavi, A., Varkonyi-Koczy, A.: Advances in machine learning modeling reviewing hybrid and ensemble methods. Preprints (2019)
73. Ardabili, S., Mosavi, A., Varkonyi-Koczy, A.: Building Energy information: demand and consumption prediction with Machine Learning models for sustainable and smart cities. Preprints (2019)
74. Ardabili, S., Mosavi, A., Dehghani, M., Varkonyi-Koczy, A.: Deep learning and machine learning in hydrological processes climate change and earth systems a systematic review. Preprints (2019)
75. Mohammadzadeh, D., Karballaeezadeh, N., Mohemmi, M., Mosavi, A., Varkonyi-Koczy A.: Urban Train Soil-Structure Interaction Modeling And Analysis. Preprints (2019)
76. Mosavi, A., Ardabili, S., Varkonyi-Koczy, A.: List of deep learning models. Preprints (2019)
77. Nosratabadi, S., Mosavi, A., Keivani, R., Ardabili, S., Aram, F.: State of the art survey of deep learning and machine learning models for smart cities and urban sustainability. Preprints (2019)

Modelling Temperature Variation of Mushroom Growing Hall Using Artificial Neural Networks

Sina Ardabili[1], Amir Mosavi[2,3](✉), Asghar Mahmoudi[4], Tarahom Mesri Gundoshmian[5], Saeed Nosratabadi[6], and Annamária R. Várkonyi-Kóczy[2,7]

[1] Institute of Advanced Studies Koszeg, Koszeg, Hungary
[2] Institute of Automation, Kalman Kando Faculty of Electrical Engineering, Obuda University, Budapest, Hungary
amir.mosavi@kvk.uni-obuda.hu
[3] School of the Built Environment, Oxford Brookes University, Oxford OX3 0BP, UK
[4] Department of Biosystem Engineering, University of Tabriz, Tabriz, Iran
[5] Department of Biosystem Engineering, University of Mohaghegh Ardabili, Ardabil, Iran
[6] Institute of Business Studies, Szent Istvan University, Godollo 2100, Hungary
[7] Department of Mathematics and Informatics, J. Selye University, Komarno, Slovakia

Abstract. The recent developments of computer and electronic systems have made the use of intelligent systems for the automation of agricultural industries. In this study, the temperature variation of the mushroom growing room was modeled by multi-layered perceptron and radial basis function networks based on independent parameters including ambient temperature, water temperature, fresh air and circulation air dampers, and water tap. According to the obtained results from the networks, the best network for MLP was in the second repetition with 12 neurons in the hidden layer and in 20 neurons in the hidden layer for radial basis function network. The obtained results from comparative parameters for two networks showed the highest correlation coefficient (0.966), the lowest root mean square error (RMSE) (0.787) and the lowest mean absolute error (MAE) (0.02746) for radial basis function. Therefore, the neural network with radial basis function was selected as a predictor of the behavior of the system for the temperature of mushroom growing halls controlling system.

Keywords: Agricultural production · Environmental parameters · Mushroom growth prediction · Machine learning · Artificial neural networks (ANN) · Food production · Food security

A. R. Várkonyi-Kóczy (Ed.): INTER-ACADEMIA 2019, LNNS 101, pp. 33–45, 2020.
https://doi.org/10.1007/978-3-030-36841-8_3

1 Introduction

Nowadays, Due to issues such as population growth and limited agricultural resources including land and freshwater, the necessity of attention to new methods and efficiency in agricultural production, is quite evident [1]. A number of clinical studies in Japan and the United States of America have shown that a certain percentage of polysaccharides against breast cancer, lung, liver, prostate and brain tumors is effective [2, 3]. The benefits of this product is promising to use this product in the diet. The growth period of this product consists of several stages and each of these stages requires different controlling condition [1]. The use of intelligent systems for automation in agriculture industries has been due to the development of computer systems and electronics in recent decades. With these systems, we can control the environmental parameters involved in mushroom production halls. Temperature is one of the parameters that shows a high impact on mushroom growth, and chemical reactions are intensive at higher temperatures. In biological processes such as growth, the effect of temperature can be easily observed where vast quantities of chemical reactions occur. The optimum temperature for mushrooms, depending on the stage and type of race, is 17 to 30 °C. The metabolism of consumed food by microorganisms in the compost contributes to their growth and activity, and as a result, it produces the heat. For example, rising compost temperature decreases crop production. Figure 1 shows the lack of mushroom production in the middle of the bed when the compost temperature is higher than the standard value [1].

Fig. 1. The effect of temperature value on mushroom production

The successful cultivation of mushroom is possible when parameters such as temperature, humidity, and carbon dioxide concentration, pests and diseases and also preparing compost have been controlled and inspected properly. Environmental factors have the most influence on the quality of the product on the growth stage [4]. Problems in the field of parameters controlling on mushroom cultivation halls forced us to do studies on controlling these parameters. Manually and traditional controlling methods are under the influence of factors such as human, measurement and environment errors [1]. To resolve this problem, several studies were carried out with different control methods. Ardabili et al. [4] presented controlling system using fuzzy and digital controllers to control the environmental parameters of the mushroom production hall. Previous studies with presented methods have the complexity of calculation in control strategies. Today, predictive control is used in industrial applications to develop control strategies [5]. Among the systems that have the capability to model and predict the behavior of systems, can point to artificial neural networks. A neural network consists of a number of processing elements or computing nodes that are very simple and interconnected. This network is an algorithm information processing that is processing by dynamic response related to processing elements and their connections to lateral inputs [6]. The most common neural networks are Multi-Layered Perceptron (MLP) and Radial Basis Function (RBF) networks.

The main aim of this study is to present a predictive model of the temperature variation of the mushroom growing room by artificial neural networks based on the variables that are affecting on room temperature (ambient temperature, water temperature, fresh air dampers, circulation air dampers, and water tap). To reach this purpose, the study consists of three phases. The first stage is analyzing the required data. The second stage presents MLP and RBF models and the last stage presents the results and a comparison of networks and introduces the best model.

2 Materials and Methods

2.1 Data Collecting

This research was studied in one of the mushroom production halls of Sabalan agro-industry company (Sabalan Mushroom) in Ardabil province of Iran. The target hall has dimensions of 22, 5/6, 5/4 m (length, width, height, respectively) and has 1850 compost with a weight of approximately 15 kg for each compost. In order to data collecting operation, three PT-100 sensors were used. The location of temperature sensors in terms of height and width was in the middle of the hall height and width and in term of length were located on three points including the beginning of hall, the middle of the hall and the end of the hall. This method of the arrangement of sensors was used because the longitudinal air circulators in two top corners of the hall make airflow in transverse and height directions of Hall and provide thermal equilibrium on transverse and height directions of the hall, and if there is a temperature difference, this difference will be in the longitudinal direction of hall.

Compost generates heat. There is a need to maintain and stabilize the temperature of compost at every stage of the growth cycle and since the volume of compost is lower than the indoor air volume of the growing hall, the operation of stabilizing of compost temperature should be done by changing hall temperature.

Data collecting operations were done in the winter season and due to variations of ambient temperature, an external temperature logger was used to record temperature changes. Data collecting operations were performed at different reps. In order to record the required data, Autonics temperature controller TKM-B4RC was used that was equipped with RS485 output and related DAQ-master software. Ambient temperature, variations of circulation air dampers, variations of fresh air dampers, hot water tap and hot water temperature were as independent variables and indoor temperature of hall was as the only dependent variable. To adjust the room temperature and variation of input variables, it was using air conditioning systems (Fig. 2).

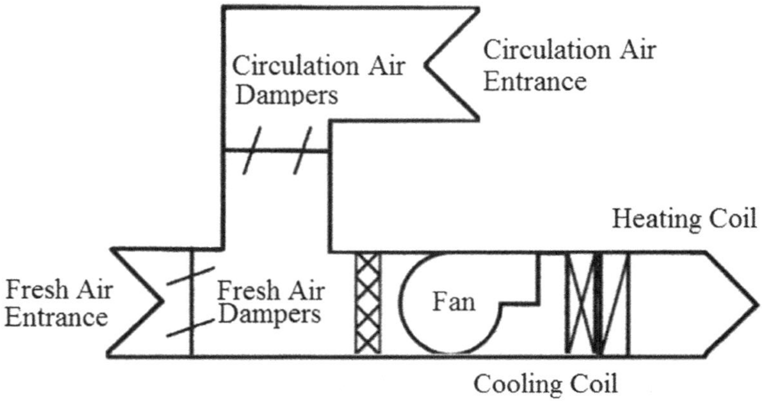

Fig. 2. Air conditioning system

This system is capable of cooling and heating, generating indoor air circulation by air ducts, providing required relative humidity and reducing the carbon dioxide concentration using fresh air dampers. Operation of temperature control in this system is performing by variating hot and cold water Debbie as well as opening and closing the air dampers. This system has two air inputs (Circulation and fresh air dampers) and an air output that the circulation air damper is for circulating the hall atmosphere and the fresh air damper is controlling temperature, humidity or carbon dioxide concentration by entering fresh air to hall [1]. Measurement of independent variables such as the air dampers and hot water tap were performed at three levels including minimum, medium and maximum value of actuators openness. According to the coolness of the air in the operation season, the outdoor air was used for cooling operations instead of cold water. Because the cold water would freeze and damage the coils in this season. Data were collected in 3 treatments and different repetitions to achieve high accuracy. Table 1 shows the different treatments for independent variables.

2.2 Artificial Neural Networks (ANNs)

Literature includes a vast number of machine learning methods used for the purpose of the modeling and prediction [7–26]. Machine learning models generally out-perform most of the statistical and mathematical models in term of computation cost, efficiency and accuracy [27–40]. ANNs are considered as an efficient methods for developing reliable models. This study employs two types of neural networks including multilinear

Table 1. Treatments of independent variables for data collecting

Parameter	Treatment		
	Maximum	Medium	Minimum
Ambient temperature (°C)	−10	0	+10
Water temperature (°C)	30	40	50
Fresh air damper (Openness)	1/3	2/3	3/3
Circulation air damper (Openness)	1/3	2/3	3/3
Water tap (Openness)	1/3	2/3	3/3

perceptron (MLP) and radial basis function (RBF). Before starting the training process, data have been divided into two categories of training data (with a share of 70%) and testing data (with a share of 30%), randomly. The training process was started with the different number of neurons in the hidden layer and the function of each parameter was measured with respect to the base parameter. To determine the optimal number of neurons in the hidden layer and to obtain the best predictor network, in first stage the network was trained with one neuron on a hidden layer.

Figure 3 presents the structure of the RBF network.

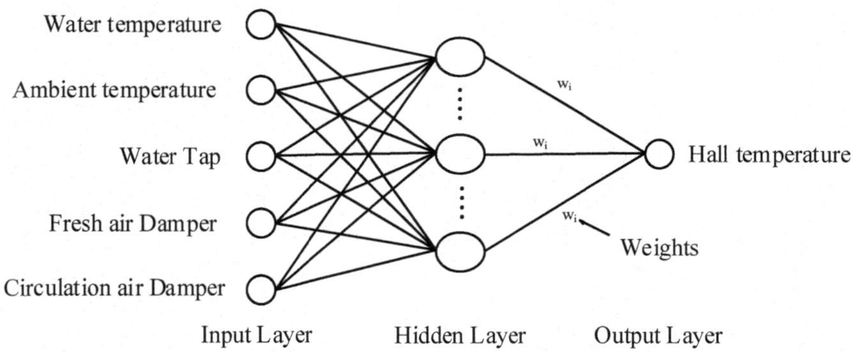

Fig. 3. The structure of RBF network

Evaluating the results have been conducted by employing Root Mean Square Error (RMSE), correlation coefficient (R) and mean absolute error (MAE) were used [41] to analyze the output of networks and target values.

$$MSE = \frac{1}{N}\sum_{i=1}^{N}(A - P)^2 \tag{1}$$

$$RMSE = \sqrt{\frac{1}{N}\sum_{i=1}^{N}(A - P)^2} \tag{2}$$

$$R = \left(1 - \left(\frac{\sum_{i=1}^{n} (A - P)^2}{\sum_{i=1}^{n} A_i^2} \right) \right)^{1/2} \tag{3}$$

$$MAE = \frac{\sum_{i=1}^{N} |A - P|}{N} \tag{4}$$

That A is the target value, P is the predicted values and N is the numbers of data.

3 Results

In this study, the temperature variation of the mushroom growing hall, as an critical factor of mushroom production, was modeled based on dependent variables including ambient temperature, fresh air damper, circulation air damper, water tap and water temperature using MLP and RBF networks. To perform modeling operations, there is a

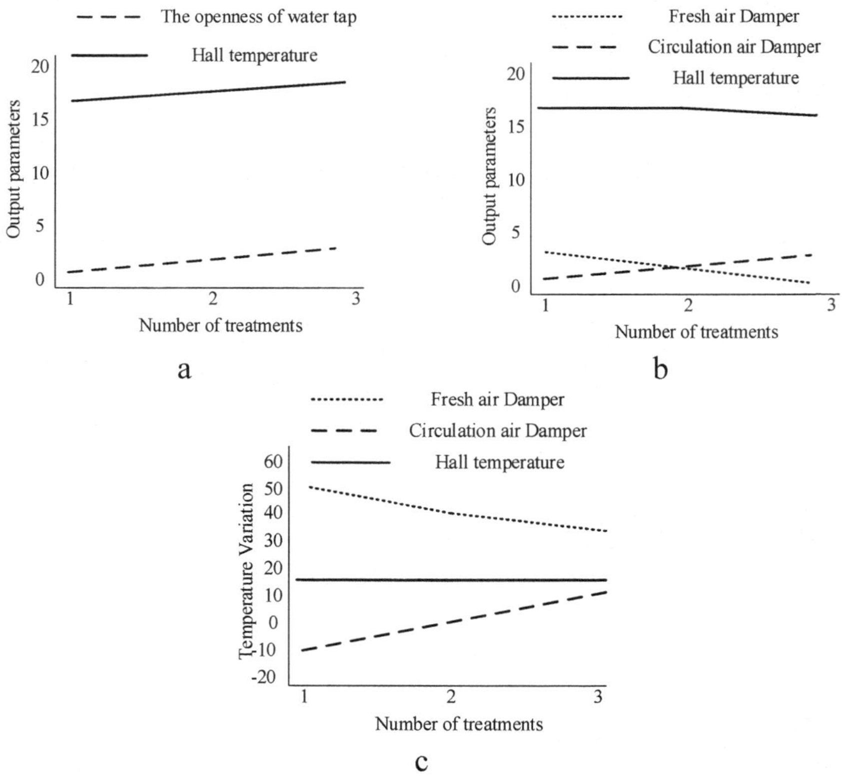

Fig. 4. The results of experimental data and the relation of dependent and independent variables

need to be aware of the general nature of the system that this is carried out by experimental data related to the system. For this purpose, the experimental data were obtained from the studied system using the introduced strategy in Table 1. Figure 4 indicates the results of data collecting and the relationship among the actuators and the related parameter.

According to Fig. 4(a) by considering the fixed value of other parameters, opening hot water tap, increases the growing hall temperature. Figure 4(b) shows the variation of growing hall temperature by opening and closing air dampers when other parameters are fixed. Accordingly, if the rate of opening and closing of circulation and fresh air dampers be equal, respectively, the hall temperature almost will be fixed. In Fig. 4(c) by reducing the temperature of the water and by increasing the ambient temperature during the day, the growing hall temperature has undergone a constant trend.

3.1 Training Process

This section presents the results of choosing the best network for the training process based on the performance functions for RBF and MLP separately. Table 2 is related to MLP network, and Table 3 is related to RBF network.

Table 2. The result of selecting the best network for MLP model

Numbers of the neuron on hidden layer/repetition	Value of performance function for validation data	Value of performance function for training data	Value of performance function for testing data
12/1	0.55064	0.42202	0.84431
12/2	0.53321	0.41001	*0.82541*
12/3	0.52248	0.41056	0.84522
Minimum value	*0.52248*	*0.41001*	*0.82541*
Maximum value	0.55064	0.42202	0.84522
Average	0.53544	0.41419	0.83831

According to Table 2, the best result (lowest values of the performance function for testing data) is obtained in the second repetition of the training process. So the second network was selected as the best prediction network.

Table 3 was prepared to choose the best number of neurons in the hidden layer for RBF network.

Table 3. The result of selecting the best number of neurons on hidden layer for RBF network

Number of neurons on hidden layer	RMSE	r	MAE
4	0.2897	0.69	0.1016
8	0.1925	0.79	0.0675
12	0.1589	0.85	0.05575
16	0.1205	0.91	0.0422
20	*0.0787*	*0.996*	*0.02746*
24	0.0787	0.996	0.02745

As shown in Table 3, by increasing the number of neurons in the hidden layer, the correlation coefficient is increased and mean absolute error and root mean square error values are reduced.

After 20 neurons, these parameters remained stable and have not changed, so the number of 20 neurons in the hidden layer were selected as the optimal number of neurons. The networks were trained after selecting the optimal number of neurons to neural network Multilayer Perceptron and Radial Basis Function networks were trained. After the training process, test data were imported to developed networks and output data were generated to compare with the target values. The results of the target and the predicted values are shown in Figs. 5 and 6.

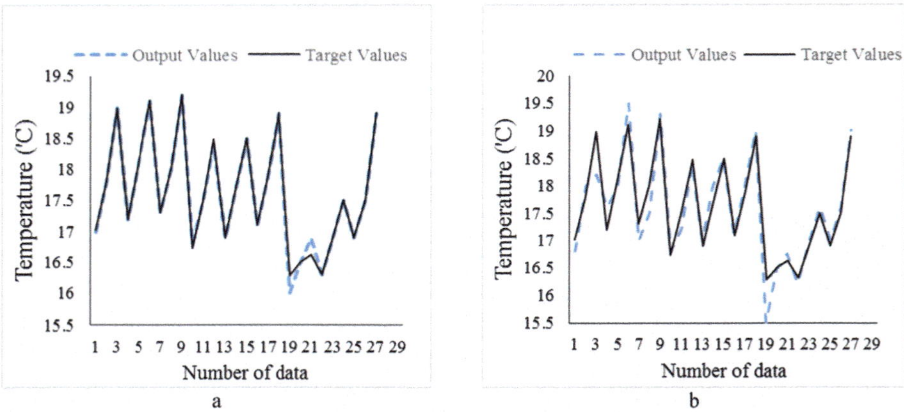

Fig. 5. Results of predicted and target values compared to target values. a) RBF network b) MLP network

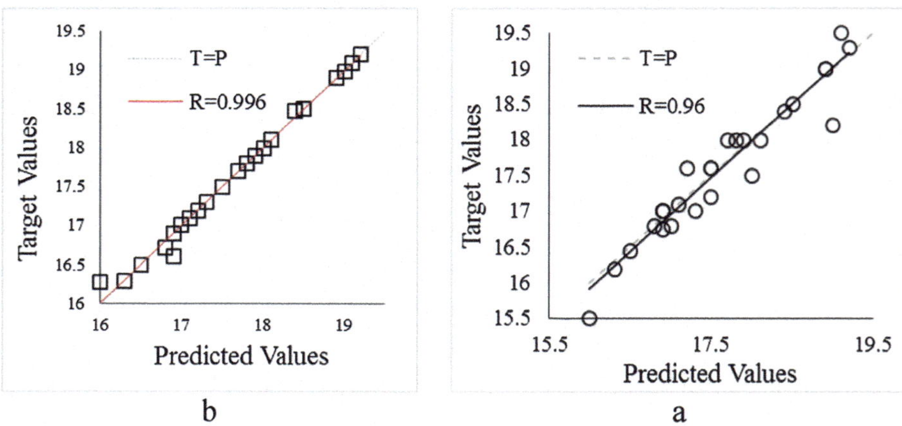

Fig. 6. Scatter plot of predicted and target values a) RBF network b) MLP network

According to Fig. 5.a, the predicted values of RBF network are following target values well and have less deviation and error from the target value, but predicted values of MLP network (Fig. 5.b) have a large error and deviation from target values and has lower compliance with target values compared to RBF network.

Based on the results of Fig. 6 and according to the description mentioned about correlation coefficient, it can be said that the output values of RBF network have 99.6% of linearship and the output values of MLP network have 96% linearship with target values. To display these results as a statistical factor, the output of models were compared with target values using the comparison parameters that were mentioned in Materials and methods. The obtained results were tabulated in Table 4.

Table 4. The results of comparison parameter for two types of networks

Network type	MAE	RMSE	r
MLP	0.137	0.9085	0.9612
RBF	0.02746	0.787	0.996

According Table 4, the results of comparison parameters indicate that the results of RBF network have high correlation coefficient (0.996) and low RMSE and MAE values (0.787 and 0.02746, respectively) compared to MLP network. Due to the high correlation coefficient for RBF network (0.996), It can be said that process modeling, compliance and linear correlation predicted by the RBF network is higher than MLP network that is confirming the obtained results from Figs. 5 and 6.

On the other hand, RBF network with the lowest root mean square error (0.787) and the lowest mean absolute error (0.02746) generated closest predicted values with minimal errors compared to the MLP neural network and it can be said that RBF has high ability to model the temperature variations compared to MLP network in this study. Therefore the designed model based on RBF is predicting the temperature value more accurate with low deviation to target values compared to the MLP network.

Ardabili et al. [42] developed a fuzzy modeling system in order to predict the temperature of the mushroom growing hall that the correlation coefficient and mean absolute error between the predicted and target values were calculated 0.67 and 0.232, respectively. The present study indicates the improvements in the prediction of temperature variations using artificial neural networks. One of the reasons that led to this happening, is that the fuzzy systems unlike the artificial neural networks, is operating by the defined laws. These rules can be affected by the accuracy of laws defining and can have a negative effect on system precision.

Figure 7 presents the error values for the predicted values of each network from the desired values. The zero value of deviation is related to the Target value. The blue line indicates the deviations related to RBF network and the red line indicates the deviations related to MLP network. Based on Fig. 7, MLP network has the maximum deviation from the target value.

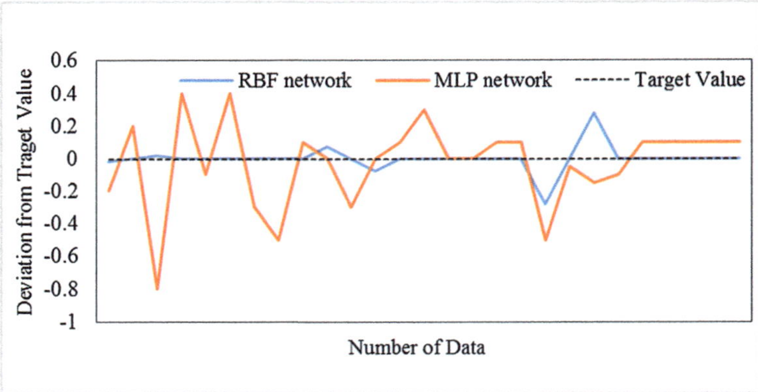

Fig. 7. Deviation Of predicted values of networks from the target value

According to Fig. 7, if the output of networks is compared in the same input values, it can be said that the deviation of temperature from target values in RBF network is higher than MLP network. This means that the energy losses of MLP network are higher than RBF network. This energy losses on MLP network can be reduced by the changes that can be applied in network inputs. This losses of energy is equal to increasing the failure risk of the system on MLP model compared to RBF model.

4 Conclusion

This study is performed in a mushroom growing hall with the aim of modeling of temperature variations. Accordingly, modeling systems including MLP and RBF networks was used. The results of the data collecting process reflected the dependence of temperature value to independent variables. Therefore results were prepared after the modelling process and extracting the output values of networks and comparing them with target values. This results showed that the RBF network has high accuracy and better performance compared to MLP network and also using RBS network will reduce energy consumption, system failure, and costs. Thus, the neural network with radial basis function was chosen as a predictive network of hall temperature in this study. For the future works, more sophisticated machine learning methods must come to consideration, e.g., [42–51].

Acknowledgments. This publication has been supported by the Project: "Support of research and development activities of the J. Selye University in the field of Digital Slovakia and creative industry" of the Research & Innovation Operational Programme (ITMS code: NFP313010T504) co-funded by the European Regional Development Fund.

References

1. Ardabili, S.F.: Simulation and comparison of control system in mushroom growing rooms environment. Thesis of Master science. Department of mechanic of agricultural machinery engineering. University of Tabriz, Tabriz, Iran (2014)
2. Kurashige, S., Akuzawa, Y., Endo, F.: Effects of Lentinus edodes, Grifola frondosa and Pleurotus ostreatus administration on cancer outbreak, and activities of macrophages and lymphocytes in mice treated with a carcinogen, N-butyl-N-butanolnitrosoamine. Immunopharmacol. Immunotoxicol. **19**(2), 175–183 (1997)
3. Zhang, H., et al.: Flammulin Purified from the Fruit Bodies of Flammulina velutipes (Curt.: Fr.) P. Karst. Int. J. Med. Mushrooms **1**(1) (1999)
4. Ardabili, S.F., et al.: Modeling and comparison of fuzzy and on/off controller in a mushroom growing hall. Measurement **90**, 127–134 (2016)
5. Fang, Y., Fei, J., Ma, K.: Model reference adaptive sliding mode control using RBF neural network for active power filter. Int. J. Electr. Power Energy Syst. **73**, 249–258 (2015)
6. Ardabili, S.F., Mahmoudi, A., Mesri Gundoshmian, T.: Modeling and simulation controlling system of HVAC using fuzzy and predictive (radial basis function, RBF) controllers. J. Build. Eng. **6**, 301–308 (2016)
7. Aram, F., et al.: Design and validation of a computational program for analysing mental maps: Aram mental map analyzer. Sustainability (Switzerland) **11**(14), 3790 (2019)
8. Asadi, E., et al.: Groundwater Quality Assessment for Drinking and Agricultural Purposes in Tabriz Aquifer, Iran (2019)
9. Asghar, M.Z., Subhan, F., Imran, M., Kundi, F.M., Shamshirband, S., Mosavi, A., Csiba, P., Várkonyi-Kóczy, A.R.: Performance Evaluation of Supervised Machine Learning Techniques for Efficient Detection of Emotions from Online Content. Preprints 2019, 2019080019. https://doi.org/10.20944/preprints201908.0019.v1
10. Bemani, A., Baghban, A., Shamshirband, S., Mosavi, A., Csiba, P., Várkonyi-Kóczy, A.R.: Applying ANN, ANFIS, and LSSVM Models for Estimation of Acid Solvent Solubility in Supercritical CO_2. Preprints 2019, 2019060055. https://doi.org/10.20944/preprints201906.0055.v2
11. Choubin, B., et al.: Snow avalanche hazard prediction using machine learning methods. J. Hydrol. **577**, 123929 (2019)
12. Choubin, B., et al.: An ensemble prediction of flood susceptibility using multivariate discriminant analysis, classification and regression trees, and support vector machines. Sci. Total Environ. **651**, 2087–2096 (2019)
13. Dehghani, M., et al.: Prediction of hydropower generation using Grey wolf optimization adaptive neuro-fuzzy inference system. Energies **12**(2), 289 (2019)
14. Dineva, A., et al.: Review of soft computing models in design and control of rotating electrical machines. Energies **12**(6), 1049 (2019)
15. Dineva, A., et al.: Multi-Label Classification for Fault Diagnosis of Rotating Electrical Machines (2019)
16. Farzaneh-Gord, M., et al.: Numerical simulation of pressure pulsation effects of a snubber in a CNG station for increasing measurement accuracy. Eng. Appl. Comput. Fluid Mech. **13**(1), 642–663 (2019)
17. Ghalandari, M., et al.: Investigation of submerged structures' flexibility on sloshing frequency using a boundary element method and finite element analysis. Eng. Appl. Comput. Fluid Mech. **13**(1), 519–528 (2019)
18. Ghalandari, M., et al.: Flutter speed estimation using presented differential quadrature method formulation. Eng. Appl. Comput. Fluid Mech. **13**(1), 804–810 (2019)

19. Karballaeezadeh, N., et al.: Prediction of remaining service life of pavement using an optimized support vector machine (case study of Semnan-Firuzkuh road). Eng. Appl. Comput. Fluid Mech. **13**(1), 188–198 (2019)
20. Menad, N.A., et al.: Modeling temperature dependency of oil - water relative permeability in thermal enhanced oil recovery processes using group method of data handling and gene expression programming. Eng. Appl. Comput. Fluid Mech. **13**(1), 724–743 (2019)
21. Mohammadzadeh, S., et al.: Prediction of compression index of fine-grained soils using a gene expression programming model. Infrastructures **4**(2), 26 (2019)
22. Mosavi, A., Edalatifar, M.: A hybrid neuro-fuzzy algorithm for prediction of reference evapotranspiration. Lecture Notes in Networks and Systems, pp. 235–243. Springer (2019)
23. Mosavi, A., Lopez, A., Varkonyi-Koczy, A.R.: Industrial applications of big data: State of the art survey. In: Luca, D., Sirghi, L., Costin, C. (eds.) pp. 225–232. Springer (2018)
24. Mosavi, A., Ozturk, P., Chau, K.W.: Flood prediction using machine learning models: Literature review. Water (Switzerland) **10**(11), 1536 (2018)
25. Mosavi, A., Rabczuk, T.: Learning and intelligent optimization for material design innovation. In: Kvasov, D.E., et al. (eds.) pp. 358–363. Springer (2017)
26. Mosavi, A., Rabczuk, T., Varkonyi-Koczy, A.R.: Reviewing the novel machine learning tools for materials design. In: Luca, D., Sirghi, L., Costin, C. (eds.) pp. 50–58. Springer (2018)
27. Mosavi, A., et al.: State of the art of machine learning models in energy systems, a systematic review. Energies **12**(7), 1301 (2019)
28. Mosavi, A., et al.: Prediction of multi-inputs bubble column reactor using a novel hybrid model of computational fluid dynamics and machine learning. Eng. Appl. Comput. Fluid Mech. **13**(1), 482–492 (2019)
29. Mosavi, A., Varkonyi-Koczy, A.R.: Integration of machine learning and optimization for robot learning. In: Jablonski, R., Szewczyk, R. (eds.) pp. 349–355. Springer (2017)
30. Nosratabadi, S., et al.: Sustainable business models: a review. Sustainability (Switzerland) **11**(6), 1663 (2019)
31. Qasem, S.N., et al.: Estimating daily dew point temperature using machine learning algorithms. Water (Switzerland) **11**(3), 582 (2019)
32. Rezakazemi, M., Mosavi, A., Shirazian, S.: ANFIS pattern for molecular membranes separation optimization. J. Mol. Liq. **274**, 470–476 (2019)
33. Riahi-Madvar, H., et al.: Comparative analysis of soft computing techniques RBF, MLP, and ANFIS with MLR and MNLR for predicting grade-control scour hole geometry. Eng. Appl. Comput. Fluid Mech. **13**(1), 529–550 (2019)
34. Shabani, S., Samadianfard, S., Taghi Sattari, M., Shamshirband, S., Mosavi, A., Kmet, T., Várkonyi-Kóczy, A.R.: Modeling Daily Pan Evaporation in Humid Climates Using Gaussian Process Regression. Preprints 2019, 2019070351. https://doi.org/10.20944/preprints201907.0351.v1
35. Shamshirband, S., Hadipoor, M., Baghban, A., Mosavi, A., Bukor J., Varkonyi-Koczy, A. R.: Developing an ANFIS-PSO Model to predict mercury emissions in Combustion Flue Gases. Preprints 2019, 2019070165 https://doi.org/10.20944/preprints201907.0165.v1
36. Shamshirband, S., et al.: Ensemble models with uncertainty analysis for multi-day ahead forecasting of chlorophyll a concentration in coastal waters. Eng. Appl. Comput. Fluid Mech. **13**(1), 91–101 (2019)
37. Shamshirband, S., Mosavi, A., Rabczuk, T.: Particle swarm optimization model to predict scour depth around bridge pier. arXiv:1906.08863 (2019)
38. Taherei Ghazvinei, P., et al.: Sugarcane growth prediction based on meteorological parameters using extreme learning machine and artificial neural network. Eng. Appl. Comput. Fluid Mech. **12**(1), 738–749 (2018)

39. Torabi, M., et al.: A Hybrid clustering and classification technique for forecasting short-term energy consumption. Environ. Prog. Sustain. Energy **38**(1), 66–76 (2019)
40. Torabi, M., et al.: A hybrid machine learning approach for daily prediction of solar radiation. Lecture Notes in Networks and Systems, pp. 266–274. Springer (2019)
41. Gundoshmian, T.M., et al.: Application of artificial neural network in prediction of the combine harvester performance. J. Food Agric. Environ. **8**(2), 721–724 (2010)
42. Ardabili, S.F., Mahmoudi, A., Mesri Gundoshmian, T.: Modeling of controlling system on temperature controlling of mushroom growing halls, in Ninth National Congress of biosystem engineering and mechanization. Karaj, Iran (2015)
43. Ardabili, S.F., Mosavi, A., Mesri Gundoshmian, M., Nosratabadi, T., Var-konyi-Koczy, A. S.: Modelling temperature variation of mushroom growing hall using artificial neural networks. Preprints 2019
44. Mesri Gundoshmian, T., Ardabili, S.F., Mosavi, A., Varkonyi-Koczy, A.: Prediction of combine harvester performance using hybrid machine learning modeling and response surface methodology. Preprints 2019
45. Ardabili, S.F., Mosavi, A., Varkonyi-Koczy, A.: Systematic review of deep learning and machine learning models in biofuels research. Preprints 2019
46. Ardabili, S.F., Mosavi, A., Varkonyi-Koczy, A.: Advances in machine learning model-ing reviewing hybrid and ensemble methods. Preprints 2019
47. Ardabili, S.F., Mosavi, A., Varkonyi-Koczy, A.: Building Energy information: demand and consumption prediction with Machine Learning models for sustainable and smart cities. Preprints 2019
48. Ardabili, S.F., Mosavi, A., Dehghani, M., Varkonyi-Koczy, A.: Deep learning and machine learning in hydrological processes climate change and earth systems a systematic review. Preprints 2019
49. Mohammadzadeh, D., Karballaeezadeh, N., Mohemmi, M., Mosavi, A., Varkonyi-Koczy, A.: Urban Train Soil-Structure Interaction Modeling and Analysis. Preprints 2019
50. Mosavi, A., Ardabili, S.F., Varkonyi-Koczy, A.: List of deep learning models. Preprints 2019
51. Nosratabadi, S., Mosavi, A., Keivani, R., Ardabili, S.F., Aram, F.: State of the art survey of deep learning and machine learning models for smart cities and urban sustainability. Preprints 2019

Formation and Research of Properties of Photocatalytic Materials on the Basis of TiO_2 for Water Treatment

D. L. Kovalenko[1(✉)], V. E. Gaishun[1], V. V. Vaskevich[1], S. A. Khakhomov[2], S. K. Khudaverdyan[2], and G. Y. Ayvazyan[2]

[1] F. Skorina, Gomel State University, Sovetskaya 104, 246019 Gomel, Belarus
dkov@gsu.by
[2] National Polytechnic University of Armenia, 9/105 Teryan, Yerevan, Armenia

Abstract. The paper describes the sol-gel method for producing photocatalytic materials based on titanium oxide. The temperature-time regimes of heat treatment of the obtained materials are determined. The surface properties of the resulting coatings were studied by scanning electron microscopy and asthmatic force microscopy. The photocatalytic properties of the synthesized coatings and bulk samples were studied on destruction of methylene blue molecules in the developed water purification systems.

Keywords: Sol-gel process · TiO_2 coating · Photocatalytic materials · Photocatalytic properties

1 Introduction

The getter, destructive and dividing technologies existing today only partially solve a water pollution problem as they provide additional neutralization and waste disposal, the sorbents or filters accumulating on a surface in the course of cleaning. The photocatalytic method stands out against the background of these technologies and has a set of positive properties. The application of this method reduces the amount of pollution oxidizers such as chlorine and ozone, eliminates the need for waste neutralization as its implementation is not associated with the accumulation of pollutants that require subsequent disposal [1, 2].

The method, that allows the formation of high-purity nanostructured materials, differs from the most promising technological approaches to the creation of photocatalytic fissile materials on the basis of sol-gel oxides of refractory metals. Based on the deposition of metal oxyhydroxides or ethoxides with the subsequent crystallization of oxides during heat treatment, this method allows varying the physical and chemical properties of the obtained materials over a wide range.

A. R. Várkonyi-Kóczy (Ed.): INTER-ACADEMIA 2019, LNNS 101, pp. 46–51, 2020.
https://doi.org/10.1007/978-3-030-36841-8_4

2 Preparation of Materials

When carrying out the hydrolysis of various titanium compounds (mainly ethoxides and propoxides) in aqueous and anhydrous solutions, the primary products at low pH values are the main salts of variable composition. At higher pH values, hydrated forms of titanium dioxide are formed. They are consistent with the formula $Ti(OH)_2$ or $TiO_2 \cdot nH_2O$ where n depends on the aging and drying conditions [3].

Freshly precipitated hydrated titanium dioxide has a high adsorption capacity with respect to cations and anions. This paper suggests using titanium ethoxide as starting alcohol compounds to prepare film-forming solutions. The hydrolysis of these compounds in an organic solvent with the addition of a well-defined amount of ultrapure water and a catalyst under mild conditions leads to the formation of a stable solution.

To prepare film-forming solutions, the required amount of titanium ethoxide was mixed with anhydrous isopropyl alcohol. Then an aqueous hydrochloric acid solution was added to the mixture and stirred. The hydrolysis reaction in systems based on titanium alkoxides takes a long time, therefore the prepared solution was kept at room temperature for 2–3 days. The finished film-forming solutions were coated with brass mesh structures by dipping. The immersion process is optimally suitable for the formation of thin coatings on substrates with a complex structure. The immersion in the solution should be smooth and uniform, without the formation of bubbles. The sample must be completely coated with the solution and maintain the same position, which helps to obtain a uniform layer on the complex mesh structure of the substrate. To study the surface properties of the materials obtained by immersion, coatings were formed on the surface of silicon wafers.

The distinctive features of the formation of titanium dioxide films are polymorphism and the parallel flow of dehydration and polymerization during heat treatment. Therefore, the composition of film-forming sols in the synthesis of powders and the deposition of functional coatings based on highly dispersed titanium dioxide are very important. In the sols based on hydrated titanium dioxide, the nuclei of nanoparticles are formed of only one polymorphic modification of TiO_2 – anatase in the form of spherical nanoparticles of hydrated titanium dioxide. The removal of alkoxititanium hydrolysis products is carried out only at temperatures above 300 °C [4].

Heat treatment of the resulting coatings was carried out step by step in air at temperatures from 100 °C to 600 °C. The heat treatment process during the gel-oxide transition can be divided into three stages:

- complete removal of physically adsorbed water and organic solvent residues (up to 100–200 °C);
- final hydrolysis of film-forming solutions, decomposition of intermediate hydrolysis products and organic residues (200–400 °C);
- complete dehydration and final oxide formation, glass formation in multicomponent films (above 400 °C) (Fig. 1 a).

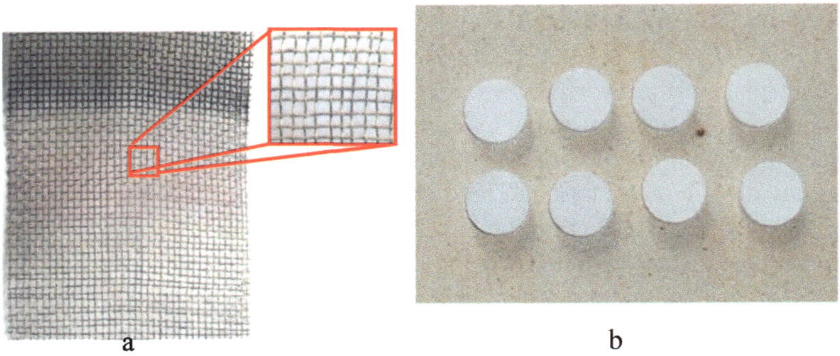

Fig. 1. Samples of materials for use in the water purification systems: **a** coatings on mesh structures, **b** volume photocatalytic materials

To get bulk photocatalytic materials the solution, used to obtain the coating, was evaporated to a gel state. Then the gel was dried in an oven at a temperature of 80 °C for 5 days. The round billets were formed from the obtained fine powder using a manual hydraulic press. The initial solution was taken as a binder. As a result, the round billets (diameter 15 mm, height 7 mm) were obtained (Fig. 1 b). A series of samples was made at annealing temperatures of 500 °C, 600 °C, 700 °C and 800 °C to study the effect of annealing temperature on the photocatalytic properties of the obtained materials.

3 Results and Discussion

To get a fuller picture of the obtained coatings structure, the studies were carried out using scanning electron microscopy and atomic force microscopy. Figure 2 presents the research results of the coating surface using a TESCAN MIRA 3 LM Scanning Electron Microscope. The surface of the coating is characterized by the presence of the chains consisting of partially immersed spherical formations with a base diameter of 150–200 nm. It should be noted that an increase in the annealing temperature of the obtained samples leads to the coating compaction and a slight decrease in spherical formations by 5–7% in the surface volume. The effect of the spherical structures formation is observed due to the crystallization of the amorphous phase and the formation of TiO2 in the form of a metastable polymorphic modification - anatase [4].

The presence of spherical formations on the coating surface is confirmed by the research of the surface using an NT-MDT SOLVER P47-PRO atomic force microscope (Fig. 3).

The research results of the surface by the AFM method confirm the presence of spherical particles on the coating surface. The data analysis confirms that the diameter of the particles at the base is 150–200 nm, and their height varies from 70 to 100 nm.

The installation for water treatment was made to study the photocatalytic properties. The results of a comparative quantitative assessment of catalytic activity on the photo degradation reaction of dye – methylene blue showed that volume samples have

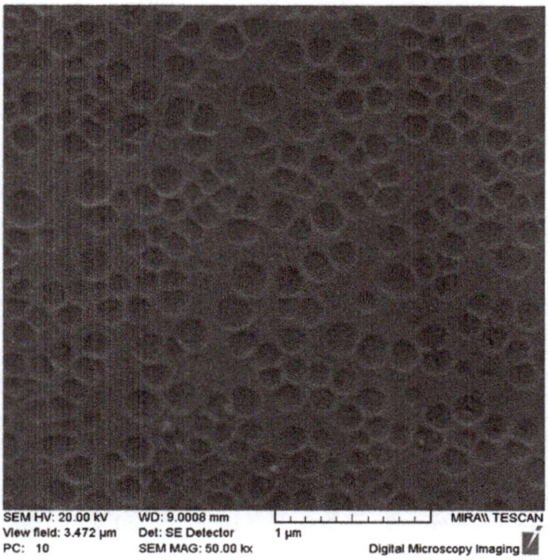

Fig. 2. SEM surface image of TiO2 coatings with a treatment temperature of 400 °C.

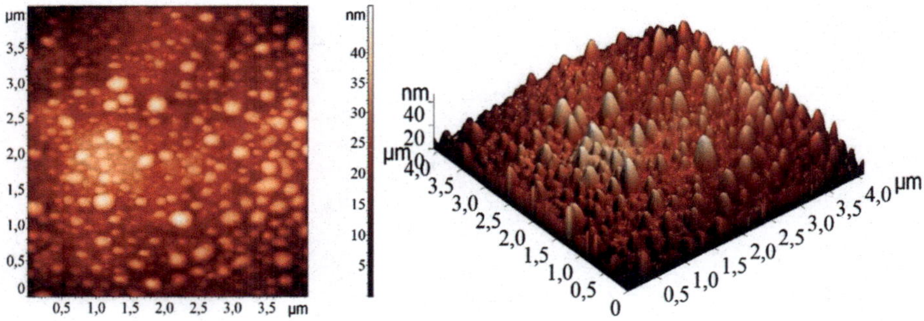

Fig. 3. 2D and 3D AFM surface image of TiO2 coatings with a treatment temperature of 400 °C.

a higher photocatalytic activity (Fig. 4), compared to the coatings (Fig. 5). This is due to the fact that volume samples have a larger superficial area in comparison with film samples on a mesh structure.

Having analyzed the research results of the photocatalytic properties of the resulting materials, it is possible to show that the samples, heat-treated at temperatures of 700 °C and 800 °C, have less pronounced photocatalytic properties in comparison with the samples manufactured at temperatures of 400–600 °C.

Such a result largely depends on the content of titanium dioxide in the anatase phase, which has more pronounced photocatalytic properties. At higher temperatures (about 700 °C), more rutile phase is formed and photocatalytic properties become worse.

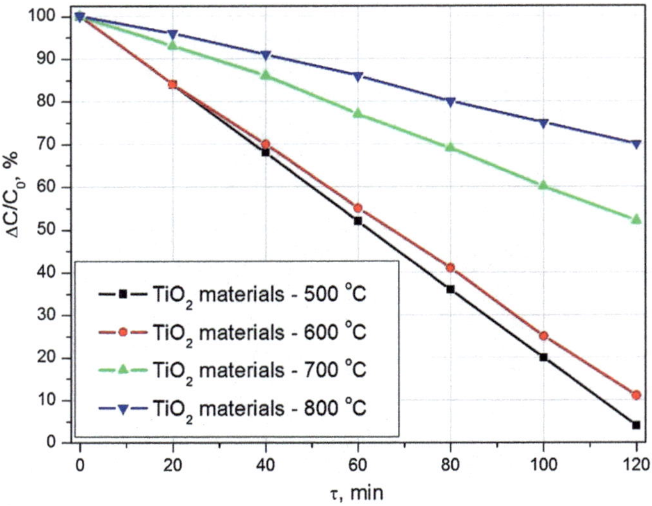

Fig. 4. Change in dye concentration - methylene blue with photo catalytic degradation on TiO2 volume materials.

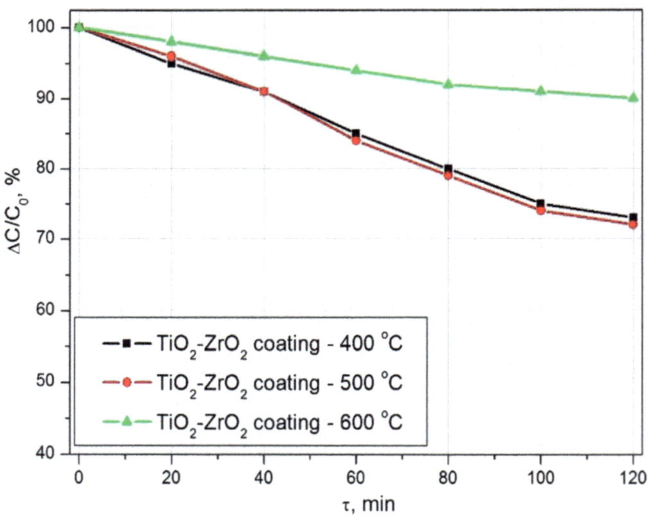

Fig. 5. Change in dye concentration - methylene blue with photo catalytic degradation on TiO2 coatings.

4 Conclusion

The paper presents a sol-gel method for producing the coatings and bulk materials based on titanium oxide. The surface morphology study indicates the formation of the ordered chains of spherical particles, based on hydrated titanium dioxide, on the coating surface. During heating, the crystallization of amorphous films occurs with the formation of anatase (at a temperature of 400–600 °C), followed by transformation to rutile (at a temperature above 650 °C).

The photocatalytic properties of the obtained samples have been studied, and it has been established that volume materials with a processing temperature of 500–600 °C have the optimal photocatalytic properties. The developed coatings can be used in small cyclic water purification systems as oxidative photocatalytic materials.

Also, the coatings possess good self-cleaning properties due to their photocatalytic ones.

Acknowledgements. The study is partially supported by the Ministry of Education of the Republic of Belarus and Belarusian Republican Foundation for Fundamental Research (project F19ARMG-001).

References

1. Fujishima, A., Zhang, X., Tryk, D.A.: TiO photocatalysis and related surface phenomena. Surf. Sci. Rep. **63**(12), 515–582 (2018)
2. Kočí, K., Matějová, L., Reli, M., Čapek, L., Matějka, V., Lacný, Z., Kustrowski, P., Obalová, L.: Sol–gel derived Pd supported TiO_2–ZrO_2 and TiO_2 photocatalysts; their examination in photocatalytic reduction of carbon dioxide. Catal. Today **230**, 20–26 (2014)
3. Vaskevich, V.V., Kovalenko, D.L., Semchenko, A.V., Sidsky, V.V., Kovalevich, M.A.: Nanostructured photocatalytic sol-gel coating based on titanium. PFMT **4**(25), 7–10 (2015)
4. Ivicheva, S.N., Kargin, Y.F, Kutsev, S.V., Volkov, V.V.: Ordered TiO2 nH2O-based films and coatings. Inorg. Mater. **7**(51), 668–672 (2015)

Deep Learning and Machine Learning in Hydrological Processes Climate Change and Earth Systems a Systematic Review

Sina Ardabili[1], Amir Mosavi[2,3](✉), Majid Dehghani[4], and Annamária R. Várkonyi-Kóczy[2,5]

[1] Institute of Advanced Studies Koszeg, Koszeg, Hungary
[2] Kalman Kando, Faculty of Electrical Engineering, Obuda University, Budapest, Hungary
amir.mosavi@kvk.uni-obuda.hu
[3] School of the Built Environment, Oxford Brookes University, Oxford OX3 0BP, UK
[4] Technical and Engineering Department, Faculty of Civil Engineering, Vali-e-Asr University of Rafsanjan, Rafsanjan, Iran
[5] Department of Mathematics and Informatics, J. Selye University, Komarno, Slovakia

Abstract. Artificial intelligence methods and application have recently shown great contribution in modeling and prediction of the hydrological processes, climate change, and earth systems. Among them, deep learning and machine learning methods mainly have reported being essential for achieving higher accuracy, robustness, efficiency, computation cost, and overall model performance. This paper presents the state of the art of machine learning and deep learning methods and applications in this realm and the current state, and future trends are discussed. The survey of the advances in machine learning and deep learning are presented through a novel classification of methods. The paper concludes that deep learning is still in the first stages of development, and the research is still progressing. On the other hand, machine learning methods are already established in the fields, and novel methods with higher performance are emerging through ensemble techniques and hybridization.

Keywords: Machine learning · Deep learning · Big data · Hydrology · Climate change · Global warming · Hydrological model · Earth systems

Nomenclatures

ANN	Artificial neural network
ELM	Extreme learning machine
ML	Machine learning
SVM	Support vector machine
WNN	Wavelet neural networks
DL	Deep learning
ARIMA	Autoregressive integrated moving average
FFNN	Feed-forward neural networks

A. R. Várkonyi-Kóczy (Ed.): INTER-ACADEMIA 2019, LNNS 101, pp. 52–62, 2020.
https://doi.org/10.1007/978-3-030-36841-8_5

MLP	Multi layered perceptron
DT	Decision tree
RSM	Response surface methodology
BPNN	Back propagation neural network
GBDT	Gradient boosting decision tree
ANFIS	Adaptive neuro fuzzy inference system
CPU	Central processing unit
FA	Fire-fly algorithm
DNN	Deep neural network
RF	Random forest
DFNN	Deep feedforward neural network
RNN	Recurrent neural network
PLS	Partial least squares
DA	Discriminant analysis
PCA	Principal component analysis
LDA	Linear discriminant analysis
SVR	Support vector regression
LS	Least-squares
SB	Sparse Bayesian
SPEI	Standard precipitation evapotranspiration index
GP	Genetic programming
MLR	Multi linear regression
MODIS	Moderate Resolution Imaging Spectroradiometer
ROM	Reduced order model
WSSFA	Wise step fire-fly algorithm
DBN	Deep belief networks.

1 Introduction

Studying the hydrological processes, climate change and earth systems are of utmost importance to expand knowledge and insight into the universe [1]. Thus, advancing the accurate models of the earth's various phenomena and systems have been the center of attention [2]. Physical models have a long tradition in simulation, understanding, and prediction of the hydrological processes, climate change, and earth systems [3–7]. Physical models are used worldwide as the trustworthy systems to study the environmental phenomena, climate behavior, atmospheric and hydrological systems, and further study of the natural hazards, extreme events, and ecosystems [8–11]. Statistical models, including the time series analysis form another major popular group of modeling techniques widely used by scientists for studying the earth systems and deliver insight on climate change and hydrological related events [12–18].

Various drawbacks are associated with physical and statistical models [19–21]. Among them, the accuracy, weakness in uncertainty analysis, high computation cost, and the need for a comprehensive amount of data, have been highlighted in the literature [22, 23]. Machine learning and deep learning methods have seen to tackle these shortcomings very well through their efficient computation and intelligence [24–29]. Only during the past few years, these methods have become very popular among the research communities [30–46]. Figure 1 represents the rapid progress of machine learning and deep learning in hydrological processes, climate change, and earth systems research and their subfields. The progressing domination of these intelligent methods is apparent. Thus, studying the novel methods and identifying the trend in using and advancement of these methods would be essential.

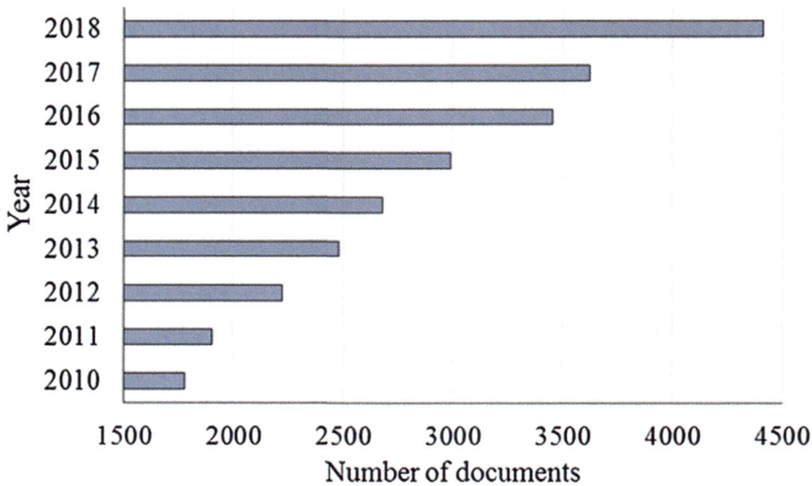

Fig. 1. The rapid growth of using machine learning and deep learning for modeling and prediction of the hydrological processes, climate change and earth systems (source: web of science)

Literature includes a number of review papers on machine learning and deep learning methods [30–46]. There exists a number of papers where the applications domains of the ML methods have been evaluated [47–62]. However, there is a gap in investigating the algorithmic advancements and application domains considering the hydrological processes, climate change, and earth systems. Consequently, the contribution of this paper is set to present the state of the art of machine learning and deep learning methods used for modeling the above-mentioned systems and identify the application areas.

2 Machine Learning Methods

In this section, the machine learning methods have been classified into the following popular subsections, i.e., tree-based, support vector-based, neural network-based, and hybrids and ensembles. Further, there are investigated according to their popularity and applications domains. The summary of the methods are provided in the tables below and methods are reviewed according to their efficiency and accuracy (Table 1).

Table 1. Top studies developed by ML methods in hydrology

References	Contribution	Method	Research domain
[63]	To estimate the hydrologic disturbance index for streams by the use of Random forest	RF	-Machine learning -Watershed management
[64]	To develop different machine learning methods for the estimation of daily reference evapotranspiration	SVM, ELM, RF, M5Tree, and GBDT	-Machine learning -Reference evapotranspiration
[65]	To develop ML methods for the prediction of the wheat production yield in the presence of satellite and climate data	RF, ANN, and SVM	-Machine learning -Hydrological data
[66]	To develop ML methods for the estimation of the atmospheric daily pollen concentration	RF and ANN	-Machine learning -Daily pollen concentration
[67]	To develop ML methods for the prediction of water flow forecasting as a hydrological parameter	MLP, SVR, RBF, and ANFIS	-Machine learning -Hydrologic parameter
[68]	To develop an innovative ML method for the prediction of daily solar radiation (monthly average values)	ELM-MODIS	-Machine learning -Daily solar radiation

Fox and Magoulick [63] developed a study to estimate the hydrologic disturbance index for streams by the use of Random forest in the presence of fish community and hydrologic data and landscape metrics for gaged streams as the training dataset. RF has been introduced as one of the effective methods for this task. RF also has been proposed in a study by Fan et al. [64], who developed SVM and ELM methods in comparison with different tree-based ensemble methods for the estimation of daily reference evapotranspiration in the presence of meteorological data. Tree-based ensemble methods included RF, M5tree, gradient boosting decision tree, and extreme gradient boosting. Evaluations have been performed using determination coefficient, root means a square error and means absolute error. Based on comparisons, RF could provide a higher accuracy compared with that of other methods.

In another study, Cai et al. [65] developed different ML methods for the prediction of wheat production yield by integrating hydrological data, including satellite and climate data. This study also could successfully present a comparison of the performance of methods using the determination coefficient. Methods included the least absolute shrinkage and selection operator, RF, ANN, and SVM. In all datasets, the highest determination coefficient value was owned by SVM followed by RF. Zewdie et al. [66] developed ML methods including RF and ANN for the prediction of atmospheric daily pollen concentration for the comparison with next-generation weather radar. Comparisons have been performed using correlation coefficient values. Based on results, RF and ANN could provide similar performance in the prediction of target values with a high correlation coefficient value.

Kovačević et al. [67] developed a study for the prediction of water flow using MLP, SVR, RBF, and ANFIS methodologies. Evaluation of results and comparison of the performance of methods have been performed using the employment of root mean square error, mean absolute error, and determination coefficient. The best method was ANFIS, followed by SVR with a linear kernel. Ghimire et al. [68] developed a novel ML method entitled the integrated ELM-MODIS. The proposed method has been compared with basic ELM, GP, ANN-GA, ANN-PSO, GA-SVR, and online sequential ELM methods in terms of determination coefficient, root mean square error and mean absolute error. Based on the results, the proposed method could significantly increase the estimation performance followed by hybrid GA-SVR and hybrid ANN-GA methods. The overall detailed results in terms of accuracy, reliability, and sustainability have been presented in Table 2 for further considerations and applications.

Table 2. the comparison results of ML-based methods in hydrology

Method	Application	Accuracy	Reliability	Sustainability	References
RF	Estimation	++	++	++	[63]
RF	Estimation	+++	++	++	[64]
M5Tree	Estimation	++	+	+	[64]
ELM	Estimation	+	+	+	[64]
SVM	Estimation	+++	+++	++	[65]
RF	Estimation	++	++	+	[65]
NN	Estimation	++	+	+	[65]
RF	Estimation	++	++	+	[66]
ANN	Estimation	++	++	+	[66]
ANFIS	Estimation	+++	++	++	[67]
SVR-linear kernel	Estimation	++	++	++	[67]
RBF	Estimation	++	++	+	[67]
MLP	Estimation	++	+	+	[67]
ELM-MODIS	Estimation	+++	+++	+++	[68]
GA-SVR	Estimation	++	++	++	[68]
ANN-GA	Estimation	++	++	+	[68]

3 Deep Learning Methods

Deep learning techniques are considered as a significant part of ML methods based on ANN. DL techniques have been widely applied in analyzing, estimating, designing, filtering, processing, recognition, and detection tasks. The most popular DL methods are DNN, DBN, RNN, and CNN techniques (Table 3).

Table 3. Top studies developed by DL techniques in hydrology

References	Contribution	Method	Research domain
[69]	To develop a DL method for the prediction and estimation of flood	LSTM-ROM	-Deep learning -Flood prediction
[70]	To develop a DL technique for making a model for monitoring drought accurately.	DFNN	-Deep learning -Drought prediction
[71]	To develop DL technique for the analysis of atmospheric imaging of Cherenkov Telescopes	CNN	-CNN -Atmospheric imaging
[72]	To develop a DL technique for the estimation of tropical cyclones and their precursors	CNN	-Deep learning -Tropical cyclones
[73]	To develop an innovative DL method for the prediction of hydrological processes.	DL-FA	-Deep learning -Hydrological processes

Hu et al. [69] developed a novel LSTM integrated by ROM method as an innovative DL method for the prediction of time-series flooding. This integrated method could successfully cop with the prediction of Spatio-temporal distribution of floods because it can use the advantage of both ROM and LSTM methods. The evaluation of results has been performed using root mean square error in the presence of different predicted periods. The validation data was included in the Okushiri tsunami test datasets. This method presented a high accuracy as well as reducing the cost of CPU. This can increase the sustainability of the proposed method significantly.

Shen et al. [70] developed DL technique architecture entitled deep feed-forward neural network for the prediction of SPEI as one of the main factors of drought in the presence of multi-source remote sensing data. The proposed method provided a good correlation with meteorological and agricultural droughts. Evaluating data included SPEI in Henan Province, China, which indicated a high-performance value.

Shilon et al. [71] developed a convolutional neural network method for the analysis of aerial imaging of Cherenkov Telescopes. This imaging system has a significant role in finding very high energy γ-ray emitters. The training phase was performed using datasets generated from Monte-Carlo simulated events and testing phase was performed on both measured and simulations data. CNN could successfully cope with the task with high accuracy. CNN also has been employed by Matsuoka et al. [72] who developed a CNN technique for the estimation of tropical cyclones. The training process was performed in the presence of longwave radiation outgoing during twenty-year simulation which has been calculated by employing a cloud-resolving global atmospheric simulation. The evaluation has been performed using the probability of detection factor.

Xu et al. [73] integrated deep learning and fire-fly algorithm for training the SVR method with optimized parameters. Also, the performance of the developed methods has been compared with other methods in term of determination coefficient. Based on results, the highest correlation coefficient was owned by wise step fire-fly based SVR algorithm followed by DLFA based SVR algorithm in training step, but in testing step the accuracy of DLFA based SVR method has been significantly reduced. This reduction caused the lowest sustainability index for this method (Table 4).

Table 4. the comparison results of DL based methods

Method	Application	Accuracy	Reliability	Sustainability	References
LSTM-ROM	Estimation	+++	+++	+++	[69]
DFNN	Estimation	+++	+++	++	[70]
CNN	Detection	+++	+++	+++	[71]
CNN	Estimation	+++	+++	+++	[72]
WSSFA-SVR	Estimation	+++	+++	+++	[73]
DLFA-SVR	Estimation	++	+	+	[73]
FA-SVR	Estimation	++	++	++	[73]

4 Conclusions

The survey of the advances in machine learning and deep learning are presented through a novel classification of methods. The paper concludes that deep learning is still in the first stages of development, and the research is still progressing. On the other hand, the machine learning methods are already established in the fields, and novel methods with higher performance are emerging through ensemble techniques and hybridization. Similar trends have also been reported in the other application domains [74–82].

Acknowledgments. This publication has been supported by the Project: “Support of research and development activities of the J. Selye University in the field of Digital Slovakia and creative industry” of the Research & Innovation Operational Programme (ITMS code: NFP313010T504) co-funded by the European Regional Development Fund.

References

1. Taylor, R.G., et al.: Ground water and climate change. Nat. Clim. Chang. **3**(4), 322 (2013)
2. Wang, C., et al.: Most of the northern hemisphere permafrost remains under climate change. Sci. Rep. **9**(1), 3295 (2019)

3. Baynes, E.R.C., et al.: Beyond equilibrium: re-evaluating physical modelling of fluvial systems to represent climate changes. Earth Sci. Rev. **181**, 82–97 (2018)
4. Bouhal, T., et al.: Technical feasibility of a sustainable concentrated solar power in morocco through an energy analysis. Renew. Sustain. Energy Rev. **81**, 1087–1095 (2018)
5. Carrassi, A., et al.: Data assimilation in the geosciences: An overview of methods, issues, and perspectives. Wiley Interdiscip. Rev. Clim. Chang. **9**(5), (2018)
6. He, C., et al.: Review and comparison of empirical thermospheric mass density models. Prog. Aerosp. Sci. **103**, 31–51 (2018)
7. Hill, J., Buddenbaum, H., Townsend, P.A.: Imaging spectroscopy of forest ecosystems: perspectives for the use of space-borne hyperspectral earth observation systems. Surv. Geophys. **40**(3), 553–588 (2019)
8. Qin, W., et al.: Comparison of deterministic and data-driven models for solar radiation estimation in China. Renew. Sustain. Energy Rev. **81**, 579–594 (2018)
9. Raseman, W.J., et al.: Emerging investigators series: A critical review of decision support systems for water treatment: Making the case for incorporating climate change and climate extremes. Environ. Sci.: Water Res. Technol. **3**(1), 18–36 (2017)
10. Schenato, L.: A review of distributed fibre optic sensors for geo-hydrological applications. Appl. Sci. **7**(9), (2017) (Switzerland)
11. Seixas, M., et al.: Active bending and tensile pantographic bamboo hybrid amphitheater structure. J. Int. Assoc. Shell Spat. Struct. **58**(3), 239–252 (2017)
12. Aslam, R.A., Shrestha, S., Pandey, V.P.: Groundwater vulnerability to climate change: A review of the assessment methodology. Sci. Total Environ. **612**, 853–875 (2018)
13. Devi, R.M., et al.: Understanding the linkages between climate change and forest. Curr. Sci. **114**(5), 987–996 (2018)
14. Estévez, J., et al.: Introduction to the special issue on "hydro-meteorological time series analysis and their relation to climate change". Acta Geophys. **66**(3), 317–318 (2018)
15. Flowers, G.E.: Hydrology and the future of the greenland ice sheet. Nat. Commun. **9**(1) (2018)
16. Mudelsee, M.: Trend analysis of climate time series: a review of methods. Earth Sci. Rev. **190**, 310–322 (2019)
17. Murray, N.J., et al.: The role of satellite remote sensing in structured ecosystem risk assessments. Sci. Total Environ. **619–620**, 249–257 (2018)
18. Rhein, M.: Taking a close look at ocean circulation: Ocean circulation patterns in the North Atlantic provide a benchmark for climate models. Science **363**(6426), 456–457 (2019)
19. Newton, R.J., McClary, J.S.: The flux and impact of wastewater infrastructure microorganisms on human and ecosystem health. Curr. Opin. Biotechnol. **57**, 145–150 (2019)
20. Royapoor, M., Antony, A., Roskilly, T.: A review of building climate and plant controls, and a survey of industry perspectives. Energy Build. **158**, 453–465 (2018)
21. Sun, Z., et al.: Evaluating and comparing remote sensing terrestrial GPP models for their response to climate variability and CO 2 trends. Sci. Total Environ. **668**, 696–713 (2019)
22. Akpoti, K., Kabo-bah, A.T., Zwart, S.J.: Agricultural land suitability analysis: State-of-the-art and outlooks for integration of climate change analysis. Agric. Syst. **173**, 172–208 (2019)
23. Lyubchich, V., et al.: Insurance risk assessment in the face of climate change: Integrating data science and statistics. Wiley Interdiscip. Rev. Comput. Stat. **11**(4) (2019)
24. Akhter, M.N., et al.: Review on forecasting of photovoltaic power generation based on machine learning and metaheuristic techniques. IET Renew. Power Gener. **13**(7), 1009–1023 (2019)
25. Al Tarhuni, B., et al.: Large scale residential energy efficiency prioritization enabled by machine learning. Energy Effic. (2019)

26. Mosavi, A., Ozturk, P., Chau, K.W.: Flood prediction using machine learning models: Literature review. Water **10**(11) (2018) (Switzerland)
27. Ponsero, A.J., Hurwitz, B.L., The promises and pitfalls of machine learning for detecting viruses in aquatic metagenomes. Front. Microbiol. **10**(MAR) (2019)
28. Shen, C.: A Transdisciplinary Review of Deep Learning Research and Its Relevance for Water Resources Scientists. Water Resour. Res. **54**(11), 8558–8593 (2018)
29. Zhu, X.X., et al.: Deep Learning in Remote Sensing: A Comprehensive Review and List of Resources. IEEE Geosci. Remote. Sens. Mag. **5**(4), 8–36 (2017)
30. Aram, F., et al.: Design and validation of a computational program for analysing mental maps: Aram mental map analyzer. Sustainability **11**(14) (2019) (Switzerland)
31. Asadi, E., et al.: Groundwater quality assessment for drinking and agricultural purposes in Tabriz Aquifer, Iran. (2019)
32. Asghar, M. Z., Subhan, F., Imran, M., Kundi, F.M., Shamshirband, S., Mosavi, A., Csiba, P. R., Várkonyi-Kóczy, A.: Performance evaluation of supervised machine learning techniques for efficient detection of emotions from online content. Pre-prints (2019), 2019080019 https://doi.org/10.20944/preprints201908.0019.v1
33. Bemani, A., Baghban, A., Shamshirband, S., Mosavi, A., Csiba, P., Várkonyi-Kóczy, A.R.: Applying ANN, ANFIS, and LSSVM models for estimation of acid sol-vent solubility in supercritical CO2. Preprints (2019), 2019060055 https://doi.org/10.20944/preprints201906.0055.v2
34. Choubin, B., et al.: Snow avalanche hazard prediction using machine learning methods. J. Hydrol. **577** (2019)
35. Choubin, B., et al.: An ensemble prediction of flood susceptibility using multivariate discriminant analysis, classification and regression trees, and support vector machines. Sci. Total Environ. **651**, 2087–2096 (2019)
36. Dehghani, M., et al.: Prediction of hydropower generation using Grey wolf optimization adaptive neuro-fuzzy inference system. Energies **12**(2) (2019)
37. Dineva, A., et al.: Review of soft computing models in design and control of rotating electrical machines. Energies **12**(6) (2019)
38. Dineva, A., et al.: Multi-label classification for fault diagnosis of rotating electrical machines. (2019)
39. Farzaneh-Gord, M., et al.: Numerical simulation of pressure pulsation effects of a snubber in a CNG station for increasing measurement accuracy. Eng. Appl. Comput. Fluid Mech. **13**(1), 642–663 (2019)
40. Ghalandari, M., et al.: Investigation of submerged structures' flexibility on sloshing frequency using a boundary element method and finite element analysis. Eng. Appl. Comput. Fluid Mech. **13**(1), 519–528 (2019)
41. Ghalandari, M., et al.: Flutter speed estimation using presented differential quadrature method formulation. Eng. Appl. Comput. Fluid Mech. **13**(1), 804–810 (2019)
42. Karballaeezadeh, N., et al.: Prediction of remaining service life of pavement using an optimized support vector machine (case study of Semnan-Firuzkuh road). Eng. Appl. Comput. Fluid Mech. **13**(1), 188–198 (2019)
43. Menad, N.A., et al.: Modeling temperature dependency of oil-water relative permeability in thermal enhanced oil recovery processes using group method of data handling and gene expression programming. Eng. Appl. Comput. Fluid Mech. **13**(1), 724–743 (2019)
44. Mohammadzadeh, S., et al.: Prediction of compression index of fine-grained soils using a gene expression programming model. Infrastructures **4**(2), 26 (2019)
45. Mosavi, A. Edalatifar, M.: A hybrid neuro-fuzzy algorithm for prediction of reference evapotranspiration, in lecture notes in networks and systems, p. 235–243. Springer (2019)

46. Mosavi, A., Lopez, A., Varkonyi-Koczy A.R.: Industrial applications of big data: State of the art survey. In: Luca, D., Sirghi, L., Costin, C. (eds.) p. 225–232. Springer (2018)
47. Mosavi, A., Rabczuk, T., Learning and intelligent optimization for material design innovation. Kvasov, D.E., et al. (eds.) p. 358–363. Springer (2017)
48. Mosavi, A., Rabczuk, T., Varkonyi-Koczy, A.R.: Reviewing the novel machine learning tools for materials design. Luca, D., Sirghi, L., Costin, C. (eds.) p. 50–58. Springer (2018)
49. Mosavi, A., et al.: State of the art of machine learning models in energy systems, a systematic review. Energies **12**(7) 2019
50. Mosavi, A., et al.: Prediction of multi-inputs bubble column reactor using a novel hybrid model of computational fluid dynamics and machine learning. Eng. Appl. Comput. Fluid Mech. **13**(1), 482–492 (2019)
51. Mosavi, A., Varkonyi-Koczy, A.R.: Integration of machine learning and optimization for robot learning. In: Jablonski, R., Szewczyk, R. (eds). p. 349–355. Springer (2017)
52. Nosratabadi, S., et al.: Sustainable business models: A review. Sustainability **11**(6) (2019) (Switzerland)
53. Qasem, S.N., et al.: Estimating daily dew point temperature using machine learning algorithms. Water **11**(3) (2019) (Switzerland)
54. Rezakazemi, M., Mosavi, A., Shirazian, S.: ANFIS pattern for molecular membranes separation optimization. J. Mol. Liq. **274**, 470–476 (2019)
55. Riahi-Madvar, H., et al.: Comparative analysis of soft computing techniques RBF, MLP, and ANFIS with MLR and MNLR for predicting grade-control scour hole geometry. Eng. Appl. Comput. Fluid Mech. **13**(1), 529–550 (2019)
56. Shabani, S., Samadianfard, S., Taghi Sattari, M., Shamshirband, S., Mosavi, A., Kmet, T. R., Várkonyi-Kóczy, A.: Modeling daily pan evaporation in humid climates using gaussian process regression. Preprints (2019), 2019070351 https://doi.org/10.20944/preprints201907.0351.v1
57. Shamshirband, S., Hadipoor, M., Baghban, A., Mosavi, A., Bukor J., Annamaria, R., Varkonyi-Koczy, A.: Developing an ANFIS-PSO model to predict mercury emissions in combustion flue gases. Preprints (2019), 2019070165 https://doi.org/10.20944/preprints201907.0165.v1
58. Shamshirband, S., et al.: Ensemble models with uncertainty analysis for multi-day ahead forecasting of chlorophyll a concentration in coastal waters. Eng. Appl. Comput. Fluid Mech. **13**(1), 91–101 (2019)
59. Shamshirband, S., Mosavi, A., Rabczuk, T., Particle swarm optimization model to predict scour depth around bridge pier (2014). arXiv preprint arXiv:1906.08863
60. Taherei Ghazvinei, P., et al.: Sugarcane growth prediction based on meteorological parameters using extreme learning machine and artificial neural network. Eng. Appl. Comput. Fluid Mech. **12**(1), 738–749 (2018)
61. Torabi, M., et al.: A Hybrid clustering and classification technique for forecasting short-term energy consumption. Environ. Prog. Sustain. Energy **38**(1), 66–76 (2019)
62. Torabi, M., et al.: A Hybrid Machine Learning Approach for Daily Prediction of Solar Radiation, in Lecture Notes in Networks and Systems, p. 266–274. Springer (2019)
63. Fox, J.T., Magoulick, D.D.: Predicting hydrologic disturbance of streams using species occurrence data. Sci. Total Environ. **686**, 254–263 (2019)
64. Fan, J., et al.: Evaluation of SVM, ELM and four tree-based ensemble models for predicting daily reference evapotranspiration using limited meteorological data in different climates of China. Agric. For. Meteorol. **263**, 225–241 (2018)
65. Cai, Y., et al.: Integrating satellite and climate data to predict wheat yield in Australia using machine learning approaches. Agric. For. Meteorol. **274**, 144–159 (2019)

66. Zewdie, G.K., et al.: Estimating the daily pollen concentration in the atmosphere using machine learning and NEXRAD weather radar data. Environ. Monit. Assess **191**(7) (2019)
67. Kovačević, M., et al.: Application of artificial neural networks for hydrological modelling in karst. Gradjevinar **70**(1), 1–10 (2018)
68. Ghimire, S., et al.: Self-adaptive differential evolutionary extreme learning machines for long-term solar radiation prediction with remotely-sensed MODIS satellite and Reanalysis atmospheric products in solar-rich cities. Remote Sens. Environ. **212**, 176–198 (2018)
69. Hu, R., et al.: Rapid spatio-temporal flood prediction and uncertainty quantification using a deep learning method. J. Hydrol. **575**, 911–920 (2019)
70. Shen, R., et al.: Construction of a drought monitoring model using deep learning based on multi-source remote sensing data. Int. J. Appl. Earth Obs. Geoinformation **79**, 48–57 (2019)
71. Shilon, I., et al.: Application of deep learning methods to analysis of imaging atmospheric Cherenkov telescopes data. Astropart. Phys. **105**, 44–53 (2019)
72. Matsuoka, D., et al.: Deep learning approach for detecting tropical cyclones and their precursors in the simulation by a cloud-resolving global nonhydrostatic atmospheric model. Prog. Earth Planet. Sci. **5**(1) (2018)
73. Xu, L., et al.: Simulation and prediction of hydrological processes based on firefly algorithm with deep learning and support vector for regression. Int. J. Parallel Emergent Distrib. Syst. (2019)
74. Ardabili, S., Mosavi, A., Mahmoudi, Mesri Gundoshmian, T, Nosratabadi, S., Var-konyi-Koczy, A.: Modelling temperature variation of mushroom growing hall using artificial neural networks. Preprints (2019)
75. Mesri Gundoshmian, T., Ardabili, S., Mosavi, A., Varkonyi-Koczy, A., Prediction of combine harvester performance using hybrid machine learning modeling and re-sponse surface methodology. Preprints (2019)
76. Ardabili, S., Mosavi, A., Varkonyi-Koczy, A.: Systematic review of deep learning and machine learning models in biofuels research. Preprints (2019)
77. Ardabili, S., Mosavi, A., Varkonyi-Koczy, A.: Advances in machine learning modeling reviewing hybrid and ensemble methods. Preprints (2019)
78. Ardabili, S., Mosavi, A., Varkonyi-Koczy, A.: Building Energy information: demand and consumption prediction with Machine Learning models for sustainable and smart cities. Preprints (2019)
79. Ardabili, S., Mosavi, A., Dehghani, M., Varkonyi-Koczy, A.: Deep learning and machine learning in hydrological processes climate change and earth systems a systematic review. Preprints (2019)
80. Mohammadzadeh D., Karballaeezadeh, N., Mohemmi, M., Mosavi, A., Varkonyi-Koczy A.: Urban train soil-structure interaction modeling and analysis. Preprints (2019)
81. Mosavi, A., Ardabili, S., Varkonyi-Koczy, A.: List of deep learning models. Preprints (2019)
82. Nosratabadi, S., Mosavi, A., Keivani, R., Ardabili, S., Aram, F.: State of the art survey of deep learning and machine learning models for smart cities and urban sustainability. Preprints (2019)

Chemical Processes and Materials

Color Control of the Mechanoluminescent Material Through a Combination of Color Centers

Kenji Murakami(✉), Keita Suzuki, Yoshiki Iwai, Masayuki Okuya, and Masaru Shimomura

Shizuoka University, Johoku Naka-Ku Hamamatsu, 432-8561, Japan
murakami.kenji@shizuoka.ac.jp

Abstract. We have synthesized the new organic mechanoluminescent material based on a 1,10-phenanthroline and an acetylacetone. The synthesized material shows both the photoluminescence and the mechanoluminescence with a green-color (main peak at 545 nm) by doping of a terbium (Tb). We have also investigated an effect of co-doping of Tb and europium (Eu) or dysprosium (Dy) on the luminescence property. Single doping gives a luminescence corresponding to each dopant with the main peak at 545, 612 or 573 nm for Tb, Eu or Dy, respectively, but the co-doping of Tb and Eu gives the luminescence corresponding to their doping amount ratio. As a result, the visual color can be controlled by the ratio. On the other hand, the co-doping of Tb and Dy keeps the luminescence corresponding to Tb single doping, but enhances its intensity. The results suggest that the electron transition process is different between the Tb-Eu and Tb-Dy co-doping. In the case of Tb/Eu co-doping, each dopant induced each own luminescence due to a significant difference in the electron energy state level between Tb and Eu. In contrary, the energy state of Dy becomes an extra electron supplier to Tb for the Tb/Dy co-doping. The results can expand application fields of the mechanoluminescence.

Keywords: Mechanoluminescence · Organic materials · Co-doping

1 Introduction

Mechanoluminescence (ML) is a phenomenon of light emission originated from a mechanical action on a solid [1]. The ML is classified into fracto-, plastico- and elastico-MLs according to an excitation mode of the electrons. The typical example of fracto-ML is a light emission induced by the plate force during and prior to an earthquake [2–4]. The plastico-ML can be observed by peeling an adhesive tape in the vacuum [5]. Additionally, the elastico-ML resulted from a mechanical stress produces a piezoelectric field on the surface of crystals [6–8]. The electric field near the color center is high due to a change in the local structure. The change may reduce the carrier trap depth or effective band bending [9–11]. When the material structure is fractured, the electrons are excited to the higher energy levels followed by the relaxation process of electrons to lower energy levels [12, 13]. The energy difference is released as a light emission with a different wavelength [14].

A. R. Várkonyi-Kóczy (Ed.): INTER-ACADEMIA 2019, LNNS 101, pp. 65–73, 2020.
https://doi.org/10.1007/978-3-030-36841-8_6

Many inorganic mechanoluminescent materials have been synthesized so far with various dopants [15–17]. Difficulty, however, to use the inorganic materials is that they are typically synthesized at high temperatures over 1000 °C [18]. In order to realize a low temperature synthesis of the ML materials, we have succeeded to synthesize the europium doped dibenzoylmethide triethylammonium (EuD_4TEA) that has been well known an organic triboluminescence material [19]. The synthesis is completed at very low temperature of 70 °C by a controlled slow cooling method. The synthesized material shows a very strong ML at 612 nm in the visible region. Furthermore, we have found that an addition of 1-ethenylpyrrolidin-2-one [(polyvinylpyrrolidone) (PVP) enhances the ML intensity of EuD_4TEA through the modification of ligand frameworks [20].

In order to realize a light emission from terbium (Tb) color center for the organic ML material, we have newly synthesized a 1,10-phenanthroline and an acetyl acetone-based material. In this study we have described luminescence properties of the newly synthesized material and investigated an effect of the co-doping with europium (Eu) or dysprosium (Dy) for a color control of the ML. When each color center is doped separately, the synthesized material shows both the photoluminescence (PL) and the ML induced by each dopant whose main peak at 545 nm (Tb), 612 nm (Eu) or 573 nm (Dy). On the other hand, Co-doping of Tb and Eu leads to a change in the visible color of luminescence depending on an amount ratio of Tb/Eu, while a co-doping of Tb and Dy just enhances the luminescence intensity originated from the Tb color center.

2 Experimental

2.1 Synthesis

The synthesis of new organic ML material was followed by our previously reported procedures and methods [20, 21]. All chemicals used were the reagent grades.

First ethyl alcohol was heated till 70 °C and then desired dopant nitrate(s) ($Tb(NO_3)_3{\cdot}5H_2O$, $Eu(NO_3)_3{\cdot}6H_2O$ and/or $Dy(NO_3)_3{\cdot}6H_2O$) was(were) added and stirred. After 1,10-phenanthroline (phen) and acetyl acetone (acac) were added in the solution, it became cloudy. Finally, trimethylamine (TEA) was added till the solute was completely dissolved. The solution container was capped tightly and inserted in the thermos for 24 h. Controlled slow cooling in the thermos is critical for optimal crystal formation. Formed crystal in the solution was collected through the filtration followed by natural drying. The synthesized materials are summarized in Tables 1, 2 and 3.

Table 1. Composition of single doped synthesized materials.

Comp. Specimen	Tb [mmol]	Eu [mmol]	Dy [mmol	acac [mmol]	phen [mlmol]	TEA [mmol]	ethanol [mL]
Tb cmplex	1.0	–	–	3.0	1.0	3.0	40
Eu complex	–	1.0	–	3.0	1.0	3.0	40
Dy complex	–	–	1.0	3.0	1.0	3.0	40

Table 2. Composition of Tb-Eu co-doped synthesized materials.

Ratio comp.	5:1	5:2	2:1	5:3	5:5
Tb $(NO_3)_3$ [mmol]	0.5	0.5	0.5	0.5	0.5
Eu $(NO_3)_3$ [mmol]	0.1	0.2	0.25	0.3	0.5
acac [mmol]	1.5	1.5	1.5	1.5	1.5
phen [mmol]	0.5	0.5	0.5	0.5	0.5
TEA [mmol]	1.5	1.5	1.5	1.5	1.5
ethanol [mL]	40	40	40	40	40

Table 3. Composition of Tb-Dy co-doped synthesized materials.

Ratio Comp.	5:1	5:2	5:3	5:4
Tb $(NO_3)_3$ [mmol]	0.5	1.0	0.5	0.5
Eu $(NO_3)_3$ [mmol]	0.1	0.4	0.3	0.4
acac [mmol]	1.5	3.0	1.5	1.5
phen [mmol]	0.5	1.0	0.5	0.5
TEA [mmol]	1.5	3.0	1.5	1.5
ethanol [mL]	40	50	40	40

2.2 Characterization

The mechanoluminescence and photoluminescence intensities were characterized by using the photonic multichannel analyzer (Hamamatsu Photonics, PMA-12). Results and discussion.

3 Results and Discussion

3.1 Luminescence Property for the Single Doping

Both the photoluminescence and the mechanoluminescense correspond to the electron energy transitions of each single dopant for the single doped synthesized materials. Figure 1 shows the ML spectrum of Tb-doped material (Tb Complex). The result indicates the peaks at 488, 545, 584 and 620 nm corresponding to the electron energy transitions of terbium, from $^5D_4$ to ^7F_n (n = 3, 4, 5 and 6).

Since the main electron energy transitions of Eu and Dy are from $^5D_0$ to ^7F_i (i = 1,2 and 3) and from $^4F_{9/2}$ to $^6H_{j/2}$ (j = 11, 13 and 15), respectively, both the PL and ML spectra show the peaks at 591, 612 and 652 nm for Eu Complex and at 482, 573 and 660 nm for Dy Complex. It is noted, however, that the ML intensity is very weak for Dy Complex.

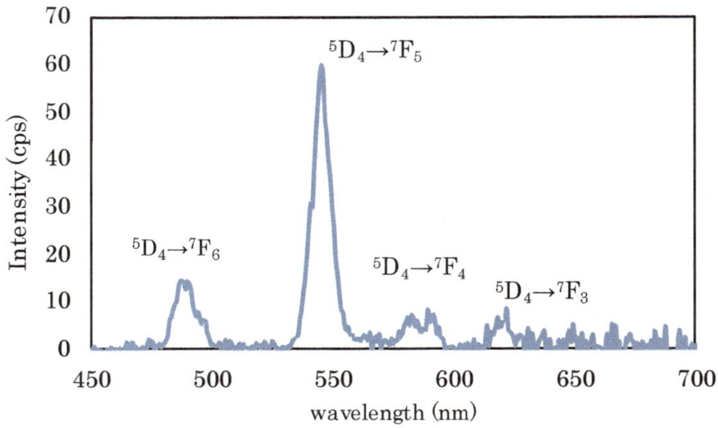

Fig. 1. Mechanoluminescence spectrum of the Tb-doped synthesized material.

We have succeeded for the first time to achieve a control of color emitted from the same ML material just by changing a dopant (color center). Energy diagrams of ligands, phen and acac which are discussed later are strong responsible for the success.

3.2 Luminescence Property for the Co-Doping

In order to realize an mixed light emission based on two color centers for the ML material, we have doped two kinds of dopant simultaneously to the synthesized ML material. Figure 2 shows the ML spectra when Tb and Eu are co-doped with the different molar ratios of Tb/Eu (Tb/Eu Complex). As shown in Fig. 1, the peaks originated from Tb Complex disappear beyond the ratio of Tb:Eu = 2:1. In contrast, the peaks originated from Eu Complex appear in conjunction with the co-doping. Property of the PL is also very similar for the Tb/Eu Comlex).

The results suggest that the co-doping ranging between Tb:Eu 5:2 and 2:1 might lead to a mixed visual light emission originated from two dopants, Tb and Eu. Actually, a chromaticity diagram for the PL of Tb/Eu Complex is obtained as shown in Fig. 3. The figure clearly indicates that the chromaticity of emitted light is changed by the amount ratio of Tb/Eu. The ML property also shows the same result.

On the other hand, the co-doping of Tb and Dy to the synthesized material (Tb/Dy Complex) shows no ML derived from the Dy dopant for any amount ratio of Tb/Dy but enhances intensities corresponding to Tb Complex up to Tb:Dy = 5:3 as shown in Fig. 4. The PL property of Tb/Dy complex is also similar. Figure 5 shows a variation of the ML intensity detected at 545 nm in Fig. 4 with the amount ratio of Dy dopant. The maximum ML intensity is obtained by the Dy doping molar ratio of 29% to Tb dopant as shown in Fig. 5.

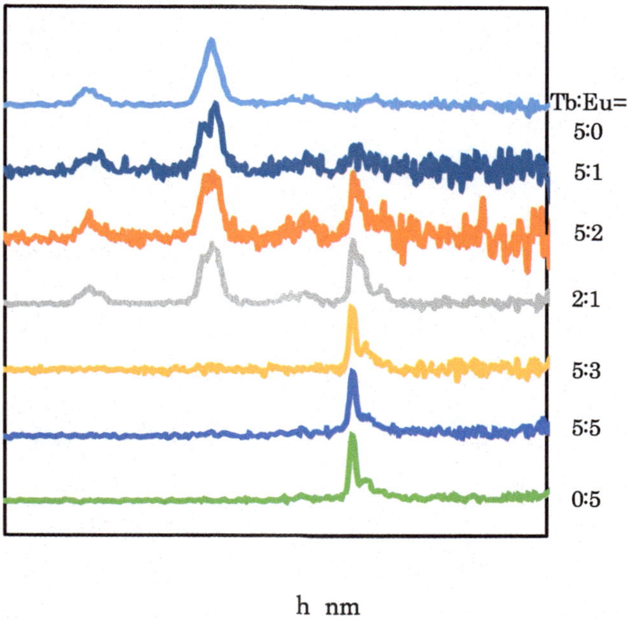

Fig. 2. Mechanoluminescence spectra of the Tb/Eu-co-doped synthesized material. The ratios indicated on right side are the molar ratios of Tb/Eu.

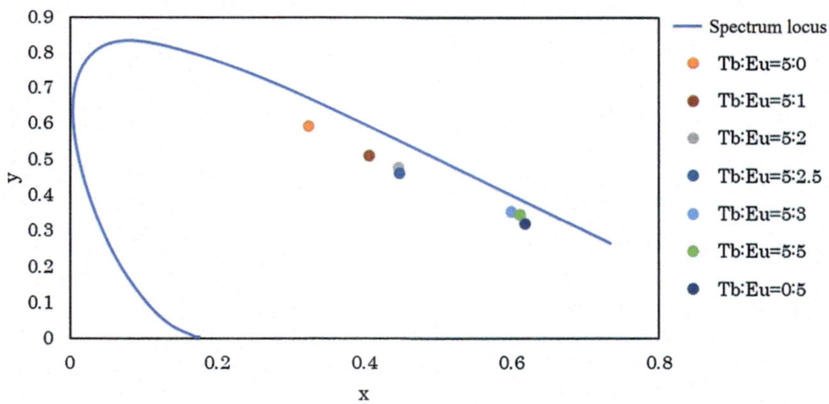

Fig. 3. Chromaticity diagram for the PL of Tb/Eu-co-doped synthesized material. The ratios indicated on right side are the molar ratios of Tb/Eu.

Difference in the effect of co-doping between the Tb/Eu Complex and the Tb/Dy Complex should be related to the difference in energy diagram of the co-doped synthesized materials as shown in Figs. 6 and 7. S and T in the figures denote the excited singlet states and the excited triplet states for the ligands, phenanthroline (phen) and acetyl acetone (acac).

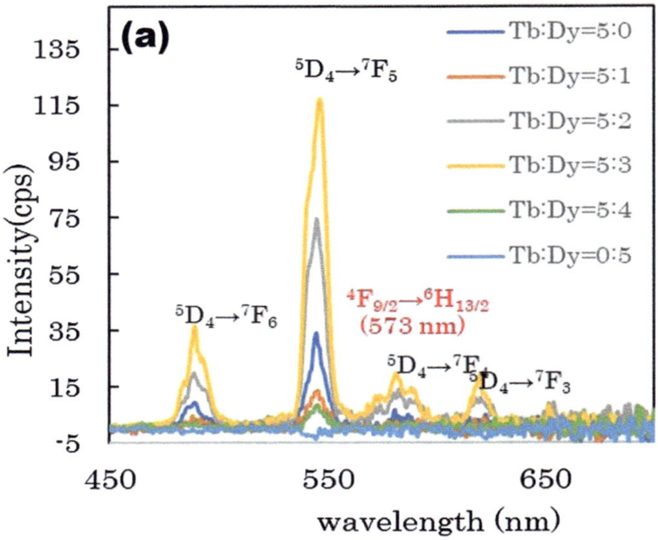

Fig. 4. Mechanoluminescence spectra of the Tb/Dy-co-doped synthesized material. The ratios inserted in the graph are the molar ratios of Tb/Dy.

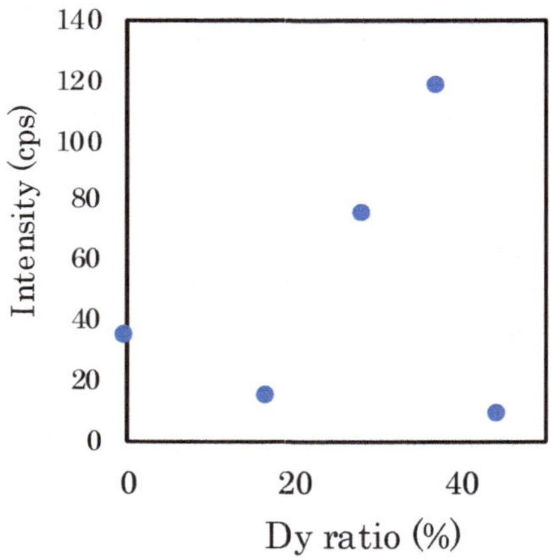

Fig. 5. Variation of the ML Intensity of the main peak at 545 m in Fig. 4 with the amount ratio of Dy dopant.

Figure 6 depicts that in the case of Tb/Eu Complex, mechanical stresses on the material excite the electrons to S(acac) and S(phen) followed by the transitions to T(acac) and T(phen) through an intersystem crossing. Collected electrons in the triplet states can transit to $^5D_4$ of Tb^{3+} and $^5D_0$ of Eu^{3+} followed by the transitions to ^7F_n (n = 3, 4, 5 and 6) of Tb^{3+} and ^7F_i (i = 1, 2 and 3) of Eu^{3+}, respectively because the energy difference between two states is significantly large. There is no communication between $^5D_4$ of Tb^{3+} and $^5D_0$ of Eu^{3+}. Thus, the luminescence from Tb/Eu Complex is

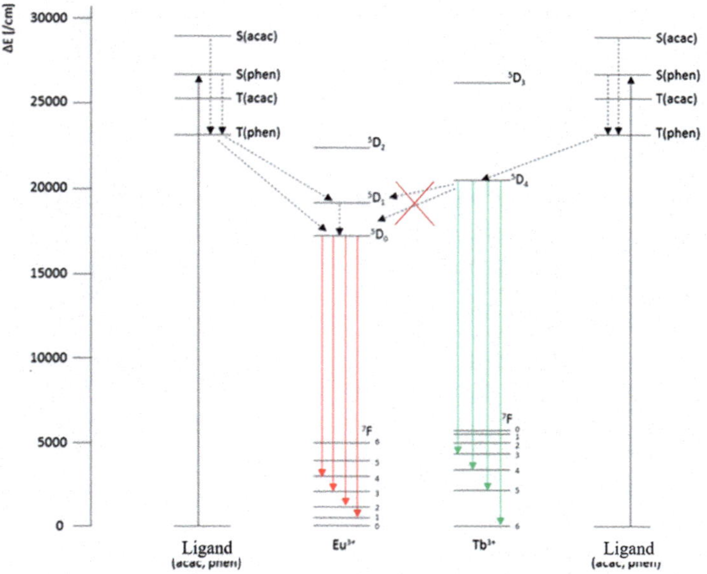

Fig. 6. Energy diagram of the Tb/Eu-co-doped synthesized material.

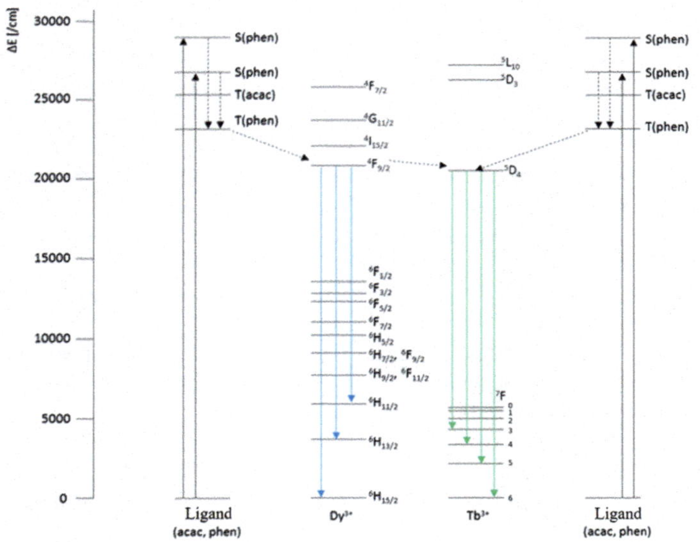

Fig. 7. Energy diagram of the Tb/Dy-co-doped synthesized material.

changed by the amount ratio of Tb and Eu. On the other hand, since the energy of $^4F_{9/2}$ of Dy^{3+} is very close to that of $^5D_4$ of Tb^{3+} for the Tb/Dy Complex as shown in Fig. 7, the collected electrons in $^4F_{9/2}$ of Dy^{3+} prefer the transition to $^5D_4$ of Tb^{3+} to that to $^6H_{j/2}$ (j = 11, 13 and 15). Thus, the electron transitions within the Tb dopant are enhanced by an increase in the amount ratio of Dy dopant. Excess amount ratio of the

Dy dopant, however, induces the electron transitions within the Dy^{3+} electron energy states as well resulting in a decrease in the transitions within the Tb dopant.

4 Conclusion

We have newly synthesized the 1,10-phenanthroline and the acetyl acetone-based organic ML material to realize a light emission induced by the Tb color center. It is found that the synthesized material also shows a scalability to accept the different color centers such as Eu and Dy, and emits the ML induced by the different single dopant. Additionally, the material has a capability to control a chromaticity of the ML by changing an amount ratio of the co-doped Tb/Eu and to enhance the ML intensity induced by Tb through a co-doping with Dy. The results might expand the application fields of ML not only to the stress sensings but also to displays.

References

1. Tiwari, N., Dubey, V., Kuraria, R.K.: Mechanoluminescence study of europium doped $CaZrO_3$ phosphor. J. Fluores. **26**(4), 1309–1315 (2016)
2. Chandra, B.P.: Luminescence of solids. Springer, Boston, MA, USA (1998)
3. Jha, P., Chandra, B.P.: Survey of the literature on mechanoluminescence from 1605 to 2013. Luminescence **29**(8), 977–993 (2014)
4. Chandra, B.P., Parganiha, S., Chandra, V.K., Jha, P., Baghel, R.N.: Sensing of shock-wave velocity and pressure using shock-wave induced mechanoluminescence of crystals. Sens. Actuators A: Phys. **235**, 203–209 (2015)
5. Teotonio, E.E.S., Faustino, W.M., Brito, H.F., Felinto, M.C.F.C., Moura, J.L., Costa, I.F., Santos, P.R.S.: Triboluminescence: theory, synthesis, and application. Springer International Publishing, Cham (2016)
6. Chandra, B.P., Chandra, V.K., Jha, P.: Microscopic theory of elastico-mechanoluminescence smart materials. Appl. Phys. Lett. **104**(3), 031102 (2014)
7. Chandra, V.K., Chandra, B.P.: Suitable materials for elastico mechanoluminescence-based stress sensors. Opt. Mater. **34**(1), 194–200 (2011)
8. Jeong, S.M., Song, S., Kim, H., Joo, K.I., Takezoe, H.: Mechanoluminescence color conversion by spontaneous fluorescent-dye-diffusion in elastomeric zinc sulfide composite. Adv. Func. Mater. **26**(7), 4848–4858 (2016)
9. Botterman, J., den Eeckkhout, K.V., Baere, I.D., Poelman, D., Smet, P.F.: Mechanoluminescence in $BaSi_2O_2N_2$:Eu. Acta Mater. **60**(15), 5494–5500 (2012)
10. Chandra, B.P., Chandra, V.K., Mahobia, S.K., Jha, P., Tiwari, R., Haldar, B.: Real-time mechanoluminescence sensing of the amplitude and duration of impact stress. Sens. Actuators A: Phys. **173**(1), 9–16 (2012)
11. Wang, X., Zhang, H., Yu, R., Dong, L., Peng, D., Zhang, A., Zhan, Y., Liu, H., Pan, C., Wang, Z.L.: Dynamic pressure mapping of personalized handwriting by a flexible sensor matrix based on the mechanoluminescence process. Adv. Mater. **27**(14), 2324–2331 (2015)
12. Takada, N., Hieda, S., Sugiyama, J., Katoh, R., Minami, N.: Mechanoluminescence from piezoelectric crystals of an europium complex. Synthesis Metals. **111–112**, 587–590 (2000)
13. Chandra, B.P.: Mechanoluminescence and high pressure photoluminescence of (Zn, Cd)S phosphors. Pramana **19**(5), 455–465 (1952)

14. Chandra, B.P., Chandra, V.K., Jha, P.: Models for intrinsic and extrinsic fracto-mechanoluminescence of solids. J. Luminesc. **135**, 139–153 (2013)
15. Lin, Y.-H., Dang, A., Deng, Y., Nan, C.W.: Studies on mechanoluminescence from $SrAl_2O_4$: Eu, Dy phosphor. Mater. Chem. Phys. **80**(1), 20–22 (2003)
16. Zhang, J.-C., Long, Y.-Z., Wang, X., Xu, C.-N.: Controlling elastico-mechanoluminescence in diphase $(Ba,Ca)TiO_3:Pr^{3+}$ by co-doping different rare earth ions. RSC Adv. **4**(77), 40665–40675 (2014)
17. Chandra, B.P., Baghel, R.N., Luka, A.K., Sanodiya, T.R., Kuraria, R.K., Kuraria, S.R.: Strong mechanoluminescence induced by elastic deformation of rare-earth-doped strontium aluminate phosphors. J. Luminesc. **129**(7), 760–766 (2009)
18. Pust, P., Weller, V., Hecht, C., Tuecks, A., Wochnik, A.S., Henss, A.-K., Wiechert, D., Scheu, C., Schmidt, P.J., Schnick, W.: Narrow-band red-emitting $Sr[LiAl_3N_4]:Eu^{2+}$ as a next-generation LED-phosphor material. Nature Mater. **13**, 891–896 (2014)
19. Fontenot, R.S., Hollerman, W.A., Bhat, K.N., Aggarwal, M.D.: Synthesis and characterization of highly triboluminescent doped europium tetrakis compounds. J. Luminesc. **132**(7), 1812–1818 (2012)
20. Ranasinghe, R.A.D.M., Tanaka, Y., Okuya, M., Shimomura, M., Murakami, K.: Structural characterizations of organic-based materials with extensive mechanoluminescence properties. J. Luminesc. **190**, 413–423 (2017)
21. Ranasinghe, R.A.D.M., Rajapakse, R.M.G Illeperuma O.A., Okuya, M. and Murakami, K.: Effect of added polyvinylpyrrolidone on mechanoluminescent property of europium-doped dibenzoylmethide triethylammonium. JJAP Conf. Proceed. **4**, 011105(1–4) (2016)

Corrosion Resistance and Hydrophobic Properties of Gradient Coatings Based on Carbon and Alloying Elements

Ekaterina A. Kulesh[2], Alexandr V. Rogachev[1,2(✉)], Dmitry G. Piliptsou[1,2], Alexandr S. Rudenkov[1,2], Jiang X. Hong[1], and Victor A. Emel′yanov[1]

[1] International Chinese-Belarusian Scientific Laboratory on Vacuum-Plasma, Technology, Nanjing University of Science and Technology, Xiao Ling Wei 200, Nanjing 210094, China
rogachevav@mail.ru

[2] Francisk Skorina Gomel State University, Sovetskaya 102, 246019 Gomel, Belarus

Abstract. Gradient metal doped diamond-like carbon (Me/α-C(5…20 Hz):Me, where Me–Ti, Cr, Al) coatings were deposited by the synergy of pulse cathode arc and direct-current cathode arc, and C content was adjusted by changing the pulse frequency. The microstructure, composition, surface morphology, the hydrophobicity and corrosion resistance of the coatings were investigated by Ra-man, XPS, SEM, contact angle measurements and potentiodynamic polarization tests, respectively. The XPS method established the formation of carbide com-pounds in the coatings obtained. It is stated that the ratio of carbide/carbon phases in the coating is determined by the type of alloying metal. The values of contact wetting angles for all coatings do not exceed 90°, therefore, the surface is well wetted by a corrosive medium and stays hydrophilic. The analysis of the polarization curves showed that the presence of gradient coatings on the surface leads to the increase in the corrosion resistance of the steel substrate. Cr/α-C(5…20 Hz):Cr coatings are characterized by the highest corrosion resistance. This work is devoted to the studying of influences of type of metal on structure and anti-corrosion property of gradient coatings.

Keywords: Carbon coatings · Gradient coatings · Mosaic coatings · Polarization curves · Corrosion resistance · Wettability · Hydrophobicity

1 Introduction

Material corrosion, as a common phenomenon, means that material is susceptible to environment and loses original function [1]. Material corrosion occurs in all walks of life, of which steel in corrosion liquid is under the most serious corrosion, bringing about enormous waste of resources [2]. One of the areas of corrosion protection is the application of carbon coatings (α-C), characterized by high mechanical and corrosion properties [2, 3].

A. R. Várkonyi-Kóczy (Ed.): INTER-ACADEMIA 2019, LNNS 101, pp. 74–83, 2020.
https://doi.org/10.1007/978-3-030-36841-8_7

Gradient metal- carbon coatings are coatings consisting of layers of carbides of transition metals and characterized by a nanocluster structure in the form of a solid continuous frame providing the necessary set of mechanical properties, and metal layers (Cr, Ti, Al and etc.), determining the limitations of the fracture thickness of the layer, and solid-lubricating layers (a-C, a-C:H, MoS_2), allowing to reduce friction and increase wear resistance. A feature of gradient coatings is the ability to change the phase composition of the layer thickness, which determines the high strength of the adhesive compound and a unique set of volumetric mechanical properties [4, 5]. As a result, such coatings have the properties of both a solid diamond-like coating and a metal, that is, have a high hardness and high toughness.

The destruction caused by corrosion processes cannot be completely stopped, but there are several methods to minimize the rate of their occurrence. These methods include: electrochemical cathodic protection, the use of corrosion inhibitors and coating [6, 7]. High corrosion resistance of coatings is realized in the absence of high internal stresses and high adhesive strength of the connection with the base. Otherwise, there is delamination and cracking of the coating, and the corrosive medium easily reaches the substrate surface through diffusion channels formed by microcracks [8]. Therefore, in the development of protective coatings, it is necessary to use methods and techniques to reduce internal stresses and increase the strength of the adhesive compound: for example, relaxation annealing, the formation of a multi-layer structure or alloying coatings [9, 10].

It is known that metal layers deposited from metals such as Ti, Cr, Al, as well as their carbides or nitrides introduced into the volume of diamond-like coating reduce the internal stress and increase the adhesion of the coating to the substrate [11, 12]. It should be borne in mind that for multilayer coatings characterized by the formation of defects at the interface layers, leading to the destruction of the multilayer system. The solution of this problem is possible by the formation of gradient coatings with a variable concentration of the "diamond-like" phase in the coating thickness, taking into account the nature of the filler, the conditions and modes of deposition of coatings [13, 14].

The aim of this work was to study the influence of the architecture of gradient metal-carbon coatings (the nature of metals, their nitrides, the morphology of the layers, the phase composition of the carbon matrix) on their hydrophobic properties and corrosion resistance.

2 Experimental

Gradient metal-carbon coatings were formed from combined flows of carbon plasma and metal atoms. For their application, a vacuum device containing a pulsed source of carbon plasma, an electric arc source of metal plasma was used (Fig. 1). Polished silicon and steel (304) plates were used as substrates for the formed coatings.

The use of combined flows makes it possible to control the composition, structure and create different conditions for the processes of chemical interaction between metal and carbon atoms, as well as to ensure a continuous and smooth change in the concentration of elements in the thickness of the coating and to ensure the synthesis of

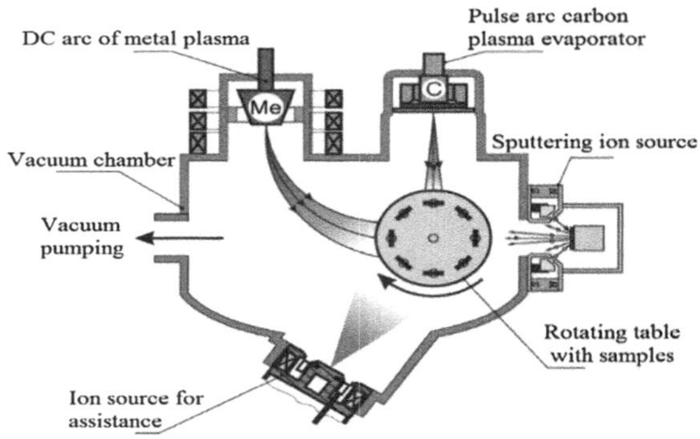

Fig. 1. Scheme of the vacuum chamber for the deposition of gradient coatings.

carbide compounds on the substrate [15]. The thickness of all obtained coatings was 459 ± 10 nm The formation of gradient metal-carbon coatings was carried out with the following stages:

1. deposition of a metal sublayer (Ti, Cr, Al) providing the required adhesion characteristics to the substrate;
2. introduction of carbon ions formed by the evaporation of the graphite cathode by a pulsed arc into the metal plasma, which, by increasing the frequency of discharge pulses in the range from 5 Hz to 20 Hz, provided a subsequent increase in the concentration of carbon atoms in the coating (Fig. 2).

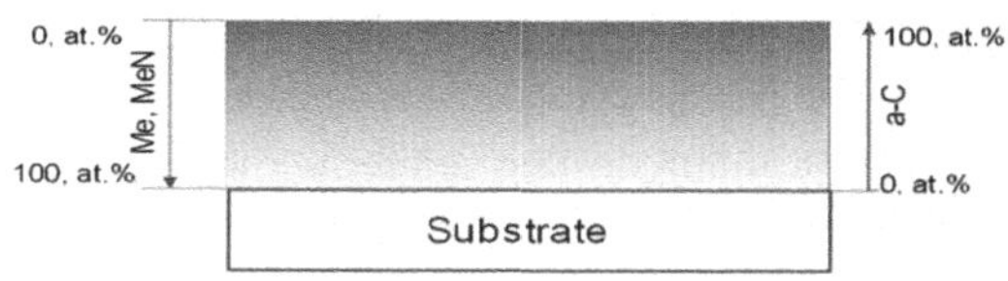

Fig. 2. Scheme of the structure of composite gradient α-C coatings.

Al, Cr and Ti were used as metal components of the coating. These metals are characterized by different chemical activity with respect to carbon. Coatings of the following structure were obtained: Ti/α-C(5…20 Hz); Ti, Cr/α-C(5…20 Hz):Cr, Al/α-C(5…20 Hz):Al, where (5…20 Hz) – shows the frequency range of the pulses of carbon plasma.

The surface structure and thickness of the coatings were studied using a scanning electron microscope Hitachi S-4800.

The chemical composition of the coatings was determined by X-ray photoelectron spectroscopy (XPS) using PHI Quanta (Japan). The obtained spectrum was calibrated along the C1s line with a binding energy of 284.6 eV.

The phase composition of the carbon matrix was determined by Raman spectroscopy, which is one of the most widely used methods of non-destructive testing of the microstructure of carbon-based coatings [16–18]. The Raman spectrum was excited by laser radiation with a power of 20 mW and a wavelength of 532 nm from the coating area with a diameter of 1 μm.

Corrosion resistance performance in artificial seawater of deposited coatings was evaluated by potentiodynamic polarization using electrochemical workstation by the standard three-electrode scheme (Fig. 3). Potential ranged from −1.5 and 1.5 V at a scanning rate of 10 mV/s. The test area of samples was 1 cm^2.

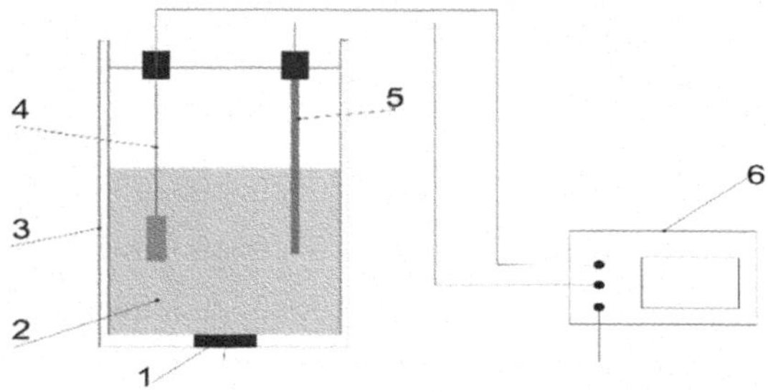

Fig. 3. Electrochemical station for corrosion test: 1-investigated electrode (coated substrate), 2-corrosive liquid, 3-electrochemical cell, 4-counter electrode (Pt sheet), 5-saturated calomel electrode, 6-potentiostat.

The wettability of coatings was investigated by static contact angle measurement using a contact angle meter with a drop volume of 0.75 ± 0.2 l at about 25 °C. The test liquid is artificial seawater. The measurement was repeated three times for each sample.

3 Result and Discussion

The results of the study of the morphology of coatings carried out by scanning electron microscopy are shown in Fig. 4.

Ti/a-C(5…20 Hz):Ti Cr/a-C(5…20 Hz):Cr Al/a-C(5…20 Hz):Al

Fig. 4. SEM images of the surface gradient Me/a-C coating with different metal.

As can be seen from the figure, the surface of the coating is fairly uniform, with a small content of particles of predominantly rounded shape. On the surface of the

Al/a-C(5…20 Hz):Al coating large droplets are recorded, which is determined by the peculiarities of the arc evaporation of the aluminum cathode, namely the presence in the flow of a large content of the droplet phase formed as a result of aluminum boiling in the cathode spot area. Analysis of cross-section view of coating showed the formation of layered structures in their volume with a thickness of not more than 20 nm. The presence of a layered structure is due to the features of the deposition of the coating: the location of the evaporators and the rotation of the substrates in the deposition process.

Regardless of the nature of the metal, the Raman spectra of gradient coatings represent an asymmetric peak with a center localized at 1580 cm^{-1} (Fig. 5). This form of the spectrum is typical for all types of amorphous carbon coatings [16].

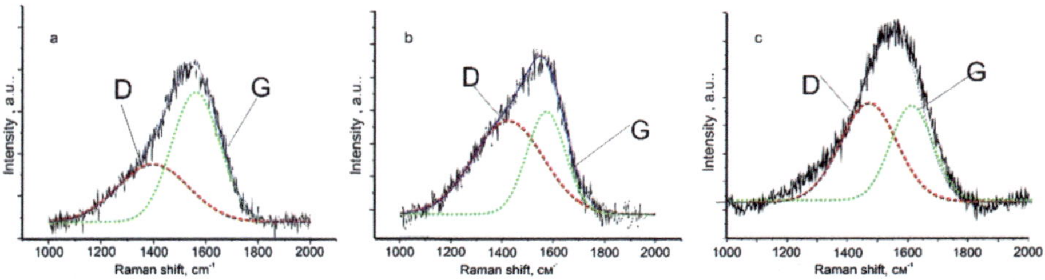

Fig. 5. Raman spectra of deposited coatings: a– Al/a-C(5…20 Hz):Al; b– Cr/a-C(5…20 Hz):Cr; c– Ti/a-C(5…20 Hz):Ti

The results of the analysis of Raman scattering spectra using Gauss functions are given in Table 1.

Table 1. Results of mathematical processing of gradient coatings.

Samples	I_D/I_G ratio	G peak shift, cm^{-1}	G peak width, cm^{-1}
Al/a-C(5…20 Hz):Al	0.51	1595.2	328.6
Cr/a-C(5…20 Hz):Cr	1.06	1585.7	201.6
Ti/a-C(5…20 Hz):Ti	0.98	1650.5	172.3

By Raman spectroscopy it was found that the ratio of I_D/I_G characterizing the phase composition of the carbon matrix depends on the nature of the metal. In this case, regardless of the nature of the metal is not only the formation of disordered structure, but also a change in the size of Csp^2 carbon clusters. The introduction of metals leads to an increase in the width of the G-peak characterizing the degree of deformation of the angles of bonds between carbon atoms in Csp^2 clusters [17].

The parameters of the Raman spectra of Al/a-C(5…20 Hz):Al coatings are close in value to the parameters characteristic of the monolayer a-C, which implies a weak effect of aluminum atoms on the microstructure of gradient Al/a-C(5…20 Hz):Al coatings. The minimum value of the I_D/I_G ratio characteristic of Al/a-C(5…20 Hz):Al coatings is determined by an increase in the number of aromatic rings in the Csp^2 cluster, as well as a decrease in the proportion of groups forming linear –C = C – chains, with an increase in the G peak width caused by a decrease in the size of Csp^2 carbon clusters.

For Ti/α-C(5…20 Hz):Ti and Cr/α-C(5…20 Hz):Cr coatings are characterized by higher values of the I_D/I_G ratio, which according to [19] is determined by a decrease in the number of Csp^2 carbon clusters. According to [20], the decrease in the number of Csp^2 carbon clusters in coatings containing titanium or chromium is determined by the higher activity of titanium and chromium with respect to carbon, which leads to the formation of TiC or CrC compounds on the surface. For Cr/α-C(5…20 Hz):Cr coatings I_D/I_G ratio is the maximum, and the position of the G-peak is shifted in the direction of smaller wave numbers, which in accordance with the rule formulated in [16], indicates a decrease in sp^3 phase.

XPS analysis of coatings Cr/α-C(5…20 Hz):Cr showed that the C1s spectrum of states of carbon atoms can be represented as a sum of the peaks. It was found that during the formation of Cr/α-C(5…20 Hz):Cr coatings chromium atoms not only enter into chemical interaction with carbon atoms, but also form solid solutions with carbon. The formation of chromium carbide occurs due to the interaction of chromium atoms with carbon atoms in the state of sp^2, which are characterized by lower binding energy in comparison with the sp^3 bond of carbon atoms. Thus, the presence of a graphite component activates the processes of carbon interaction with Cr atoms, which leads to the formation of a solid carbide phase [17]. Therefore, in the structure of Cr/α-C (5.20 Hz):Cr coating chromium is present in the form of carbide, oxide, and in the metal form.

The C1s spectrum characteristic of carbon atoms in Ti/α-C(5…20 Hz):Ti coating can be represented as the sum of four peaks with maxima localized at 284.4 eV (Csp^2), 285.3 eV (Csp^3), 286.6 eV (C-O), and 283.2 eV (C-Ti). The position of the peak center for Csp^2 and Csp^3 phase is shifted to higher binding energies in comparison with the peak centers characteristic of Cr/α-C(5…20 Hz):Cr coatings, which is explained by the change in the ratio of Csp^2/Csp^3 and the percentage of C-Ti bonds in the coating. Also on the shift of the centers of the bands influences and the formation of a layered structure (Fig. 4), the formation of a layered structure leads to the formation of diffusion channels in which there is a violation of the order carbon clusters due to the interaction with titanium and the rearrangement of atoms in the volume of the coating as a result of the reaction between the nanolayers of titanium and carbon. Analysis of the XPS data showed that the coating contains carbon atoms in the sp^2 and sp^3 phase and titanium carbide TiC. Comparing the ratio of peak areas characteristic of TiC and CrC bonds, we can conclude that a more active interaction of carbon atoms with chromium.

The analysis of the XPS spectrum for the C1s state can be attributed to the states Csp^2, C-Csp^3 and C–O, respectively. As a rule, the bond which are defining the CAl interaction and appearing in the C1s spectrum is determined by the peak with the center at 281.5 eV for Al–C and 282.5 eV for Al–O–C. However, no this peaks were found in the processing of the obtained XPS spectra, indicating that the structure of Al/α-C(5… 20 Hz):Al coatings are a composite of amorphous carbon with different types of hybridization of bonds and metal phases of aluminum. the sp^2/sp^3 ratio determined on the basis of the ratio of the areas of peaks Csp^2 and Csp^3 is higher than for Ti/a-C(5… 20 Hz):Ti and Cr/a-C(5…20 Hz):Cr coatings, which confirms the data of Raman spectroscopy, indicating an increase in the graphite component of the coating. The presence of only two peaks in the Al2p spectrum indicates that aluminum does not bind to carbon and does not form compounds of the Al–C type [18].

Based on the results of the analysis of polarization curves (Fig. 6, Table 2), it was found that the deposition of gradient coatings leads to an increase in the corrosion resistance of the steel substrate made of 304 steel. In this case, Cr/a-C(5…20 Hz):Cr coatings exhibit higher corrosion resistance in comparison with other types of coatings, which is confirmed by the displacement of the corrosion potential in the positive region.

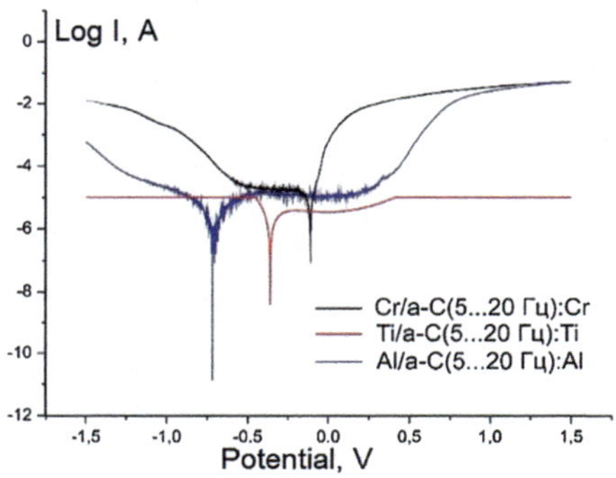

a

Fig. 6. Polarization curves of gradient coatings.

Table 2. Results of mathematical processing of electrochemical corrosion curves and determination of the wetting angle θ° of gradient coatings.

Samples	E_{corr}, V	$I_{corr} \times 10^{-5}$, A/cm²	Θ, °
Steel 304	−0.444	1.035	–
Ti/a-C(5…20 Hz):Ti	−0.372	1.253	68.3
Cr/a-C(5…20 Hz):Cr	−0.117	0.352	76.2
Al/a-C(5…20 Hz):Al	−0.715	2.865	82.8

Note that the values of the wetting angle θ for all coatings do not exceed 90°, so all coatings belong to the class of hydrophilic and, therefore, are well wetted with a corrosive environment. The values of the wetting angles are in good agreement with the results of [19], in which it is established that the coatings based on carbon and metals have an average hydrophobicity, and, according to these data, are characterized by low values of the contact wetting angle (not more than 70°). As shown in [20], the presence of hydrogen (type –C–H– bonds) in the coating volume leads to an increase in the wetting angle to 91.2°. Introduction to the volume of carbon coating of metals, changes the hydrophobic properties, the wetting angle increases to 102°, but to achieve such high values it is necessary that the particle size of the alloying material does not exceed 20 nm.

Corrosion of coatings based on amorphous carbon is determined by the ratio of sp^3/sp^2 carbon atoms with different types of bonds, the presence of inhibitory or inert phases in their volume. High corrosion resistance is shown by coatings with a high content of carbon atoms in the state with sp^3 hybridization bonds. With the increase in the concentration of graphite components in the coating, the corrosion potential is shifted to the negative value, which reduces the protective properties of the coating. Higher protective properties of gradient Cr/α-C (5…20 Hz):Cr coatings, slowing down of corrosion processes due to the formation of chromium carbide in their volume. At the same time, the formation of gradient coatings, the change in the concentration of the alloying metal in thickness allows to change the phase composition of the carbon matrix, while providing the necessary surface properties. At the same time, it can be expected that the improvement of corrosion resistance of gradient carbon coatings is possible due to an increase in the thickness of the coating, which statistically reduces the possibility of formation of through defects, as well as the formation of coatings from alternating layers characterized by different electrochemical properties.

4 Conclusions

The features of the phase and chemical composition of gradient containing chromium, aluminum, titanium carbon coatings are established. The formation of carbide compounds in coatings containing chromium and titanium was established by the XPS. The kinetic parameters of the electrochemical corrosion process are determined. It is shown that the highest corrosion resistance in the electrolyte based on NaCl is characterized by a gradient Cr/α-C(5…20 Hz):Cr coating (the change in the corrosion potential is from -0.117 to -0.715 V), which is explained by the more intensive formation of chromium carbide in its volume. It is established that gradient metal-containing carbon coatings have low wetting angle values and are hydrophilic.

Acknowledgements. This work was supported by Intergovernmental Cooperation Projects in the National Key Research and Development Plan of the Ministry of Science and Technology of PRC (projects No. 2016YFE0111800, for 2016–2019), the Belarusian Republican Foundation for Fundamental Research (project No. T18КИ-008, for 2018–2019)).

References

1. Cui, X.-J., Ning, C.-M., Shang, L.-L., Zhang, G.-A., Liu, X.-Q.: Structure and anticorrosion, friction, and wear characteristics of pure diamond-like carbon (DLC), Cr-DLC, and Cr-H-DLC films on AZ91D Mg Alloy. J. Mater. Eng. Perform. **28**, 1213–1225 (2018). https://doi.org/10.1007/s11665-019-3854-8
2. Niu, F., Ma, J., Yu, L., Zhang, P.: Electrochemical deposition and characterization of diamond-like carbon films doped with surface-capped silica nanoparticles. Surf. Coat. Technol. **221**, 77–87 (2013). https://doi.org/10.1016/j.surfcoat.2013.01.030

3. Bewilogua, K., Hofmann, D.: History of diamond-like carbon films – From first experiments to worldwide applications. Surf. Coat. Tech. **242**, 214–225 (2014). https://doi.org/10.1016/j.surfcoat.2014.01.031
4. Miao, Y.M., Jiang, X.H., Piliptsou, D.G., Zhuang, Y., Rogachev, A.V., Rudenkov, A.S., Balmakou, A.: Chromium-modified a-C films with advanced structural, mechanical and corrosive-resistant characteristics. Appl. Surf. Sci. **379**, 424–432 (2016). https://doi.org/10.1016/j.apsusc.2016.04.112
5. Zhang, L.F., Wang, F., Qiang, L., Gao, K., Zhang, B., Zhang, J.: Recent advances in the mechanical and tribological properties of fluorine-containing DLC films. RSC Adv. **5**, 9635–9649 (2015). https://doi.org/10.1039/C4RA14078H
6. Yildiz, R.: An electrochemical and theoretical evaluation of 4,6-diamino-2 pyrimidinethiol as a corrosion inhibitor for mild steel in HCl solutions. Corros. Sci. **90**, 544–553 (2015). https://doi.org/10.1016/j.corsci.2014.10.047
7. Guo, W., Li, X., Chen, M., Xu, L., Dong, L., Cao, X., Tang, W., Zhu, J., Lin, C., Pan, C., Wang, Z.L.: Electrochemical cathodic protection powered by triboelectric nano-generator. Adv. Funct. Mater. **24**, 6691–6699 (2014). https://doi.org/10.1016/j.corsci.2014.10.047
8. Son, M.J., Zhang, T.F., Jo, Y.J., Kim, K.H.: Enhanced electrochemical properties of the DLC films with an arc interlayer, nitrogen doping and annealing. Surf. Coat. Tech. **329**, 77–85 (2017). https://doi.org/10.1016/j.surfcoat.2017.09.025
9. Cui, M.J., Pu, J., Zhang, G., Wang, L., Xue, Q.: The corrosion behaviors of multilayer diamond-like carbon coatings: influence of deposition periods and corrosive medium. RSC Adv. **6**, 28570–28578 (2016). https://doi.org/10.1039/C6RA05527C
10. Zhang, L.F., Wang, F., Qiang, L., Gao, K., Zhang, B., Zhang, J.: Recent advances in the mechanical and tribological properties of fluorine-containing DLC films. RSC Adv. **5**, 9635–9649 (2015). https://doi.org/10.1039/C4RA14078H
11. Dai, W., Wua, G., Wang, A.: Preparation, characterization and properties of Cr-incorporated DLC films on magnesium alloy. Diam. Rel. Mat. **19**, 1307–1315 (2010). https://doi.org/10.1016/j.diamond.2010.06.018
12. Dai, W., Ke, P.L., Wang, A.Y.: Microstructure and property evolution of Cr-DLC films with different Cr content deposited by a hybrid beam technique. Vacuum **85**, 792–797 (2011). https://doi.org/10.1016/j.vacuum.2010.11.013
13. Rogachev, A.V., Kulesh, E.A, Piliptsou, D.G., Rudenkov, A.S., Hong, J.X.: Structure and mechanical properties of gradient metal-carbon coatings. In: Recent Advances in Technology Research and Education., vol. 53. Springer, Cham (2019). https://doi.org/10.1007/978-3-319-99834-3_1
14. Andersson, M., Högström, J., Urbonaite, S., Furlan, A., Nyholm, L., Jansson, U.: Deposition and characterization of magnetron sputtered amorphous Cr–C films. Vacuum **86**(9), 1408–1422 (2012). https://doi.org/10.1016/j.vacuum.2012.01.021
15. Cloutier, M., Harnagea, C., Hale, P., Seddiki, O., Rosei, F., Mantovani, D.: Long-term stability of hydrogenated DLC coatings: Effects of aging on the structural, chemical and mechanical properties. Diam. Relat. Mat. **48**, 65–72 (2014). https://doi.org/10.1016/j.diamond.2014.07.002
16. Robertson, J.: Diamond-like amorphous carbon. Mater. Sci. Eng. **37**, 129–281 (2002). https://doi.org/10.1016/S0927-796X(02)00005-0
17. Martínez-Martínez, D., López-Cartes, C., Fernández, A., Sánchez-López, J.C.: Influence of the microstructure on the mechanical and tribological behavior of TiC/a-C nanocomposite coating. Thin Solid Films **517**, 1662–1671 (2009). https://doi.org/10.1016/j.tsf.2008.09.091

18. Wang, Q., Zhou, F., Zhou, Z., Yang, Y., Yan, C., Wang, C., Zhang, W., Li, L.K.-Y., Bello, I., Lee, S.-T.: Influence of Ti content on the structure and tribological properties of Ti-DLC coatings in water lubrication. Diam. Relat. Mat. **25**, 163–175 (2012). https://doi.org/10.1016/j.diamond.2012.03.005
19. Viswanathan, S., Mohan, L., Bera, P., Kumar, V.P., Barshilia, H.C., Anandan, C.: Corrosion and Wear Behaviors of Cr-Doped Diamond-Like Carbon Coatings. J. Mater. Engin. Perform. **28**, 3633–3647 (2017). https://doi.org/10.1007/s11665-017-2783-7
20. Wei, X.: Hydrophobicity of hydrogen DLC films. Surf. Technol. **45**(5), 154–161 (2016)

Polarization Properties of a Rectangular Balanced Omega Element in the THz Range

Igor Semchenko(✉), Sergei Khakhomov, Maxim Podalov, and Andrey Samofalov

Francisk Skorina Gomel State University, Sovyetskaya Str. 104, 246019 Gomel, Belarus
isemchenko@gsu.by

Abstract. The paper aims to create a new type of polarizers in the THz range of electromagnetic waves, comprising an array of micro-dimensional planar rectangular omega elements. The metal omega elements under consideration have a well-balanced shape, since the incident electromagnetic wave induces in them an electric dipole moment and a magnetic moment, which are equally significant. Such an optimal shape of the omega elements allows their use in the absorbers of microwave and THz waves. The paper illustrates that this shape of omega resonators is also universal for their use in THz polarizers.

Keywords: Metamaterials · Omega-element · Polarization · Terahertz range

1 Introduction

Currently, the development of metamaterial technology is experiencing rapid growth. At the same time, despite the successful studies, the broadband polarization converter in the THz range has not been implemented yet. So the idea of creating such a polarizer is fairly new and relevant. The devices for producing circularly polarized THz waves are of great interest. The electromagnetic waves with this polarization type can be used to study and diagnose various objects, including biological ones, in the THz frequency range. Chiral structures, which can exhibit polarization-selective properties, stand out among these objects. The study of such properties requires THz waves with right and left circular polarization. These waves can interact differently with chiral objects and thereby reveal their structure and distinctive features.

The polarizers of electromagnetic waves in different spectral bands can have elements of various shapes, e.g. helices with different numbers of turns, open rings directed in different positions, as well as classical or rectangular omega elements [1–9]. The main range of these polarizers is microwave, and, to a lesser extent, the THz band of the spectrum. Paper [1] presents numerical studies on a multi-functional polarization convertor for terahertz light, composed of a bilayer wire-split-ring structure chiral metamaterial. Paper [2] studies a highly sensitive refractive index sensor based on the conjugated bilayer Γ-shaped chiral metamaterial. Paper [3] provides the studies on the bilayer structure with microstrip connecting line to achieve giant circular dichroism. Paper [4] presents a bi-layered chiral metamaterial to realize high–efficiency broadband

A. R. Várkonyi-Kóczy (Ed.): INTER-ACADEMIA 2019, LNNS 101, pp. 84–93, 2020.
https://doi.org/10.1007/978-3-030-36841-8_8

asymmetric transmission of linearly polarized electromagnetic wave in the microwave region. An asymmetric chiral metamaterial circular polarizer based on bilayer twisted split-ring resonator structure is proposed and investigated in [5]. The plasmonic version of a 3D chiral meta-atom which consists of a loop-wire structure, namely the so-called twisted omega particle, is implemented in [6]. Paper [7] presents systematical studies of the Ω elements of classical shape as well as considers the impact of geometric parameters on chiral effects, which occur in the metamaterial. Paper [8] studies the chiral metamaterial consisting of double L resonators on two sides of the dielectric substrate. Paper [9] considers the multi-layered structure comprising two sub-wavelength gratings and split-ring resonator, printed on two sides of dielectric substrate.

This paper presents the study of the properties of a rectangular omega element for potential use of such elements in the polarizers of THz waves. Since the element being examined is flat, this facilitates the production of metamaterials and metasurfaces by using planar technologies. The second advantage of the omega element is the fact that the electric current, arising in it, induces simultaneously an electric dipole moment and a magnetic moment. This contrasts with the coupled elements of metamaterials, e.g. straight wires and split rings. The first element in this couple is necessary for producing the dielectric properties of the metamaterial, and the other one is for producing the magnetic properties. Compared to the omega element, such coupled elements are more difficult to tune, as different modes of electromagnetic waves are activated in each of them. The third advantage of the rectangular omega-element is that the conversion of the THz wave polarization will occur in the designed metamaterial only when the wave is reflected, but not when it passes through the metamaterial. This will reduce the losses of wave intensity for the absorbing metamaterial.

The implementation of the elements to convert wave polarization in two-dimensional and three-dimensional arrays of metamaterials is associated with certain difficulties. For example, helix elements have quite a complicated spatial form, and their production has a number of limitations in the size and material of the substrate. Omega elements and split rings have a planar geometry and can be easily manufactured using modern photolithography techniques.

The production of a new type of the omega-element for the THz range is due to the use of the balance property of this element. The optimality condition for the omega-element as part of a two-dimensional array was previously formulated and experimentally investigated. This was done primarily to produce an absorber of microwave and THz waves. The optimality condition determines the equality of the induced magnetic and electric dipole moments. They are equal in absolute value in the optimal element. The length of the metallized strip, forming the polarizing element, is approximately equal to half the wavelength of the incident radiation, which corresponds to the main condition of frequency resonance [10–13].

The aim of our research paper is to obtain and study a balanced rectangular omega-element and eventually produce two-dimensional and three-dimensional arrays on its basis in the THz range for frequency filtration and wave polarization conversion. In the future, a reflective polarizer of the THz wave is expected to be produced. This polarizer will allow changing the wave polarization from linear to circular when the wave is reflected from the metamaterial.

2 Simulation

The research methodology consists in numerical simulation of the electric current, arising in a rectangular omega-element under the influence of an incident electromagnetic wave. The electric dipole moment and the magnetic moment of the omega element, which have a mutually perpendicular direction, are calculated. Then the ellipticity coefficient of the reflected wave is determined, which must be close to 1 under the balance of moments condition. The influence of the omega-element geometric parameters on the possibility of obtaining a circular polarization of the reflected wave is taken into account. The change in the frequency of the incident wave near the resonance and two different directions of its electric vector are considered as well.

The study uses a traditional bianisotropic particle with an omega coupling, which is a metal Ω shape strip. It is possible to find a balance condition for the polarizability for a single uniaxial omega particle, made from a conducting wire. It is also known that some sort of restriction on the electromagnetic properties of the wire omega particle can be found in the approximation of electrically small dimensions.

All the polarizabilities of the omega particle are determined on the basis of the finite element method (FEM). Computer simulation and an analytical approach are used as well to determine the polarizability of the rectangular omega element.

The parameters of the element can be calculated approximately considering the model of a quasi-stationary current. It is expected that the current intensity does not change depending on the coordinate measured along the omega element. The design of the rectangular shape of the omega element is selected on the basis of the balance condition, since the induced electric and magnetic dipole moments are equally significant. It should be noted that electromagnetic and magnetoelectric polarizabilities are equal as well.

The polarizability balance is achieved by changing the geometric parameters of the omega element. The polarization properties of the overall metamaterial can be enhanced using the array of balanced micro-resonators on the metasurface.

Figure 1(a) and (b) shows Ω-resonator design with new structural parameters and its polarizability. As can be seen, this omega resonator is balanced, since the induced electrical and magnetic dipole moments are equally significant. This property is reflected in the formula in Fig. 1(b). It should be noted that electromagnetic and magnetoelectric polarizabilities are also equal. Consequently, this Ω-resonator fully satisfies the polarizability balance condition. Using the array of these resonators, it is possible to achieve complete polarization of electromagnetic waves. At the same time the polarizer will be inactive away from resonance. However, additional optimization of the particles position in the array is required, since their interaction in the array also significantly affects the properties of the polarizer.

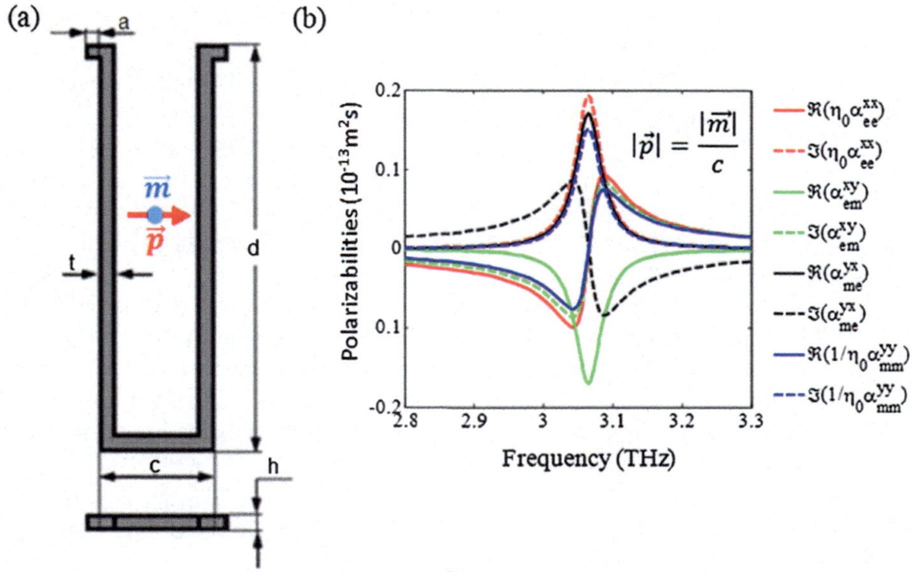

Fig. 1. (a) The Ω-resonator design with structural parameters; (b) calculated polarizability of a balanced Ω-resonator

The main initial parameters are shown in Fig. 1. The width of the element is approximately equal to c = 10 μm, the width of the metallized strip is t = 1 μm, the length of the element is d = 70 μm, the thickness of the metallized strip is h = 1 μm, the arm length is a = 0 μm. Figure 1 shows the arms for illustrative purposes, since in parametric simulation the parameter **a** can change upwards, but it is 0 (zero) when calculating a balanced rectangular Ω-element. The metallization material is copper; the element is in a vacuum in the initial conditions.

2.1 Boundary Conditions

The Ω-elements in the microwave range (2.55–3.8 GHz) were previously calculated and investigated in [10, 11]. These particles were also considered for the terahertz range as elements of effective polarizers or absorbers of electromagnetic waves [12–14]. At this stage of research, a rectangular Ω-element is being developed, the parameters of which are optimized for the terahertz radiation. It is shown that the metamaterial, based on an array of omega elements, can perform the functions of an effective polarization transducer of an electromagnetic wave.

Designing a balanced rectangular omega element for the THz range is associated with a change in the parameters of the element in accordance with a change in the wavelength of the incident radiation. Under the main resonance condition, the length of the optimal omega-element in the rectified state is approximately equal to half the wavelength of the incident radiation. In particular, the wavelength for the 1 THz frequency is 0.3 mm, and the length of the metallized strip, forming the omega element, will be about 0.15 mm.

The optimality condition for the omega-element means that equally significant electric dipole moment and magnetic moment are induced in the element [10]. Considering this condition, the following parameters and design of the element are obtained (Fig. 1).

The balance of the rectangular omega element can be obtained, for example, by optimizing the geometrical parameters of this omega element, designed with the use of the quasi-stationary current model.

The polarization properties of the rectangular omega element in the terahertz range are simulated. The parameters of the metasurface, consisting of omega elements, are determined. The omega-elements are necessary for the effective conversion of the incident linearly polarized wave into the reflected wave with circular polarization in the THz frequency range. The studies are carried out with different directions of the incident wave vector $\vec{E}$. In the first case, the vector of the electric field of the incident wave oscillates in the plane of incidence, i.e. in the plane that passes through the ends of the omega element. In this case, an electric dipole moment is induced by the electric field, and a magnetic moment arises as a result of the electric current in the omega element. In the second case, the magnetic field vector of the incident wave oscillates in the plane of incidence and has a component which is perpendicular to the plane of the omega element. In this case, a magnetic moment is induced by the magnetic field, and an electric dipole moment arises as a result of the electric current in the omega-element. The moments moduli are equal in both cases, which leads to the appearance of a reflected wave with a circular polarization.

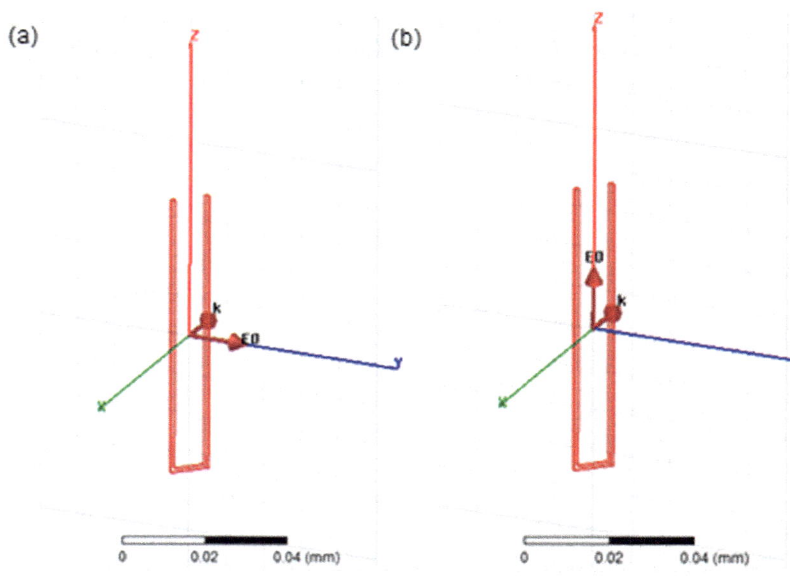

Fig. 2. The direction of the vector $\vec{k}$ of the incident EM wave relative to the structural elements of the omega cell, (a) $\vec{E}$ vector oscillates in the XY plane, (b) $\vec{E}$ vector is directed along the Z axis

On constructing the simulation object, the boundary conditions were introduced, and the parameters of the incident electromagnetic wave were set. When addressing the current task, an incident plane wave is used. The vector of the incident wave $\vec{E}$ is oriented either horizontally in the XY plane, i.e. in the plane parallel to the arms of the element, or vertically, along the Z axis, i.e. orthogonal to the arms of the element. The wave vector $\vec{k}$ of the incident wave is directed along the X axis, i.e., at an angle of 45 degrees to the omega-element plane (Fig. 2).

3 Results of the Simulation and Their Analysis

The current paper considers the case with the oblique incidence of an electromagnetic wave; $\vec{k}$ vector of the incident wave is directed at 45° to the axis of the rectangular omega element. The calculated wavelength is 0.3 mm, and the frequency is 1 THz for the optimal balanced omega element.

The oblique incidence of the EM wave (when the vector is directed at an angle of 45 degrees to the plane of the omega element turn) allows to activate both the electric dipole moment and the magnetic moment of the omega elements, so the ellipticity coefficient of the reflected EM wave will reach maximum values at the calculated frequency. In this case, the condition of half-wave resonance is satisfied, therefore the reflection coefficient takes the maximum value as well.

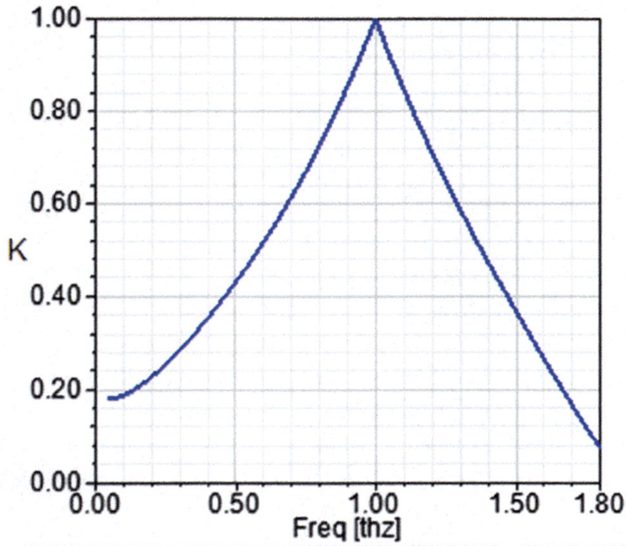

Fig. 3. The graph of the frequency dependence of the ellipticity coefficient of the reflected wave for the element with the parameters: a = 0.3 μm, h = 1.5 μm, d = 69 μm, $\vec{E}$ vector oscillates in the plane parallel to the arms of the Ω-element

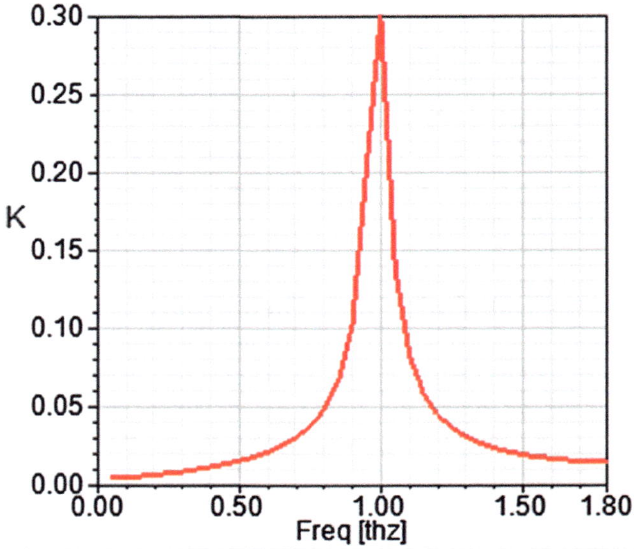

Fig. 4. The graph of the frequency dependence of the ellipticity coefficient of the reflected wave for the element with the parameters: a = 0.1 μm, h = 1.5 μm, d = 69 μm, $\vec{E}$ vector oscillates in the plane orthogonal to the arms of the Ω-element

Consider the first case with the direction of $\vec{E}$ vector of the incident wave in the plane parallel to the arms of the rectangular Ω-element. Parametric simulation is carried out using mainly three parameters of the element: the length, the thickness of the metallized strip and the length of the arm. The Ω-element with the parameters a = 0.3 μm, h = 1.5 μm, d = 69 μm, c = 10 μm, t = 1 μm prove to be the best polarizer of all the parameters under study. The Ω-element with such parameters shows the ellipticity coefficient of the reflected wave K_{Max} = 0.997 at the calculated frequency equal to 1 THz (Fig. 3). In this case, the Ω-element converts well enough the polarization of an electromagnetic wave, when it is reflected from a plane-polarized incident wave, to a circularly polarized reflected wave in a wide frequency interval. The peak of the electric field intensity of the reflected wave is observed at a frequency of 1 THz, which indicates a good excitation of the Ω-element by the electromagnetic wave at the calculated frequency. Figure 5 shows the electromagnetic wave activation of the rectangular omega element for this case.

While analyzing the other case, when $\vec{E}$ vector is orthogonal to the arms of the rectangular omega element, it can be concluded that the Ω element with the following parameters has the best polarization properties for this case: a = 0.1 μm, h = 1.5 μm, d = 69 μm, s = 10 μm, t = 1 μm. The Ω-element with such parameters showed the ellipticity coefficient of the reflected wave K_{Max} = 0.299 at the calculated frequency of 1 THz, which corresponds to the elliptically polarized reflected wave (Fig. 4). In this case, the "turn" of the omega element is penetrated by the magnetic field vector of the incident wave, i.e. the element is activated by $\vec{H}$ vector. A small peak of the electric field intensity of the reflected wave is recorded at a frequency of 0.95 THz, which

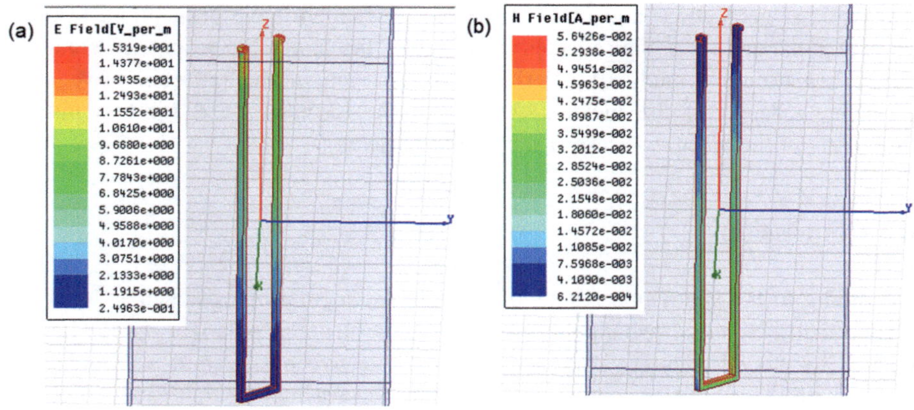

Fig. 5. Field overlay: a – $\vec{E}$ vector; b – $\vec{H}$ vector, the Ω-element with the parameters: a = 0.3 μm, h = 1.5 μm, d = 69 μm, $\vec{E}$ vector oscillates in the plane parallel to the arms of the Ω-element

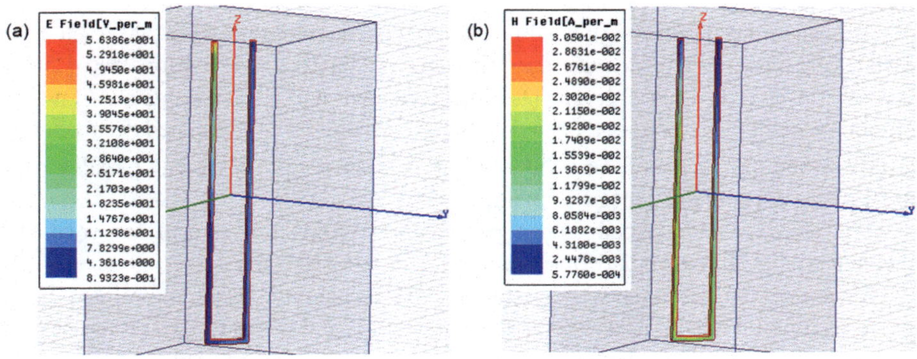

Fig. 6. Field overlay: a – $\vec{E}$ vector; b – $\vec{H}$ vector, Ω-element with the parameters: a = 0.1 μm, h = 1.5 μm, d = 69 μm, $\vec{E}$ vector oscillates in the plane orthogonal to the arms of the Ω-element

generally corresponds to theoretical calculations. A small ellipticity coefficient is explained by the insufficiently effective electric component of the incident wave, since the wave mainly activates the front leg of the rectangular element (Fig. 6).

This paper provides a numerical simulation of the electric current arising in a rectangular omega-element under the influence of an incident electromagnetic wave. The excited mode of electromagnetic oscillations produces both an electric dipole moment and a magnetic moment in the omega-element. These moments have a mutually perpendicular direction. The arising moments and polarizabilities of the omega-element as a bianisotropic particle have been calculated. The ellipticity coefficient of the reflected wave has been determined. This coefficient must be close to 1 under the balance of moments condition. Consequently, it becomes possible to convert the incident linearly polarized wave into a reflected wave with circular polarization.

The optimal geometric parameters of the omega element have been found to obtain circular polarization of the reflected wave under the half-wave resonance condition. There has been the study of polarization conversion of the reflected wave at the frequency change of the incident wave near the resonance and for two different directions of the electric vector of the incident wave.

4 Conclusions

According to the simulation results of a single rectangular omega element, it is concluded that a balanced rectangular omega element has high polarizing properties calculated for the THz range. This element is an effective polarizer of electromagnetic waves with the ellipticity coefficient of the reflected wave close to 1, with an oblique incidence of the electromagnetic wave at an angle of 45 degrees to the element plane.

One of the advantages of the rectangular omega element under consideration is that the conversion of the THz wave polarization in the designed metamaterial will become possible for the reflected wave, and not for the wave passing through the metamaterial. This will allow using the absorbing metamaterials and reducing losses in the wave intensity with converted polarization.

The maximum value of the ellipticity coefficient, close to 1, is observed for a rectangular omega element with the following parameters: $a = 0.3$ μm, $h = 1.5$ μm, $d = 69$ μm, $c = 10$ μm, $t = 1$ μm. The omega element with such parameters shows the ellipticity coefficient of the reflected wave $K = 0.997$ at a calculated frequency of 1 THz, when the vector of the incident wave $\vec{E}$ is directed in the plane parallel to the arms of the rectangular omega element.

The rectangular balanced omega particle is a promising element for creating a metamaterial with high polarization properties in the THz range.

Since the omega elements under study have a rectangular shape, the printed-circuit board techniques can be used to create metamaterials and metasurfaces on the basis of these elements. Vacuum-plasma technologies can also be applied to produce the omega-structured metamaterials and metasurfaces containing rectangular elements.

The research has been performed as a part of the government research program "Photonics, opto- and microelectronics" of the "Photonics" subprogram.

References

1. Cheng, Z., Cheng, Y.: A multi-functional polarization convertor based on chiral metamaterial for terahertz waves. Opt. Comm. **435**, 178–182 (2019)
2. Sun, B., Yu, Y.: Optical refractive index sensor based on the conjugated bilayer Γ-shaped chiral metamaterials Optik **182**, 587–593 (2019)
3. Yingying, Yu., Sun, B.: Analysis of giant circular dichroism metamaterial based on conductive coupling. Optik **182**, 1046–1052 (2019)
4. Mirzamohammadi, F., Nourinia, J., Ghobadi, C., Majidzadeh, M.: A bi-layered chiral metamaterial with high-performance broadband asymmetric transmission of linearly polarized wave. AEU—Int. J. Electron. Commun. **98**, 58–67 (2019)

5. Cheng, Y.Z., Nie, Y., Cheng, Z.Z., Wang, X., Gong, R.Z.: Asymmetric chiral metamaterial circular polarizer based on twisted split-ring resonator. Appl. Phys. B **116**, 129–134 (2014)
6. Sakellari, I., Yin, X., Nesterov, M.L., Terzaki, K., Xomalis, A., Farsari, M.: 3D chiral plasmonic metamaterials fabricated by direct laser writing: the twisted omega particle. Adv. Opt. Mater. **5**(16), 1700200 (2017)
7. Stojanović, D.B., Beličev, P.P., Radovanović, J., Milanović, V.: Numerical parametric study of chiral effects and group delays in Ω element based terahertz metamaterial. Phys. Lett. A **383**, 1816–1820 (2019)
8. Liu, D.-j., Xiao, Z.-y., Ma, X.-l., Kai-Kai, X., Tang, J.-y., Wang, Z.-h.: Broadband asymmetric transmission and polarization conversion of a linearly polarized wave based on chiral metamaterial in terahertz region. Wave Motion **66**, 1–9 (2016)
9. Liu, D.-j., Xiao, Z.-y., Ma, X.-l., Wang, Z.-h.: Broadband asymmetric transmission and multi-band 90° polarization rotator of linearly polarized wave based on multi-layered metamaterial. Opt. Commun. **354**, 272–276 (2015)
10. Semchenko, I.V., Khakhomov, S.A., Podalov, M.A., Tretyakov, S.A.: Radiation of circularly polarized microwaves by a plane periodic structure of Ω elements. J. Commun. Technol. Electron. **52**(9), 1002–1005 (2007)
11. Semchenko, I.V., Khakhomov, S.A., Samofalov, A.L., Podalov, M.A., Songsong, Q.: The effective optimal parameters of metamaterial on the base of omega-elements. In: Jablonski, R., Szewczyk, R. (eds.) Recent Global Research and Education: Technological Challenges, INTER-ACADEMIA 2016. Advances in Intelligent Systems and Computing, vol. 519, pp. 3–9 (2017)
12. Semchenko, I.V., Khakhomov, S.A., Samofalov, A.L., Podalov, M.A., Solodukha, V.A., Pyatlitski, A.N., Kovalchuk, N.S.: Omega-structured substrate-supported metamaterial for the transformation of wave polarization in THz frequency range. In: Luca, D., Sirghi, L., Costin, C. (eds.) Recent Advances in Technology Research and Education, INTER-ACADEMIA 2017. Advances in Intelligent Systems and Computing, vol. 660, pp. 72–80 (2018)
13. Semchenko, I., Khakhomov, S., Samofalov, A., Faniayeu, I., Slepiankou, D., Solodukha, V., Pyatlitski, A., Kovalchuk, N., Goncharenko, A., Sinitsyn, G.: Design and Creation of Metal-Polymer Absorbing Metamaterials Using the Vacuum-Plasma Technologies. Springer Nature Switzerland AG 2019, G. Laukaitis (Ed.): INTER-ACADEMIA 2018, LNNS 53, pp. 105–112 (2019)
14. Balmakou, A., Podalov, M., Khakhomov, S., Stavenga, D., Semchenko, I.: Ground-plane-less bidirectional terahertz absorber based on omega resonators. Opt. Lett. **40**(9), 2084–2087 (2015)

Real-Time Analysis and Characterization of Dynamic Performances for Coloration-Bleaching Processes in Electrochromic Devices Based on WO_3 Layers Prepared by Sol-Gel Synthesis

E. V. Sokhovich(✉), V. I. Khalimon, O. V. Prostitenko, S. V. Mjakin, M. M. Sychov, V. F. Borodzyulya, V. S. Zemko, and S. O. Lebedev

Saint-Petersburg State Institute of Technology (Technical University), 26 Moskovsky pr., Saint-Petersburg 190013, Russia
svmjakin@technolog.edu.ru

Abstract. A comparative characterization of multiple coloration-bleaching cycles under various modes of power (voltage and current supply) is performed for electrochromic devices (ECDs) based on WO_3 layers prepared from metal tungsten or tungsten carbide and annealed at different temperatures from 150 to 350° C. The results indicate that optimal voltage/current values and exposure time strongly depend on WO_3 synthesis and annealing conditions. Particularly, a sustainable functioning of ECD involving WO_3 prepared from metal tungsten and annealed at 150° C is achieved upon a stepwise change of coloration and bleaching voltage from 1.5 to 2.5 V at the current 1 mA within 600 s, while for the sample prepared from tungsten carbide and annealed at 350° C a similar mode within 300 s is effective, probably due to certain differences in the electrochromic material structure. Deviations from optimal power supply conditions result in the destabilization of ECD performances and significant efficiency decrease after several coloration-bleaching cycles, particularly due to an incomplete bleaching.

Keywords: Electrochromism · Electrochromic device · Coloration · Bleaching · Tungsten oxide · Tungsten carbide

1 Introduction

Electrochromic devices (ECD) providing a reversible and adjustable change of transparency and color caused by electric current find a growing demand in “smart” windows, special glassing for various vehicles, such as anti-glare coatings, special displays and light filters. Generally, ECD consists of several functional layers, including supports with transparent conducting layers (ITO, FTO), cathode electrochromic layer (usually based on tungsten oxide WO_3), counter electrode (commonly TiO_2) and electrolyte separating the electrodes and containing lithium ions (Li^+) that reversibly penetrate WO_3 layer to promote the required coloration. The main performances

A. R. Várkonyi-Kóczy (Ed.): INTER-ACADEMIA 2019, LNNS 101, pp. 94–100, 2020.
https://doi.org/10.1007/978-3-030-36841-8_9

characterizing ECD are coloration contrast, coloration-bleaching rates, electrochromic efficiency (integral parameter taking into account both coloration intensity and power consumption) as well as the stability in the course of storage and multiple (up to thousands) coloration-bleaching cycles. The complexity of electrochemical processes involving different functional layers and their boundaries, particularly at voltage polarity reversion providing consecutive coloration-bleaching, requires a precise dynamic testing and analysis of ECD parameters.

In continuation of earlier studies [1] relating to the optimization of the conditions for sol-gel synthesis and treatment of WO_3 electrochromic layers, in this research we studied approaches to the improvement of electric conditions for ECD operation. For this purpose, we developed automated installations and software for real-time acquisition, processing and comparative analysis of electric and optical performances of ECDs in the course of multiple coloration-bleaching cycles.

2 Experimental

Comparative studies were performed for tungsten oxide based ECDs prepared according to the procedure involving the following stages. First, dispersed metal tungsten $WO5_{-10}$ (Wolfram Co., Germany) of 0.5–0.75 μm size was dissolved in hydrogen peroxide 30% solution to obtain peroxotungstic acid (PTA) H_2WO_5, followed by drying of the prepared solution to provide PTA polycondensation yielding a transparent yellow solution of linear poly(peroxotungstic acid) PPTA $H_2W_2O_9$, and subsequent drip drying of this solution under equilibrium conditions at 50–60°C. For comparison, in another experiment tungsten carbide was dissolved in hydrogen peroxide instead of metal tungsten at the first step of the synthesis. The resulting glassy product (hydrated α-WO_3 xerogel) was dissolved in a purified ethanol within 30 min to obtain a transparent yellow WO_3 alcosol (15%wt.). Thus prepared alcosols were spin coated onto a 60 × 60 mm glass support of 3.8 mm thickness with a transparent SnO_2: F (FTO) conducting layers of 150–200 nm thickness and surface resistance 20–25 Ω (Pilkington Co., UK).

Supports with alsosol layers were placed into Petri dishes, tightly covered and annealed within 1 h at 150–350 °C. In a similar way, TiO_2 layers were deposited onto similar supports and annealed at 350 °C.

Based on the prepared samples, ECDs were assembled by sticking supports with WO_3 and TiO_2 layers separated by 1 M $LiClO_4$ solution in propylene carbonate as an electrolyte with the conductivity $9.5 \cdot 10^{-3}$ Cm/cm. The electrolyte was injected between the supports via a syringe followed by sealing the ECD with a silicone sealant.

Thus obtained ECD samples involved WO_3 layers prepared from metal tungsten and annealed at 150 (ECD-1), 250 (ECD-2) and 350 °C (ECD-3), as well as WO_3 layer prepared from tungsten carbide and annealed at 350 °C (1-ECD-3).

According to XRD data, the prepared WO_3 and TiO_2 layers were amorphous, and the counducting FTO layer contained a tetragonal SnO_2 phase with particle size 70–100 nm.

The obtained ECDs were tested using an automated installation [1, 2] providing a programmed remote adjustment of coloration-bleaching parameters and calculation of optical and electric performances. The program of performed tests involved 20 cycles of consecutive coloration and bleaching (under reverse polarity voltage) with a stepwise change of coloration and bleaching voltage from 1.5 to 2.5 V with the step 0.5 V within 300 s at current 1.65 mA (mode 1) or within 600 s at current 1 mA (mode 2). In both modes, all the voltage change steps were separated with a 10 s pause.

The coloration intensity was measured according to the receiving photodiode signal in arbitrary units in the range from 10000 to 25000. The coloration-bleaching contrast R was calculated as the difference between the consecutive coloration (maximum) and bleaching (minimum) signals.

3 Results and Discussion

The testing results for all the studied ECD samples are presented in Fig. 1 as changes of voltage (top plot), current (middle plot) and coloration-bleaching intensity (bottom plot) in the course of 20 coloration-bleaching cycles.

The data summarized in Fig. 1 indicate that for ECD-1 annealed at 150 °C (Fig. 1a) the coloration-bleaching intensity stabilization is achieved after 10 cycles. However, the application of mode 1 leads to abrupt surges of current up to 1.65 mA that deteriorates a sustainable functioning of electrochromic layer. The increase of the pause between voltage switches to 600 s (mode 2) results in the reduction of these surges to about 1.2 mA (Fig. 1b), stabilization of coloration-bleaching performances already after 4 cycles and increase of the contrast R from ~4500 to ~6000 (the latter determined as the difference between colored and bleaching state intensities 16000 and 10000, respectively).

Similar tests for the sample ECD-2 (annealed at 250 °C) in the mode 1 (Fig. 1c) indicated a significant decrease of the contrast since the second cycle (due to increased level of the bleached state intensity, i.e. incomplete bleaching upon the reverse polarity voltage application) with the growth of current surges from 1.2 to about 2 mA. For mode 2, a similar result was observed after 5 cycles. Such an undesirable behavior is probably determined by structural features of tungsten oxide annealed at the considered temperature complicating lithium ions transfer from WO_3 layer to the electrolyte at bleaching, especially under the studied voltage-currency supply conditions inappropriate for this sample.

For ECD-3 sample, testing in both modes indicated even more prominent drop of the contrast (from 6000 to 2000 in mode 1, as shown in Fig. 1d, and from 9000 to 4000 in mode 2) since the first coloration-bleaching cycles, probably due to the WO_3 transition into a condense glassy state upon annealing at 350°C as confirmed by DSC characterization.

Experiments with 1-ECD-3 prepared from tungsten carbide in mode 1 with a reduced bleaching (reverse polarity) voltage (−0.75 V) indicated the worst result with almost no bleaching observed after 3–5 cycles and current surges growing up to 14–17 mA (Fig. 1e). However, the voltage increase to −1.5 V provided the process stabilization after 12 cycles with the reduction of current surges to about 1 mA and increase of the contrast up to about 11500 (Fig. 1f).

a

b

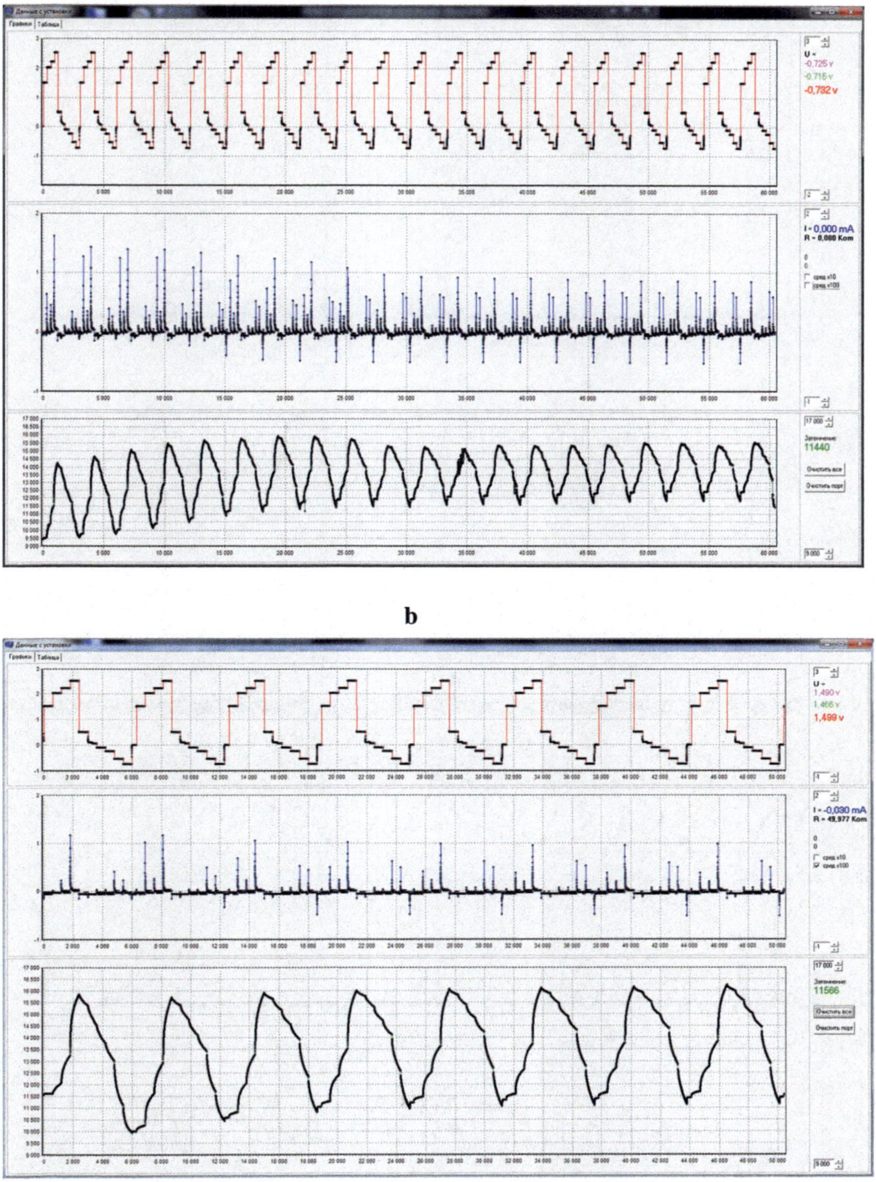

Fig. 1. Voltage (top), current (middle) and coloration-bleaching intensity (bottom) changes within 20 cycles for ECD-1 (a – mode 1, b – mode 2), ECD-2 (c – mode 1), ECD-3 (d – mode 1) and 1-ECD-3 (e - mode 1, bleaching voltage −0.75 V, f – mode 1, bleaching −1.5 V)

c

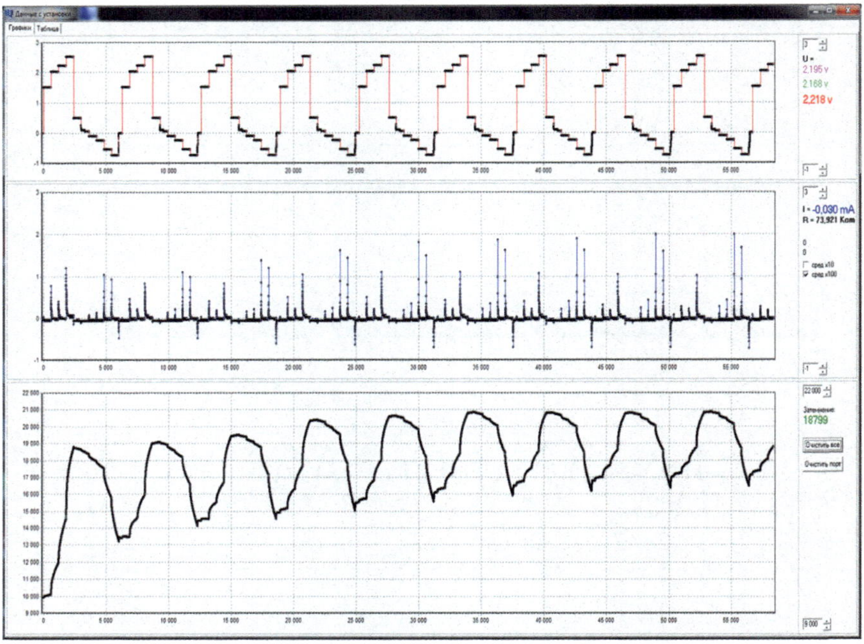

d

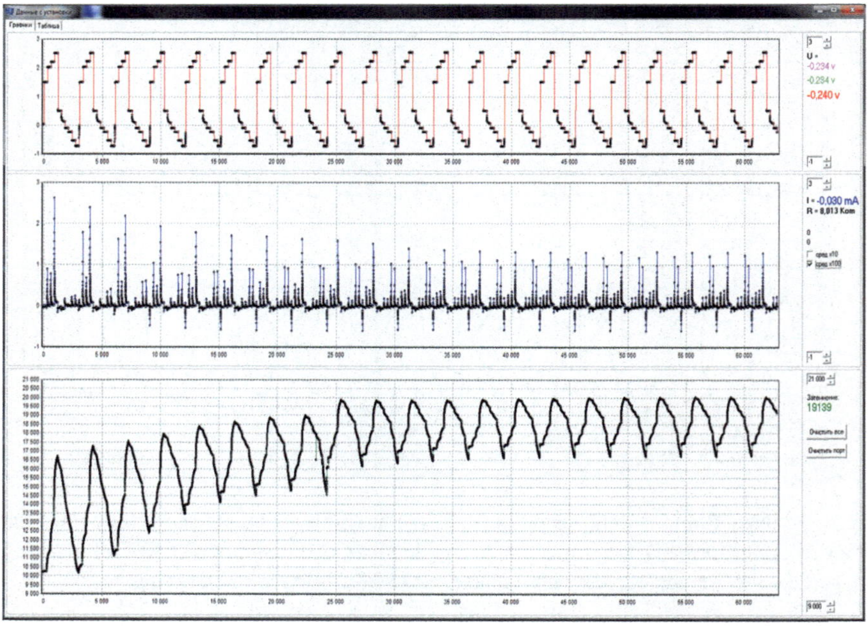

Fig. 1. (*continued*)

e

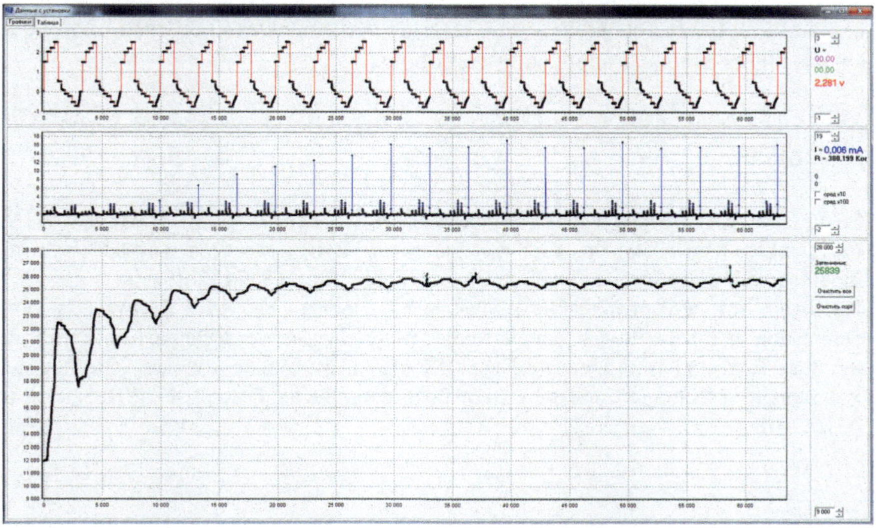

f

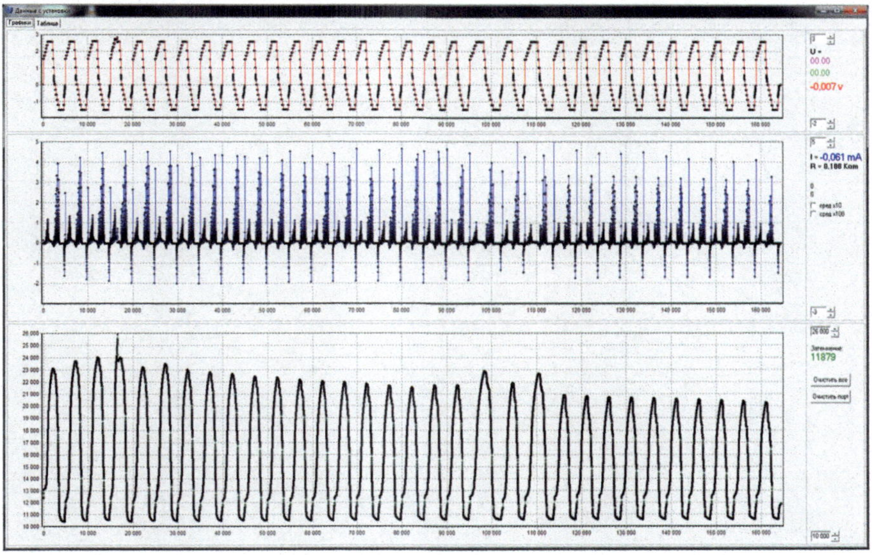

Fig. 1. (*continued*)

4 Conclusion

Generally, the obtained results suggest that values and exposure timing modes of coloration and bleaching voltage and current applied to electrochromic devices strongly depend on tungsten oxide layer synthesis and thermal processing conditions and should be precisely optimized, that determines the necessity in the further thorough characterization of all the factors affecting ECD target performances in order to improve their efficiency under practical application conditions.

References

1. Sokhovich, E.V., Mjakin, S.V., Semenova, A.A., Zemko, V.S., Bakhmetyev, V.V., Prostitenko, O.V., Khalimon, V.I.: Electrochromic devices based on tungsten oxide layers modified with polyethylene glycol. J. Opt. Technol. **1**(86), 68–74 (2019)
2. Khalimon, V.I., Prostitenko, O.V., Mjakin, S.V., Sokhovich, E.V.: Software system for automation of electrochromic sample studies using client-server technologies. Certificate of the State Registration of Software No. 2018661945, filed in the Russian State Register of Programs for Electronic Computers 24.09.2018 upon the application No. 2018617158 from 03.07.2018

Melting Threshold and Thermal Conductivity of CdTe Under Pulsed Laser Irradiation

V. P. Veleschuk[1(✉)], V. A. Gnatyuk[1,2], T. Aoki[2], Z. K. Vlasenko[1], S. N. Levytskyi[1], A. V. Shefer[3], A. G. Kuzmich[4], K. V. Dubyk[4], V. V. Kuryliuk[4], and M. V. Isaiev[4,5]

[1] V.E. Lashkaryov Institute of Semiconductor Physics of the National Academy of Sciences of Ukraine, Prospekt Nauky 41, Kyiv 03028, Ukraine
gnatyuk@ua.fm
[2] Research Institute of Electronics, Shizuoka University, 3-5-1 Johoku, Naka-Ku, Hamamatsu 432-8011, Japan
[3] Poltava National Technical Yuriy Kondratyuk University, Prospekt Pershotravnevyi 24, Poltava 36011, Ukraine
[4] Taras Shevchenko National University of Kyiv, Volodymyrska Str. 64/13, Kiev 01601, Ukraine
[5] Université de Lorraine, CNRS, LEMTA (UMR 7563), Vandoeuvre les Nancy 54500, France

Abstract. The paper is devoted to the study of the features of CdTe surface treatment under laser irradiation with both different wavelengths (λ = 300–800 nm) and pulse durations (τ_p = 7 ns–1 ms). The thermal conductivity of the semi-insulating *p*-like CdTe semiconductor crystals was evaluated using the photoacoustic gas-microphone method. Simulations of the melting threshold were performed based on the three stage model of the laser induced excitation and relaxation. In particular, the following processes were considered in frames of the model: (i) rapid interband thermalization, (ii) nonradiative interband and (iii) nonradiative surface recombination. It was revealed that in the range of pulse durations from 7 ns to 1 μs, the melting threshold of the CdTe mainly depended on the absorption coefficient $\alpha(\lambda)$. For pulse durations longer than 1 μs the threshold started to depend also on the spectra of the reflectivity coefficient $R(\lambda)$. The obtained results have been used for optimization of the laser-assisted techniques of surface processing and stimulated doping of CdTe crystals.

Keywords: CdTe · Melting threshold · Photoacoustic method · Pulsed laser irradiation

1 Introduction

Pulsed laser irradiation (PLI) is an efficient tool for investigation and modification of properties of various materials including semiconductors and device structures [1–6]. The unique characteristics of laser pulses have revolutionized surface processing that has opened great possibilities for obtaining the required parameters of semiconductors [1, 2]. For instance, PLI has been widely employed for cleaning of crystal surfaces,

A. R. Várkonyi-Kóczy (Ed.): INTER-ACADEMIA 2019, LNNS 101, pp. 101–109, 2020.
https://doi.org/10.1007/978-3-030-36841-8_10

annealing of structure imperfections, modification of the morphology, changing of carrier sheet concentration, surface recombination velocity, surface states and other characteristics of the superficial region of semiconductors [1–11].

The use of PLI in surface processing of semiconductors limited heat diffusion within the treated region, minimizes a heat-affected zone and suppresses formation of point and extended structure defects in the surroundings [1–4]. This feature of PLI plays a key role under creation of surface barrier structures, formation of electrical junctions or preliminary treatment of the semiconductor crystal surface before contact (electrode) deposition. Therefore, PLI has been intensively used for surface processing of CdTe crystals [7–9], in particular for laser-induced doping, creation of a *p-n* junction and formation of diode-type X/γ-ray detectors [10, 11].

For PLI of CdTe semiconductor, laser radiation with various pulse durations τ_p and wavelengths λ have been used. Ruby (λ = 694 nm) and YAG:Nd (1–3 harmonics with λ_1 = 1064 nm, λ_2 = 532 nm and λ_3 = 355 nm, respectively) lasers, as well as excimer KrF (λ = 248 nm), XeCl (λ = 308 nm) and XeF (λ = 351 nm) ones, are mainly employed for modification of crystal surface. The following laser pulse durations are most commonly used: 7 ns, 15 ns, 20 ns, 80 ns, 120 ns, 1 μs and 1 ms [7–11, 18–23].

It is clear, that information regarding the features of interaction of laser radiation with the semiconductor surface is necessary to modify the properties in a more desirable way. Particularly, for efficient modification of CdTe surface, the information regarding the melting threshold I_{th} of the material is crucial. The correct estimation of the melting threshold is a complicated issue, because this value depends on sequences of different parameters of the laser and semiconductor, in particular radiation wavelength and pulse duration and, thermal conductivity and parameters of nonequilibrium excess carriers (NEC). Thus, the development of the approaches and corresponding simulations as well as experimental study of photothermal transformation in the semiconductor is important both from fundamental and applied points of view.

In the present study, we have considered the features of the photothermal phenomena in bulk CdTe single crystals which were used for fabrication of diode-type X/γ-ray detectors by laser-induced doping [10, 11]. In particular, the photoacoustic (PA) set-up in the classical gas-microscope configuration was used for the evaluation of the thermal conductivity of CdTe crystals. Further, the estimated value was taken into consideration in the calculations of the melting thresholds of CdTe for lasers with different wavelengths and pulse durations.

2 Thermal Conductivity of the CdTe Measured by the Photoacoustic Technique

The use of PA techniques has a lot of advantages for the study of thermal physical properties of various materials [12–15], especially it concerns the situation when the creation of a contact is undesirable. The PA gas-microphone method in the conventional configuration [12, 15] was used for the evaluation of the thermal conductivity of semi-insulating Cl-compensated (111) oriented *p*-like CdTe single crystals grown with THM [9–11]. A typical experimental setup of the PA method is schematically shown in Fig. 1. A green laser (λ = 532 nm), modulated by a square signal generator with the output optical power of ~100 mW, was used as a heating source. The samples were alternately mounted at the bottom of the inner space of the PA cell. The informative acoustic response was recorded by an electret microphone Panasonic WM-61A, which was built into the sidewall of the PA cell. This signal was compared with the reference one by a lock-in amplifier in order to extract the amplitude and phase of the PA signal.

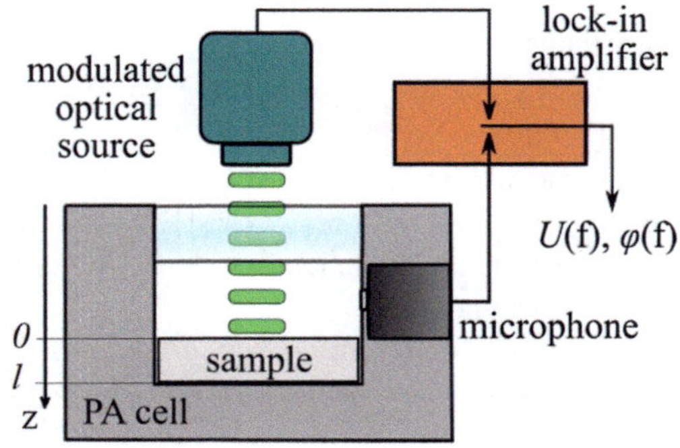

Fig. 1. Sketch view of the classical PA cell employed.

According to the Rosencsweig and Gersho approach [16], the periodical heating of the sample inside the PA cell causes the corresponding temperature variation of the adjacent to the sample`s surface gas (air). The periodically heated gas compresses the rest of the inner gas in the system, performing the role of a «thermal piston». Thus, the acoustic waves are generated in the system and propagate through the volume of the gas to the detector. The variable temperature component in the structure under the periodical irradiation in one-dimensional (z) approximation can be presented as:

$$\frac{d}{dz}\left(\chi(z)\frac{dT}{dz}\right) - 2\pi fc\rho T = I(1-R)\alpha\exp(-\alpha z), \quad (1)$$

where χ is the thermal conductivity, T is the temperature, f is the modulation frequency, c is the specific heat capacity, ρ is the mass density, I is the laser intensity, α is the optical absorption coefficient, R is the reflectivity coefficient. Considering the adiabatic boundary conditions of the studied sample, the pressure fluctuations inside the PA cell can be described by the following equation:

$$p(f) = \int_0^\infty T(z,f)dz = -T(0)\sqrt{\chi_g/(i2\pi fc_g\rho_g)}, \quad (2)$$

where the index g means that the following parameters belong to the gas.

We measured the PA signal in the frequency range from 13 Hz to 2250 Hz. In this frequency domain, the PA cell with the dimensions of 10 × 10 × 3 mm^3 operated in the «non-resonant» regime. Additionally, the calibration of the PA cell by the sample with well-established thermal and optical properties was performed. As a reference sample, we chose *p*-type boron-doped (100)-oriented monocrystalline silicon ($\chi \sim 130$ W/(m·K)) with the same thickness, as in the CdTe sample (Table 1).

Table 1. The fitting parameters of the reference Si and investigated CdTe samples.

Fitting parameter	Sample	
	Si	CdTe
Density [kg/m^3]	2330	5860
Heat capacity [J/(kg·K)]	680	209
Thickness [m]	$5 \cdot 10^{-4}$	$5 \cdot 10^{-4}$
Optical absorption [m^{-1}]	$8 \cdot 10^5$ [17]	$8 \cdot 10^6$ [18]
Reflectivity	0.36 [17]	0.32 [18]
Thermal conductivity [W/(m·K)]	130 ± 5	3.7 ± 0.7

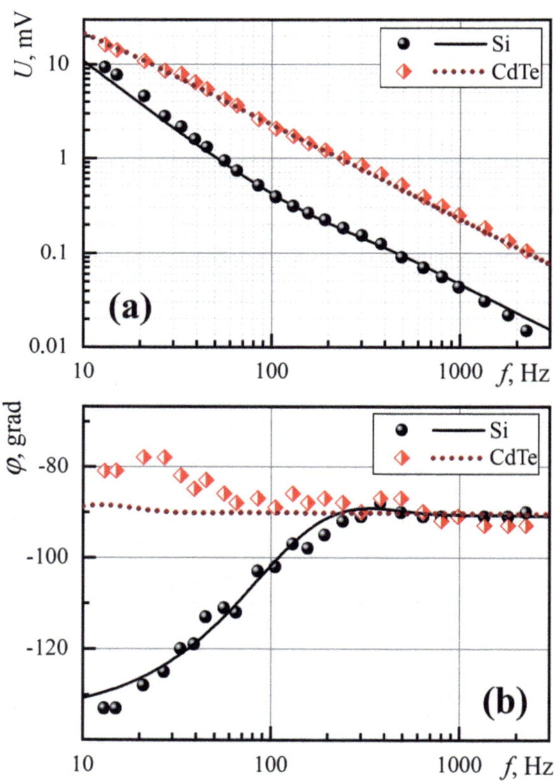

Fig. 2. The experimental amplitude-frequency (a) and phase-frequency (b) dependencies of the PA response obtained by optical irradiation of the reference Si (circles) and investigated CdTe (rhombus) samples, approximated by model calculations (solid and dot lines, respectively).

The normalization process was carried out by achieving the best correlation between the experimental results and the simulation with the finite difference method based on (1) and (2) for the reference Si sample [15]. Thus, the normalized amplitude-frequency characteristics and the phase-frequency characteristics, obtained for both the CdTe and the reference Si samples, are shown in Fig. 2.

As we can see from Fig. 2, the magnitude of the PA signal versus frequency for the CdTe sample have a constant slope of −1 and the phase is almost constant and equal to −90° for the entire experimental frequency range. Thus, the qualitative coincidence of the theoretical curve amplitude with the experimental result for the CdTe sample occurred when its thermal conductivity value as a fitting parameter is equal to be 3.7 ± 0.7 W/(m·K). This value is in good agreement with the literature data [19–21].

3 Estimation of the CdTe Melting Threshold

Certainly, the description of the processes arising under heating of the CdTe surface induced by nanosecond PLI is complicated because of the temperature dependencies of the thermal and optical parameters [18–21] and the concentration of photo-excited NEC [22]. The laser intensity I_0 required to heat the surface of the semiconductor from an initial temperature T_0 to a temperature T_f can be estimated by the following equation [18, 23]:

$$I_0 = \frac{1}{\tau_p} \int_{T_0}^{T_f} \frac{dTc\rho}{(1-R)\alpha} \left[\eta_T/(L_T\alpha + 1) + \eta_B^{NR}/(L_T\alpha + L_D\alpha + 1) + \eta_S^{NR}/L_T\alpha\right]^{-1}, \quad (3)$$

where $\Delta T = T_f - T_0$ is the temperature rise, $L_D = (D\tau_B)^{1/2}$ is the free carrier diffusion depth, D is the ambipolar carrier diffusion coefficient, τ_B is the bulk carrier recombination lifetime, $L_T(T)$ is the thermal diffusion depth. The parameters η_T, η_B^{NR} and η_S^{NR} are the fractions of the laser energy which go into the thermalization of NEC immediately (10^{-12}–10^{-15} s) after excitation (interband relaxation), into both nonradiative interband (bulk) and surface recombination with surface lifetime $\tau_S = L_D/S$, where S is the surface recombination velocity.

The temperature dependent thermophysical characteristics of CdTe and NEC parameters that used in the simulations are presented in Table 2. In the calculations, we used the typical parameters of detector-grade semi-insulating *p*-like CdTe single crystals produced by Acrorad Co. Ltd which were used for fabrication of M-*p*-*n* structured X/γ-ray diode detectors [10, 11].

Figure 3(a) demonstrates the melting thresholds calculated by (3) under PLI within the wavelength region λ = 300–800 nm for different pulse durations τ_p. As one can see, an increase in pulse duration leads to a sensitivity decrease of the melting threshold I_{th} on the excitation wavelength λ. For instance, at τ_p = 7 ns, the threshold laser intensity changes by ΔI_{th} = 5.6 MW/cm^2 (from 4.7 to 10.3 MW/cm^2) with a wavelength variation from 300 nm to 800 nm (Fig. 3(a), curve 1). While at τ_P = 1 μs, the meting threshold of CdTe changes (increased) only by ΔI_{th} = 0.045 MW/cm^2 (Fig. 3 (a), curve 5).

Table 2. Parameters of CdTe.

Parameter	Value
Melting temperature [°C]	1092
Temperature dependency of density ρ [kg/m^3]	5887–0.1165·T [20]
Temperature dependency of specific heat c [J/(kg·K)]	205 + 3.6·10^{-2}·T [19]
Ambipolar diffusion coefficient D [m^2/s] (300 K)	3.9·10^{-4} [21]
Bulk carrier recombination lifetime τ_B (p-CdTe) [s]	(10–30)·10^{-9} [21]
Surface carrier lifetime τ_S [s]	(0.5–1)·10^{-9} [21]
Auger recombination coefficient γ_{Aug} [cm^6/s]	1.7·10^{-30} [21]
Surface recombination velocity S [m/s]	10^3–10^5 [21]
Recombination energy E_R [eV]	1.5
Electron mobility μ_n [cm^2/(V·s)] (300 K)	1000–1100 [21]
Hole mobility μ_p [cm^2/(V·s)] (300 K)	80–100 [21]
Electron diffusion depth L_D (p-CdTe) [m]	(0.4–2) 10^{-6} [18]

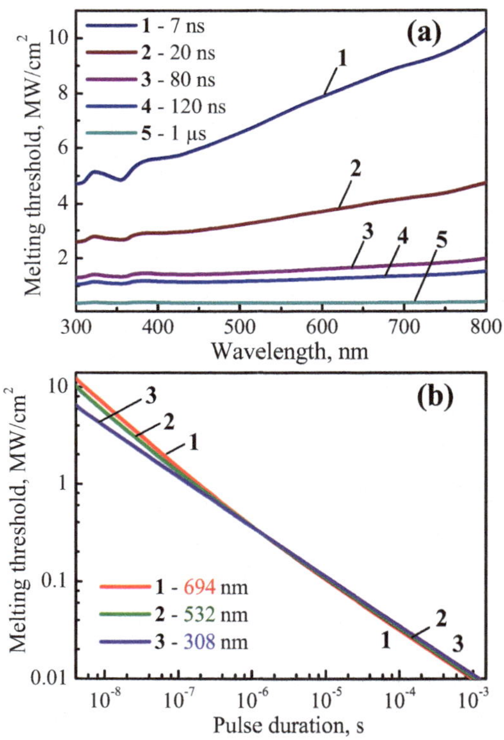

Fig. 3. The dependencies of the CdTe melting threshold on laser wavelength for different pulse durations (a) and on the pulse duration for different wavelengths (b).

Such decrease in the value ΔI_{th} with increasing τ_P can be explained by the fact that the thermal diffusion depth L_T increases with rising τ_P and becomes much larger than $1/\alpha$, and therefore the melting threshold I_{th} becomes less dependent on the optical absorption coefficient $\alpha(\lambda)$. When λ changes from 380 to 800 nm, the reflection coefficient R changes from 0.4 to 0.29, i.e. it increases by a factor of 1.38. At the same time, the absorption coefficient α changes from $4.4 \cdot 10^7\ m^{-1}$ to $2.1 \cdot 10^6\ m^{-1}$, i.e. it decreases by more than an order of magnitude [18]. It is evident that, if only the reflectivity decreases, the melting threshold I_{th} decreases according to (3); but if only α decreases, I_{th} increases because a larger volume of the CdTe surface layer is heated up. Therefore, the dependency $\alpha(\lambda)$ determines the dependency $I_{th}(\lambda)$ (Fig. 3(a)).

Also, the dependency of the melting threshold on pulse duration is important (Fig. 3(b)). The melting threshold falls drastically with an increase in τ_p in accordance with (3) because the heated-up layer volume increases. If the pulse duration becomes shorter, the rate of laser energy delivery becomes higher. With increasing τ_p by six orders of magnitude, I_{th} decreases by three orders of magnitude, for example from 8.7 MW/cm^2 to 10 kW/cm^2 at $\lambda = 532$ nm (Fig. 3(b)). This dependency is similar to the corresponding dependencies for GaAs and Si [23].

Also, one can see that at pulse duration around $\tau_p = 1$ µs, the curves intersect (Fig. 3(b)). The explanation is as follows: initially, I_{th} is determined by α, and it is obvious that the rise of α leads to a decrease of I_{th}. With increasing pulse duration, the thermal diffusion depth becomes much larger in comparison with the radiation depth penetration d, i.e. $L_T >> 1/\alpha$. Therefore, the factor $L_T\alpha$ is determined by τ_P. Then, at $\tau_P = 1$ µs, the ratio d/L_T equals 0.01, whereas the difference in the reflection coefficients, for example $\Delta R/R = (R_{308\ nm} - R_{694\ nm})/R_{308\ nm} = 0.15$. At $\tau_P = 1$ ms, the ratio d/L_T is only 0.0033. Accordingly, the value ΔR begins to affect the melting threshold significantly after $\tau_P = 1$ µs. And since $R_{308\ nm} > R_{532\ nm} > R_{694\ nm}$, the melting threshold slightly increases because with increasing R, more light is reflected. Therefore, even though $\alpha_{308\ nm} > \alpha_{532\ nm} > \alpha_{694\ nm}$, after the intersection point (Fig. 3(b)) I_{th} becomes determined by the reflectivity $R(\lambda)$.

According to our calculations, carried out for $\lambda = 694$ nm, $I_{th} = 1.7$ MW/cm^2 at $\tau_p = 80$ ns and $I_{th} = 4.1$ MW/cm^2 at $\tau_p = 20$ ns that are comparable with the results shown in [20] where $I_{th} = 1.25$ MW/cm^2 at $\tau_p = 80$ ns, and $I_{th} = 3$ MW/cm^2 at $\tau_p =$ 20 ns. As seen from Fig. 3, $I_{th} = 7$ MW/cm^2 for $\lambda = 532$ nm and $\tau_p = 7$ ns that is close to the experimental data obtained for CdTe at similar laser parameters: $I_{th} =$ 8.6 MW/cm^2 [9].

4 Conclusions

The features of photothermal transformation in semi-insulating detector-grade CdTe crystals have been studied under PLI with different wavelengths and pulse durations. The gas-microphone PA technique was applied for the evaluation of the thermal conductivity of the CdTe. The obtained value was used for the estimation of the dependencies of the melting threshold I_{th} of CdTe crystals on wavelength ($\lambda = 300$–800 nm) and pulse duration ($\tau_p = 7$ ns – 1 ms) of laser radiation. It has been shown that in the range of laser pulse duration from 7 ns to 1 µs, the CdTe melting threshold depends mainly on the absorption coefficient $\alpha(\lambda)$. For longer pulse durations ($\tau_p > 1$ µs), I_{th} starts to depend also on the spectra of the reflectivity coefficient $R(\lambda)$, because the thermal diffusion depth become much larger in comparison with the laser radiation depth penetration in CdTe. It has been demonstrated that for shorter laser pulse durations, the melting threshold changes more, when radiation wavelength varies. The obtained results have been used for optimization of the laser-assisted techniques of surface processing and stimulated doping of CdTe crystals.

Acknowledgements. This research was supported by the following research projects: "Laser pulsed photothermoacoustic formation of CdTe-based sensor structures with super-high resolution for radiation monitoring systems" (State Fund For Fundamental Research of Ukraine, grant number F85), "Features of photothermal and photoacoustic processes in low-dimensional silicon-based semiconductor systems" (Program of the Ministry of Education and Science of Ukraine, state registration number 0118U000242), "Development of Cd(Zn)Te-based X/gamma-ray detectors with high resolution for security and diagnostics instruments" (The 2019 Cooperative Research at Research Center of Biomedical Engineering, Japan, grant number 2022), "Development of perovskite single crystal X/gamma-ray detectors for environmental radioactive contamination monitoring" (The Short-term Recruitment Program of Foreign Experts in Anhui (APFEP, 2019, China), "Exploring novel perovskite single-crystal based gamma-ray detector for trace environmental radioactivity monitoring" (Program of the Joint Ukraine - The People's Republic of China R&D Projects for the period of 2019–2020 adopted by the Ministry of Science and Technology of the People's Republic of China, grant number CU03-15).

References

1. Wood, R.F., White, C.W., Young, R.T.: Pulsed laser processing of semiconductors. In: Semiconductors and Semimetals 23, 1st edn. Academic Press, (1984)
2. Misra, N., Xu, L., Rogers, M.S., Ko, S.H., Grigoropoulos, C.P.: Pulsed laser annealing of semiconductor structures for functional devices. Physica Status Solidi (c) **5**(10), 3264–3270 (2008)
3. Phillips, K.C., Gandhi, H.H., Mazur, E., Sundaram, S.K.: Ultrafast laser processing of materials: a review. Adv. Opt. Photon. **7**(4), 684–712 (2015)
4. Malinauskas, M., Žukauskas, A., Hasegawa, S., Hayasaki, Y., Mizeikis, V., Buividas, R., Juodkazis, S.: Ultrafast laser processing of materials: from science to industry. Light: Sci. Appl. **5**, e16133-1–14 (2016)
5. Burbelo, R., Andrusenko, D., Isaiev, M., Kuzmich, A.: Laser photoacoustic diagnostics of advanced materials with different structure and dimensions. Arch. Metall. Mater. **56**(4), 1157–1162 (2011)
6. Voitenko, K., Veleschuk, V., Isaiev, M., Kuzmich, A., Lyashenko, O., Melnychenko, M., Malyarenko, E., Zhelnakov, S., Lysenko, V., Burbelo, R.: Nonlinear laser ultrasound formation in silicon. AIP Adv. **6**(10), 105306-1–6 (2017)
7. Gnatyuk, V.A., Aoki, T., Nakanishi, Y., Hatanaka, Y.: Surface state of CdTe crystals irradiated by KrF excimer laser pulses near the melting threshold. Surf. Sci. **542**(1–2), 142–149 (2003)
8. Gnatyuk, V.A., Aoki, T., Hatanaka, Y., Vlasenko, O.I.: Metal-semiconductor interfaces in CdTe crystals and modification of their properties by laser pulses. Appl. Surf. Sci. **244**(1–4), 528–532 (2005)
9. Aoki, T., Gnatyuk, D.V., Odarych, V.A., Poperenko, L.V., Yurgelevych, I.V., Levytskyi, S.N.: Influence of nanosecond laser irradiation on optical properties of surface layers of CdTe crystals. Thin Solid Films **519**(9), 2834–2837 (2011)

10. Zelenska, K., Gnatyuk, D., Aoki, T.: Laser-based technique of formation of CdTe-metal diode structures for high energy radiation detectors. In: 2016 IEEE Nuclear Science Symposium, Medical Imaging Conference and Room-Temperature Semiconductor Detector Workshop (NSS/MIC/RTSD), Proceedings, pp. 8069953-1–3 (2017)
11. Gnatyuk, V.A., Levytskyi, S.N., Vlasenko, O.I., Aoki, T.: Formation of doped nano-layers in CdTe semiconductor crystals by laser irradiation with nanosecond pulses. J. Nanosci. and Nanotechnol. **1**(2), 7–16 (2016)
12. Lishchuk, P., Andrusenko, D., Isaiev, M., Lysenko, V., Burbelo, R.: Investigation of thermal transport properties of porous silicon by photoacoustic technique. Int. J. Thermophys. **36**(9), 2428–2433 (2015)
13. Neimash, V., Shepelyavyi, P., Dovbeshko, G., Goushcha, A., Isaiev, M., Melnyk, V., Didukh O., Kuzmich A.: Nanocrystals growth control during laser annealing of Sn:(α-Si) composites. J. Nanomater., 7920238-1–8 (2016)
14. Isaiev, M., Didukh, O., Nychyporuk, T., Timoshenko, V., Lysenko, V.: Anisotropic heat conduction in silicon nanowire network revealed by Raman scattering. Appl. Phys. Lett. **110** (1), 011908-1–4 (2017)
15. Lishchuk, P., Isaiev, M., Osminkina, L., Burbelo, R., Nychyporuk, T., Timoshenko, V.: Photoacoustic characterization of nanowire arrays formed by metal-assisted chemical etching of crystalline silicon substrates with different doping level. Physica E **107**, 131–136 (2019)
16. Rosencwaig, A., Gersho, A.: Theory of the photoacoustic effect with solids. J. Appl. Phys. **47**, 64–69 (1976)
17. Green, M.A.: Self-consistent optical parameters of intrinsic silicon at 300 K including temperature coefficients. Sol. Energy Mater. Sol. Cells **92**(11), 1305–1310 (2008)
18. Veleshchuk, V.P., Vlasenko, O.I., Vlasenko, Z.K., Gnatyuk, V.A., Levytskyi, S.N.: Dependence of the CdTe melting threshold on the pulse duration and wavelength of laser radiation and the parameters of non-equilibrium charge carriers. Ukrainian J. Phys. **62**(2), 159–165 (2017)
19. Gnatyuk, V.A., Aoki, T., Gorodnychenko, O.S., Hatanaka, Y.: Solid-liquid phase transitions in CdTe crystals under pulsed laser irradiation. Appl. Phys. Lett. **83**(18), 3704–3706 (2003)
20. Kovalev, A.A., Zhvavyi, S.P., Zykov, G.L.: Dynamics of laser-induced phase transitions in cadmium telluride. Semiconductors **39**(11), 1299–1303 (2005)
21. Madelung, O.: Semiconductors: Data Handbook, 3rd edn. Springer, Berlin (2004)
22. Isaiev, M., Kuryliuk, V., Kuzmich, A., Burbelo, R.: Photothermal transformation in heterogeneous semiconductors structures under its pulse laser irradiation: role of electron-hole diffusion. Arch. Metall. Mater. **58**(4), 1351–1354 (2013)
23. Meyer, J.R., Kruer, M.R., Bartoli, F.J.: Optical heating in semiconductors: Laser damage in Ge, Si, InSb, and GaAs. J. Appl. Phys. **51**(10), 5513–5521 (1980)

Electronics and Nanoelectronics

Structural Properties of $BiFeO_3$ and $Bi_{0,9}La_{0,1}FeO_3$ Powders Synthesized by Sol-Gel Process

A. V. Semchenko[1(✉)], S. A. Khakhomov[1], V. V. Sidsky[1], V. E. Gaishun[1], D. L. Kovalenko[1], W. Strek[2], and D. Hreniak[2]

[1] F. Skorina, Gomel State University, Sovetskaya 104, 246019 Gomel, Belarus
semchenko@gsu.by
[2] Institute of Low Temperature and Structures Research PAN, Okolna st. 2, Wroclaw, Poland

Abstract. The present work aims to design and study novel functional materials with multiferroic properties required in electric applications, such as magnetic and magnetoresistive sensors, actuators, microwave electronic devices, phase shifters, mechanical actuators etc. Complex oxides $BiFeO_3$ and $Bi_{0,9}La_{0,1}FeO_3$ for analysis of its structural properties were synthesized as powders by sol-gel method. The size, shape, and degree of crystallinity of the formed nanoparticles can be changed by varying the temperature and the concentrations of the initial reactants and the stabilizer. This work is devoted to interrelation between composition of sol-gel $BiFeO_3$ and $Bi_{0,9}La_{0,1}FeO_3$ nanopowders and their nanostructural properties.

Keywords: Sol-gel · Powder · Bismuth ferrite · Ferromagnets

1 Introduction

Need of efficient materials having outstanding functional properties with controllable characteristics and those which meet demanding ecological restrictions motives researchers to look at complex transition metal oxides with perovskite-like structure. It is well known that ferrites possess several unique properties such as magnetization/magnetotransport and magnetoelectric coupling being enhanced near the structural phase boundaries [1].

Multiferroics have been known as materials with ferromagnetic and ferroelectric properties at the same time, which have exhibited interesting physical properties as well as a pos possibility of practical applications. Multiferroic materials, owing to the coexistence of ferroelectricity, ferromagnetism, and even ferroelasticity in the same phase, have shown promising applications in nonvolatile information storages, spintronic devices, and magnetoelectric sensors. Among the multiferroic materials studied so far, $BiFeO_3$ (BFO) is known to have a rhombohedrally distorted perovskite structure described by space group R3c. At present, the ceramics of BFO have been extensively investigated [2, 3]. Although rhombohedral $BiFeO_3$ (BFO R-phase) has been studied since first discovery in 1960s, electrical properties of the pure BFO R-phase have been

A. R. Várkonyi-Kóczy (Ed.): INTER-ACADEMIA 2019, LNNS 101, pp. 113–118, 2020.
https://doi.org/10.1007/978-3-030-36841-8_11

rarely reported due to its high conductivity, which may originated from uncertain oxygen stoichiometry, high defect density and poor sample quality [3–5]. It is known [6, 7] that the introduction of small impurity (up to 10%) of rare-earth ion into sol composition facilitates the formation of the perovskite phase and consequently leads to an increase in the antiferromagnetic properties of $Bi_{0,9}La_{0,1}FeO_3$ nanopowders in comparison with $BiFeO_3$.

Wet chemical methods are promising routes to prepare fine and homogeneous oxide powders [8, 9]. Various wet chemical methods such as hydrothermal, co-precipitation, combustion synthesis, molten-salt method, thermal decomposition, and sol-gel process have been developed and designed to prepare $BiFeO_3$ and $Bi_{0,9}La_{0,1}FeO_3$ nanopowders. However, many of these routes do not lead to obtain a pure phase. Controllability of the functional parameters can be achieved particularly using sol-gel synthesis method favoring the superior properties via modification of chemical bond character, variation of the structural parameters, and controlling the defectiveness/stoichiometry of the compounds [10]. However many parameters of the sol-gel $BiFeO_3$ and $Bi_{0,9}La_{0,1}FeO_3$ nanopowders are not studied. This work is devoted to interrelation between composition of sol-gel $BiFeO_3$ and $Bi_{0,9}La_{0,1}FeO_3$ nanopowders and their nanostructural properties.

2 Experimental

Synthesis of $BiFeO_3$ powder used nitrate salts of Fe and Bi, HNO_3, and citric acid. The compounds were dissolved in distilled water, which was then evaporated on a hot plate at 80–90 °C to form a gel (about 4–5 h). The resulting gel was then heated in an oven at a temperature of 180 °C for 2 h. The aim is to remove water molecules. The annealing temperature for powders was 550 °C (during 3 or 10 h), 600 °C (during 3 h), 700 °C (during 3 h), 800 °C (during 3 h) (see Fig. 1). The $Bi_{0,9}La_{0,1}FeO_3$ (BLFO) powder were synthesized using the same procedure. $La(NO_3)3$ was used as lanthanum source.

X-ray diffraction was carried out on diffractometer PANalytical X'Pert MPD Pro in the reflection mode (geometry Bragg - Brentano) using Cu-Kα radiation. The identification of the diffraction peaks was performed using the data bank JCPDS affordable software company Search-Match. Processing of the diffraction patterns was made in the JANA2006 program.

The resulting powder (previously introduced into ethyl alcohol) was applied to the substrate by centrifugation. Single-crystal silicon wafer was used as a substrate. Heat treatment at a temperature of 100° C for 10 min. Surface scanning. was measured by atomic force microscopy (AFM) on 47 SOLVER-PRO and was analyzed using the program Gwyddion.

3 Results and Discussion

As can be seen from the XRD data (Fig. 1), the BFO reaction product was not monophasic (rhombohedral phase). The determining factor is associated with the peculiarities of the sol-gel synthesis technique. The increasing of the synthesis temperature leads to the decrease in the content of the perovskite phase due to the weak bond of bismuth ions in the crystalline cell which causes the removal of bismuth from the crystal lattice. Further annealing of the formed material at higher temperatures does not lead to an increase in the content of the required phase (Table 1).

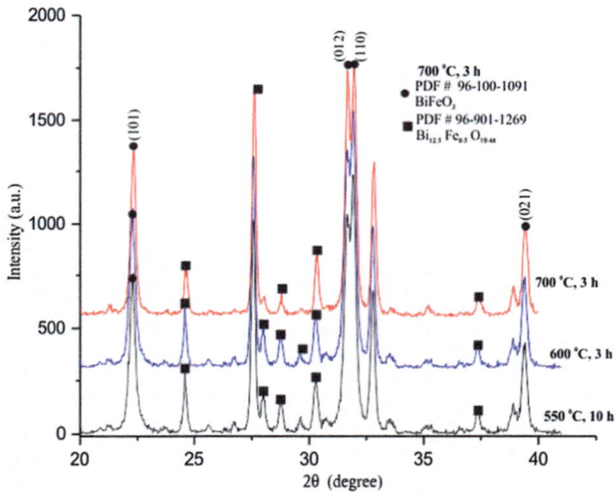

Fig. 1. XRD of $BiFeO_3$ powder after heat treatment at 550 °C during 10 h and after 600 °C and 700 °C during 3 h.

Table 1. Rhombohedra phase content of $BiFeO_3$ powder.

Powder	Temperature and processing time, °C	Phase content $BiFeO_3$, %
BFO	550-10 h	74
	700-3 h	64
	800-3 h	63

Table 2. Rhombohedral phase content of $Bi_{0,9}La_{0,1}FeO_3$ powder.

Powder	Temperature and heat treatment time, °C	Perovskite phase content, (%)
BLFO1	180-2 h	–
BLFO2	550-10 h	80.5
BLFO4	700-3 h	59.8
BLFO5	800-3 h	65.5

The heat treatment conditions of BLFO nanopowders and results of XRD analysis are presented in Table 2.

After synthesis at 180 °C, the sample does not have any crystalline structure. As can be seen from the XRD data (Fig. 1 and Table 2), the BFLO reaction product was not monophasic. The increasing of the synthesis temperature firstly leads to the decrease of the content of the required phase due to the weak bond of bismuth ions in the crystalline cell. But the total content of the required phase in doped nanopowders is higher than in powders without the addition of lanthanum annealed at similar temperatures. This is explained by the expansion of the region of the concentrations of the initial metals necessary to form the required phase by increasing the number of components. The increasing of the content of the required phase from 60% to 65% with the increasing of heat temperature from 700 °C up to 800 °C is bound with the restructuring of the crystal structure of the nanomaterial. However, the limiting phase content at high heat treatment temperatures is lower than at heat treatment of 550 °C. The low annealing temperature should be compensated by the long annealing time (see Fig. 2).

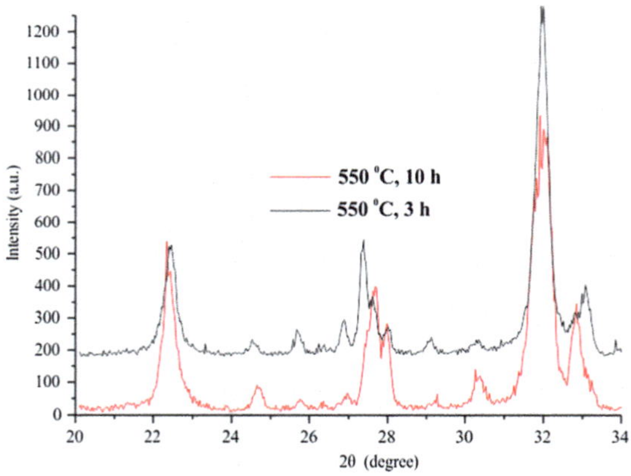

Fig. 2. XRD of $Bi_{0,9}La_{0,1}FeO_3$ powder after heat treatment at 550 °C during 3 h and during 10 h.

Table 3. Domains sizes in BFO and BFLO nanopowders heat treated in different conditions.

Sample	Annealing temperature, °C, Annealing time, hour	Domain size, nm
BFLO	550, 3 h	25
	600, 3 h	19
	700, 3 h	18
	800, 3 h	16
BFO	550, 10 h	15
	600, 3 h	15
	700, 3 h	15

The sizes of domains are also determined (see Table 3).

The results of the investigation of the surface of the synthesized powder after heat treatment at 550 °C during hours are shown in Fig. 4 (AFM image) and Table 4. It has been established that the doping by lanthanum in BFO leads to the significant (about 1.5 times) reduction of the average grain size and decrease (about 3.5 times) in the surface roughness. From the point of view of potential using in nanoelectronics, these facts increase the applicability of the material.

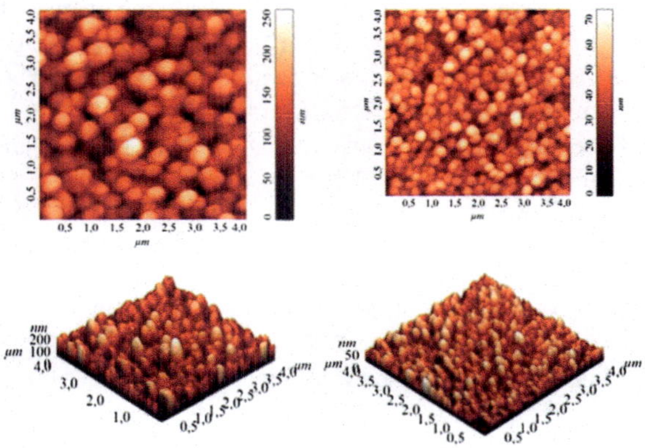

BFO BFLO

Fig. 4. AFM image of $BiFeO_3$ (left) and $Bi_{0,9}La_{0,1}FeO_3$ (right) powders.

Table 4. Surface parameters of $BiFeO_3$ and $Bi_{0,9}La_{0,1}FeO_3$ powders

Sample	Ra, nm	Average grain size, nm	Number of grains
BFO	28	127	380
BFLO	8	90	705

4 Conclusion

In conclusion, the mixture of different phases BFO with the perovskite phase content about 70% and BLFO with the perovskite phase content about 80% were synthesized using sol-gel method. The highest content of the required rhombohedral phase is observed for $Bi_{0,9}La_{0,1}FeO_3$ powder annealed for 10 h at the temperature of 550 °C.

Acknowledgements. This project has received funding from the European Union's Horizon 2020 research and innovation programme under the Marie Sklodowska-Curie grant agreement No 778070.

References

1. Damjanovic, D.: J.: A morphotropic phase boundary system based on polarization rotation and polarization extension. Appl. Phys. Lett. **97**, 062906–062911 (2010)
2. Karpinsky, D.V., et al: J. Mater. Sci. Phase Coexistence in $Bi_{1-x}Pr_xFeO_3$ Ceram. **49** (20), 6937–6943 (2014)
3. Arnold: Composition-driven structural phase transitions in rare-earth-doped $BiFeO_3$ ceramics: a review. IEEE Trans. Ultrason. Ferroelectr. Freq. Control **62**, 62–70 (2015)
4. Karpinsky, D.V., et al: Structural and magnetic phase transitions in $Bi_{1-x}Ca_xFe_{1-x}Mn_xO_3$ multiferroics. J. Alloys Compound **692**, 955–962 (2017)
5. Suastiyanti, D., Wijaya, M.: ARPN Synthesis of $BiFeO_3$ nanoparticle and single phase by sol-gel process for multiferroic material. J. Eng. Appl. Sci. **11**(2), 901–905 (2016)
6. Xu B., et al: Finite-temperature properties of rare-earth-substituted $BiFeO_3$ Multiferroic solid solutions. Adv. Funct. Mater. **23**, 552–558 (2014)
7. Karpinsky, D.V., et al: Thermodynamic potential and phase diagram for multiferroic bismuth ferrite ($BiFeO_3$) npj Comput. Mater. **3**, 20–27 (2017)
8. Sidsky, V.V., et al: Ferroelectric properties of nanostructured SBTN Sol-Gel layers. In: Recent Global Research and Education. Springer, vol. 519, pp. 103–108 (2017)
9. Khakhomov, S.A., et al: Nanostructure and Ferroelectric Properties of Sol-Gel SBTN-Films for Electronic Devices, Springer, vol. 660, pp. 144–150 (2018)
10. Khakhomov, S.A. et al: Synthesis of $BiFeO_3$-powders by Sol-Gel process. In: Recent Advances in Technology Research and Education. Springer, vol. 978, pp. 43–48 (2019)

Numerical Experiments for Studying the Structure of the Electromagnetic Field on the Surface of a Small Spherical Conductive Medium

Nugzar Gomidze, Miranda Khajisvili(✉), Izolda Jabnidze, Kakha Makharadze, and Ana Slusareva

Batumi Shota Rustaveli State University, Batumi 6010, Georgia
mirandukht@gmail.com

Abstract. The present paper proposed the results of numerical experiments based on the classical theory of electrodynamics but it has practical importance in radiolocation and radio-spectroscopy at specific laboratorian and technological conditions. There is investigation the structure of the scattered electromagnetic field of a linearly or circularly polarized incident wave with frequency ω on the surface of an ideal conductive sphere with radius a in the condition $\lambda \gg a$ $(ka \leq 1)$, where λ is the wavelength of the incident wave. The general equations for the scattered field and Poynting vector, both directly near the conductive sphere and in the far zone (Fraunhofer zone) from the scattered object are analytically received. Using computer simulation there are obtained vector diagrams for the components incident and scattered electromagnetic waves, the structure 3D Poynting vector and total electromagnetic field on the surface of a small spherical conducting object located near the antenna system. In spite of the fact that the influence on the fields at the receiving point from separate small objects is negligible, should be mentioned that under certain conditions with a lot of obstacles they can affect the summary field at the receiving point. Therefore, studying the field structure on the surface of a scattering objects is great importance.

Keywords: Conductive medium · Polarization · Scattering

1 Introduction

The purpose of this study is to investigate the role of parasitic radiation near the antenna system on the useful signal. For this reason, it is important to consider the structure of filed on surface objects. Practice shows that compactly located electronic devices influence with each other and of course on some antenna system [1]. The effective cross section (ECS) of the scattered radiation waves from source located in near zone to antenna system is increased. Electromagnetic waves carry information about the structure of the medium in which it propagates. Therefore, the determination of scattered parameters from objects are very importance for analyses the electromagnetic properties of the medium.

A. R. Várkonyi-Kóczy (Ed.): INTER-ACADEMIA 2019, LNNS 101, pp. 119–128, 2020.
https://doi.org/10.1007/978-3-030-36841-8_12

Numerous papers are devoted to these issues [2, 3] but the results obtained by us are original and allows to judge the objects from the parasitic field structure, even when they cannot exert significant influence on useful signal.

Theoretical and experimental studies of scattering electromagnetic waves near located antenna systems show that antennas (such as, lens, specular, horn, etc.) with a relatively large aperture have a large effective scattering cross section, respectively. Due to large values of the effective scattering cross section for they are characterized by large energy losses when receiving a signal, since a significant part is scattered and reflected in space [4, 5].

Many scientists prefer the "microscopic" representation of wave processes, according to which the incident wave to an object causes charge distributions to it [6–12]. In the work [13] was considered a case when the objects were placed in the proximity of the antenna system. The field in the far zone was calculated considering those barriers, but the influence between the objects was not taken into account. Poynting's vector is interpreted as usual a local radiation force at a given point, but it contradicts the Huygens principle [14], according to which wave phenomena are described as the result of interference of scattered waves [15, 16].

The structure of the full Poynting vector is of particular interest in the far zone [17]. Due to the small diffusers, it includes pulsation which is obtained by small diffusors but detection which is almost impossible with standard antennas located at the receiving point [18]. Despite the smallness of these pulsations, they nevertheless formally exist and must be taken into account in general equations. It can be done when studying the 3D structure of the Poynting vector, but in the near zone of the conducting sphere it is difficult to obtain a general analytical expression of the Poynting vector. It can be sold if we take into account the boundary conditions $\overrightarrow{E_{\parallel}} = 0$, $ka \to 0$ and taking into account the fact that the conducting sphere does not absorb and does not radiate energy. In these conditions it is possible to obtain an analytical expression for full Poynting vector near the conducting sphere.

In the conducting sphere, the distribution of currents can be easily calculated in the first approximation. The present paper shows that the distribution of the total current on the conducting sphere to some extent determines the resulting field in the far zone.

2 Problem Statement

The interest of studying the field in the near zone from the source is due to numerous objects such as the connection points of the antenna-feeder components and other obstacles that can affect the radiation field. Therefore, we set the problem of studying the field structure on the surface of such an obstacle, considering it ideal conducting sphere.

True, distortion of the near zone cannot significantly affect the field in the far zone, but the fact that the field due to objects in the near zone exists. The transmitting antenna is powered by a high-frequency generator, which is connected via a coaxial cable with an antenna-feeder system. A certain power is transmitted through the cable to the antenna power points. For Young's consideration, the transverse electromagnetic waves are scattering freely on the antenna's power point. The three-dimensional structure of the antenna limits the scattered radiation.

Let assume that the antenna feed points are spherical and small in comparison of the wavelength $a \ll \lambda$, where a – is the radius of the conducting sphere and λ – is wavelength of the incident wave. We determine the structure of the field on the surface of the sphere located in the near zone of the source. Suppose a spherical object is located in the center of the coordinate system. The fields of incident waves to sphere are denoted by $\vec{E}_{in}$ and $\vec{B}_{in}$ the fields of scattered waves are denoted $\vec{E}_{sc}$ and $\vec{B}_{sc}$, respectively.

The goal is to study the structure of the field on the surface of the sphere by numerical calculations bases analytical expressions considering the incident waves on a spherical obstacle, the scattering waves by sphere, and the interacting field between the incident and scattering waves.

3 Analytical Expressions of Field's Components on the Surface of Spherical Conductive Object

We write the effective scattering cross section in a spherical coordinate system (r, θ, φ):

$$\frac{d\sigma}{d\Omega} = \frac{\text{Scattering wave power in d}\Omega}{\text{Incident wave power in unit square}},$$

$$\frac{d\sigma}{d\Omega} = r^2 \frac{\vec{S}_{sc}(\theta, \varphi)}{\vec{S}_{in}} = r^2 \frac{\left|\vec{E}_{sc}\right|^2}{E_0^2},$$

Usually the antenna is made to good conductors, which means that the tangential component of the total electric field is exactly zero.

The total vector of Poynting can be determined by the formula:

$$\left[\vec{S}_{tot} = \frac{c}{4\pi}\left[\left[\vec{E}_{tot}\right]\vec{B}_{tot}\right] = \frac{c}{4\pi}\left[\left(\vec{E}_{in} + \vec{E}_{sc}\right)\left(\vec{B}_{in} + \vec{B}_{sc}\right)\right] = \frac{c}{4\pi}\left[\vec{E}_{in}\vec{B}_{in}\right] + \frac{c}{4\pi}\left(\left[\vec{E}_{in}\vec{B}_{sc}\right] + \left[\vec{E}_{sc}\left[\vec{B}_{in}\right]\right]\right) + + \frac{c}{4\pi}\left[\vec{E}_{sc}\vec{B}_{sc}\right] = \vec{S}_{in} + \vec{S}_{inter} + \vec{S}_{sc}\right] \tag{1}$$

$\vec{n} = \vec{r}/r$—unit vector in the direction of the observer. $\vec{S}_{in}$—is Poynting vector for incident wave on the sphere, $\vec{S}_{sc}$—Poynting vector of scattering wave from sphere, $\vec{S}_{inter}$—interacting component between the incident and scattering waves.

In the near zone the tension of the scattered electrical field can be expressed by the formula [11]:

$$\vec{E}_{sc} = k^2 \frac{e^{i(kr-\omega t)}}{r}\left\{[[\vec{n}\vec{p}_0]\vec{n}] + (3(\vec{n}\cdot\vec{p}_0)\vec{n} - \vec{p}_0)\left(\frac{1}{k^2r^2} - \frac{i}{kr}\right) - \left(1 + \frac{i}{kr}\right)[\vec{n}\vec{m}_0]\right\} \tag{2}$$

where $\vec{p}_0$ and $\vec{m}_0$ are unit vectors of electrical and magnetic dipole moments: $\vec{p}_0 e^{i\omega t}$ and $\vec{m}_0 e^{-i\omega t}$ are, the dipole moments of the electric and magnetic fields respectively, which are oriented in the conducting sphere under the action of the incident fields. As was shown in work [9], in the case sphere is conductive, these moments, are determined by the formula:

$$\vec{p}_0 = a^3 \vec{E}_0 \cdot \vec{m}_0 = -\frac{a^3}{2} \vec{B}_0 \tag{3}$$

where $\vec{E}_0$ and $\vec{B}_0$—are fields of source. Let initially electric field $\vec{E}_0$—is directed on axis x and magnetic field $\vec{B}_0$ on axis y:

$$\vec{E}_0 = E_0 \vec{x};\ \vec{B}_0 = E_0 \vec{y} \tag{4}$$

Let's go on spherical coordinates $r = (r, \theta, \phi)$ and establish the connection between spherical and Cartesian coordinates:

$$\vec{n} = \vec{r}$$

$$\begin{aligned} \vec{x} &= \sin\theta\cos\phi \cdot \vec{r} + \cos\theta\cos\phi \cdot \vec{\theta} - \sin\phi \cdot \vec{\phi} \\ \vec{y} &= \sin\theta\sin\phi \cdot \vec{r} + \cos\theta\sin\phi \cdot \vec{\theta} + \cos\phi \cdot \vec{\phi}, \\ \vec{z} &= \cos\theta \cdot \vec{r} - \sin\theta \cdot \vec{\theta}. \end{aligned} \tag{5}$$

Considering Eqs. (3), (4) and (5) in (2), after the conversion we get:

$$\begin{aligned} \vec{E}_{sc}(r,t) &= k^2 a^3 \frac{e^{i(kr-\omega t)}}{r} \Big\{ \cos\theta\cos\phi\vec{\theta} - \sin\phi \cdot \vec{\phi} + \\ &+ \left(2\sin\theta\cos\phi \cdot \vec{r} - \cos\theta\cos\phi \cdot \vec{\theta} + \sin\phi \cdot \vec{\phi}\right)\left(\frac{1}{k^2 r^2} - \frac{i}{kr}\right) \\ &- \frac{1}{2}\left(1 + \frac{i}{kr}\right)(\cos\phi \cdot \vec{\theta} - \cos\theta\sin\phi \cdot \vec{\phi} \Big\} \\ &= \left[\cos\theta\left(1 - \frac{1}{k^2 r^2} - \frac{i}{kr}\right) - \frac{1}{2}\left(1 + \frac{i}{kr}\right)\right] \cdot \vec{\theta} \\ &- \sin\phi\left[1 - \frac{1}{k^2 r^2} - \frac{i}{kr} - \frac{cos\theta}{2}\left(1 + \frac{i}{kr}\right)\right] \cdot \vec{\phi} \end{aligned} \tag{6}$$

similarly, taking into account the second formula in (4) for a scattered magnetic field, we obtain:

$$\vec{B}_{sc}(r,t) = k^2 \frac{e^{i(kr-\omega t)}}{r}\left\{[[\vec{n}\vec{m}_0]\vec{n}] + [3[\vec{n}\vec{m}_0]\vec{n} - \vec{m}_0]\left(\frac{1}{k^2r^2} - \frac{i}{kr}\right) + \left(1 + \frac{i}{kr}\right)[\vec{n}\vec{p}_0]\right\}$$
$$= -k^2a^3\frac{e^{i(kr-\omega t)}}{2r}\left\{\left[\left[\vec{n}\vec{B}_0\right]\vec{n}\right] + \left[3\left[\left[\vec{n}\vec{B}_0\right]\vec{n}\right] - \vec{B}_0\right]\left(\frac{1}{k^2r^2} - \frac{i}{kr}\right) - 2\left(1 + \frac{i}{kr}\right)\left[\vec{n}\vec{E}_0\right]\right\}$$
$$= -k^2a^3E_0\frac{e^{i(kr-\omega t)}}{2r}\left\{(\cos\theta\sin\phi\cdot\vec{\theta} + \cos\phi\cdot\vec{\phi}\right.$$
$$\left. + \left(2\sin\theta\sin\cdot\vec{r} - \cos\theta\sin\phi\cdot\vec{\theta} - \cos\phi\cdot\vec{\phi}\right)\right\}$$
$$= -k^2a^3E_0\frac{e^{i(kr-\omega t)}}{2r}\{2\sin\theta\sin\phi\left(\frac{1}{k^2r^2} - \frac{i}{kr}\right)\cdot\vec{r}$$
$$+ \sin\phi\left[\cos\theta\left(1 - \frac{1}{k^2r^2} - \frac{i}{kr}\right) - 2\left(1 + \frac{i}{kr}\right)\right]\cdot\vec{\theta}$$
$$+ \cos\phi\left[1 - \frac{1}{k^2r^2} - \frac{i}{kr} - 2\cos\theta\left(1 + \frac{i}{kr}\right)\right]\cdot\vec{\phi}\Bigg\} \tag{7}$$

On the surface scattered sphere, we have $r = a$. Thus, after a small simplification of the obtained equations for the incident fields on the sphere and scattered from it, we obtain:

$$\vec{E}_{in}(r = a) \approx E_0e^{-i\omega t}\vec{x} = E_0e^{-i\omega t}(\sin\theta\cos\phi\cdot\vec{r}$$
$$+ \cos\theta\cos\phi\cdot\vec{\theta} - \sin\phi\cdot\vec{\phi}), \tag{8}$$

$$\vec{B}_{in}(r = a) \approx -\frac{E_0}{2}e^{-i\omega t}\vec{y} =$$
$$-\frac{1}{2}E_0e^{-i\omega t}\left(\sin\theta\sin\phi\cdot\vec{r} + \cos\theta\sin\phi\cdot\vec{\theta} + + \cos\phi\cdot\vec{\phi}\right). \tag{9}$$

$$\vec{E}_{sc}(r = a) \approx E_0e^{-i\omega t}\left(2\sin\theta\cos\phi\cdot\vec{r} - \cos\theta\cos\phi\cdot\vec{\theta} + \sin\phi\cdot\vec{\phi}\right), \tag{10}$$

$$\vec{B}_{sc}(r = a) \approx -\frac{1}{2}E_0e^{-i\omega t}\left(2\sin\theta\sin\phi\cdot\vec{r} + \cos\theta\sin\phi\cdot\vec{\theta} + \cos\phi\cdot\vec{\phi}\right). \tag{11}$$

Taking into account the formula (8)–(11) in (2), for the Poynting's vector on the surface of the sphere we get:

$$\left\langle\vec{S}_{in}\right\rangle = \frac{c}{8\pi}E_0^2\cdot\vec{z} \tag{12}$$

$$\vec{S}_{sc} = \frac{c}{8\pi}Re\left(\left[\vec{E}^*_{sc}\vec{B}_{sc}\right]\right) = \frac{c}{8\pi}Re\left\{\left(E^*_{\theta,sc}B_{\varphi,sc} - E^*_{\varphi,sc}B_{\theta,sc}\right)\cdot\vec{r}\right.$$
$$\left. + \left(E^*_{\varphi,sc}B_{r,sc} - E^*_{r,sc}B_{\varphi,sc}\right)\cdot\vec{\theta} + \left(E^*_{r,sc}B_{\theta,sc} - B^*_{\theta,sc}B_{r,sc}\right)\cdot\vec{\varphi}\right\} \tag{13}$$

$$\left\langle \vec{S}_{inter} \right\rangle = \frac{c}{4\pi}\left(\left[\vec{E}_{in}\vec{B}_{sc}\right] + \left[\vec{E}_{sc}\vec{B}_{in}\right]\right) = \frac{c}{8\pi} Re\left[\left(\vec{E}^{*}_{\theta,sc}\vec{B}_{\varphi,in} + \vec{E}^{*}_{\theta,sc}\vec{B}_{\varphi,sc} - \vec{E}^{*}_{\varphi,sc}\vec{B}_{\theta,in} - \vec{E}^{*}_{\varphi,in}\vec{B}_{\theta,sc}\right)\cdot\vec{r} + \left(\vec{E}^{*}_{\varphi,sc}\vec{B}_{r,in} + \vec{E}^{*}_{\varphi,in}\vec{B}_{r,sc} - \vec{E}^{*}_{r,sc}\vec{B}_{\varphi,in} - \vec{E}^{*}_{r,in}\vec{B}_{\varphi,sc}\right)\cdot\vec{\theta} + \left(\vec{E}^{*}_{r,sc}\vec{B}_{\theta,in} + + \vec{E}^{*}_{r,in}\vec{B}_{\theta,sc} - \vec{E}^{*}_{\theta,sc}\vec{B}_{r,in} - \vec{E}^{*}_{\theta,in}\vec{B}_{r,in}\right)\cdot\vec{\varphi} \tag{14}$$

The total field on the surface of the sphere is obtained as a result of the superposition of the incident and scattered fields:

$$\vec{E}_{tot}(r = a) = \vec{E}_{in}(r = a) + \vec{E}_{sc}(r = a) \tag{15}$$

$$\vec{B}_{tot}(r = a) = \vec{B}_{in}(r = a) + \vec{B}_{sc}(r = a) \tag{16}$$

Accordingly, the components of the incident, scattered and full fields of the spherical coordinate system can be written as:

$$E_{in}(r,\theta,\phi) = \begin{pmatrix} \sin\theta\cos\phi \\ \cos\theta\cos\phi \\ -\sin\phi \end{pmatrix} \qquad B_{in}(r,\theta,\phi) = \begin{pmatrix} -\frac{1}{2}\sin\theta\sin\phi \\ -\frac{1}{2}\cos\theta\sin\phi \\ -\frac{1}{2}\cos\phi \end{pmatrix} \tag{17}$$

$$E_{sc}(r,\theta,\phi) = \begin{pmatrix} 2\sin\theta\cos\phi \\ -\cos\theta\cos\phi \\ \sin\phi \end{pmatrix} \qquad B_{sc}(r,\theta,\phi) = \begin{pmatrix} -\sin\theta\sin\phi \\ -\frac{1}{2}\cos\theta\sin\phi \\ -\frac{1}{2}\cos\phi \end{pmatrix} \tag{18}$$

$$E_{tot}(r,\theta,\phi) = \begin{pmatrix} 3\sin\theta\cos\phi \\ 0 \\ 0 \end{pmatrix} \qquad B_{tot}(r,\theta,\phi) = \begin{pmatrix} 0 \\ \frac{3}{2}\cos\theta\sin\phi \\ \frac{3}{2}\cos\phi \end{pmatrix} \tag{19}$$

4 Numerical Calculations and Discussion

Numerical experiments were performed in the Mathcad interface using the formulas (16)–(18) and (17)–(19). Some of the results are shown on Figs. 1–4.

Figure 1 shows the vector diagrams of the distribution of the radial components of the incident and scattered fields on the surface of the sphere. It need to take into account that the strength of the tension increases from blue to red. Figure 2 shows the interacting component of the Poynting vector and the result for the full Poynting vector. Figure 3 shows dependence interacting Poynting's vector on angle θ in spherical and polar coordinate system and on Fig. 4 shows the structure of Total electromagnetic wave on surface small spherical conductive medium (blue – is electric filed, red – magnetic fields).

It is possible to calculate field in observation point which is placed in far zone and study influence wave on the surface of the sphere. As mentioned, the tangential

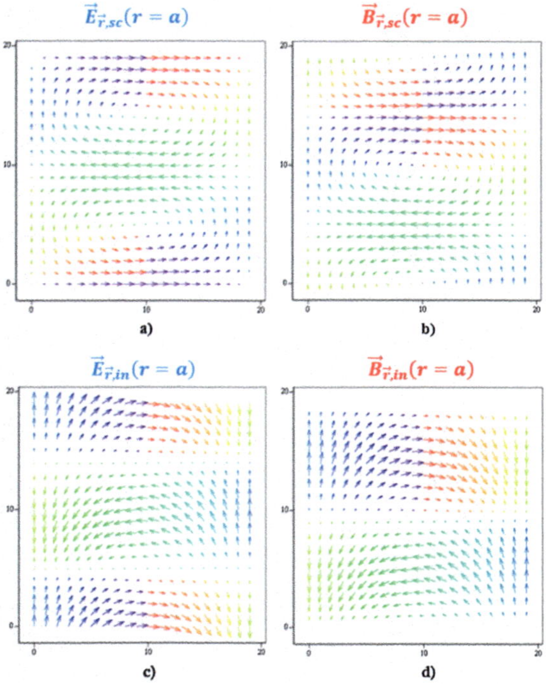

Fig. 1. Vector diagram of electromagnetic fields on the surface of small spherical conductive media. **a** radial components of scattering electric fields $\vec{E}_{\vec{r},sc}(r = a)$; **b** radial component of scattering magnetic field $\vec{B}_{\vec{r},sc}(r = a)$; **c** radial component of incident electric field $\vec{E}_{\vec{r},in}(r = a)$; **d** radial component of incident magnetic field $\vec{B}_{\vec{r},in}(r = a)$

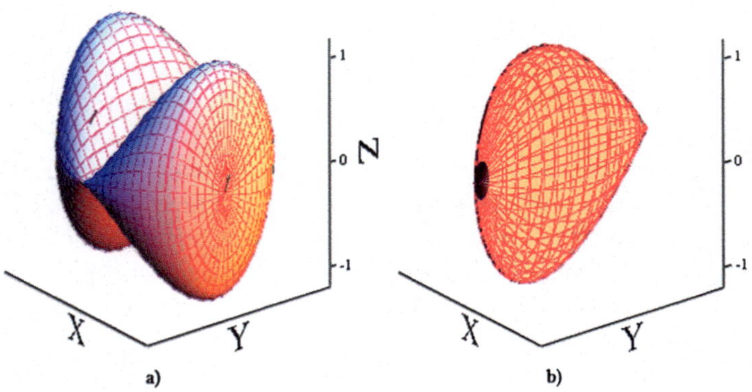

Fig. 2. 3D Poynting's vector on the surface of small spherical conductive media: **a** interacting Poynting's vector $\vec{S}_{inter}(r = a)$; **b** total Posting's vector $\vec{S}_{tot}(r = a)$;

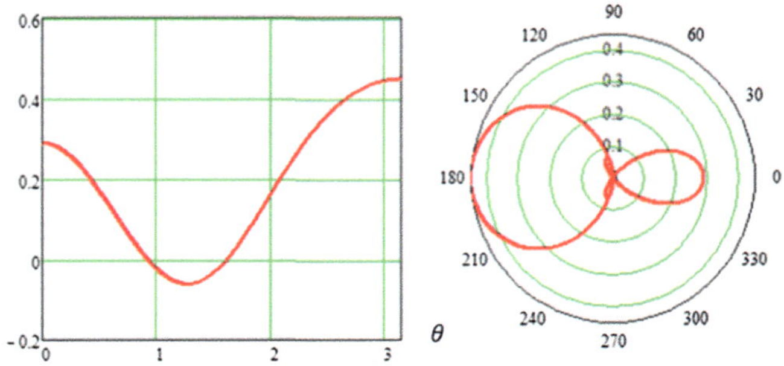

Fig. 3. Interacting Poynting's vector vs angle θ: **a** spherical coordinate system $\vec{S}_{inter}(r,\theta)$; **b** polar coordinate system $\vec{S}_{inter}(\rho)$

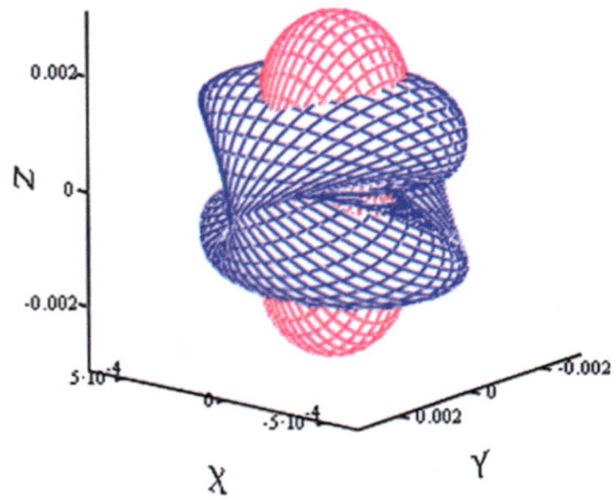

Fig. 4. Total electromagnetic wave on surface small spherical conductive

component of electric filed can be equal to zero when the angle of scattering is equal to $\theta = \pi/3$, and the transverse component never becomes zero. Based on the analytical images, it is easy to show that the ECS of scattering wave increases with the increase the radius of the sphere.

The structures of incident and scattered electric and magnetic fields components were examined via vector diagram on *xoy* plane and by the analysis of 3D images of radial and angular components of incident, scattering and interactive fields. It is shown that radial component of total electric field is not zero but angular components equal to zero. While, radial component of total magnetic field is zero, and the angular components give us the 3D effect.

The field under study in the near zone includes an incident and scattered wave and also interacting wave between those fields as well.

The structure wave in far zone on the observed point due to objects located in the near to source, does not change significantly, as it should be, but by choosing the size and number of obstacles and the wavelength we can get a tangible effect on the summary field in the receiving point.

5 Conclusion

The structure of incident and scattered fields on the surface of the conductive sphere is investigated based on a vector diagram and analysis of 3D patterns of radial and angular components of scattering and incident waves. It is shown that the angular components of the summary scattering and incident waves are equal to zero, while the radial components are different from zero. Scattering and incident electric fields cancel each other, while the summary radial magnetic component is exactly zero, and the angular components are different from zero. On the surface of the sphere, the total charge density exceeds the effective cross section scattered by 3/2.

Acknowledgement. Work was supported by the 2019 Competition for Targeted Scientific Research Projects "Diagnostics of a liquid medium based on an estimate of the correlation function of the intensity of partially coherent waves scattered from the fluid volume". Batumi Shota Rustaveli State University (2019), Senior scientists: PhD of Physics Miranda Khajishvili, Prof. Zhuzhuna Diasamidze; project manager PhD of Physics Kakha Makharadze.

References

1. Kildal, P.S.: Foundations of antenna engineering: a unified approach for line-of-sight and multipath. Chalmers University of Technology (2015)
2. Mallick, K.: Conducting sphere in an electromagnetic input field. Geophys. Prospect. **20**(2), 293–303 (2006)
3. Muller, R.S.: Electromagnetic fields of a nonprecessing and processing, spinning, permanent magnet, conducting sphere. Radio Sci. **26**(1), 137–147 (1991)
4. Mittra, R.: Computer technologies for electrodynamics. New York-Oxford-Toronto (1957)
5. Serebryakov, G.U.: Modern problems of statistical physics (2002) V.I.P. 95 (in Russian)
6. Lambert, K.M., Rudduck, R.C., Lee, T.H.: IEEE Trans. Antennas Propag. **38**, 896–903 (1990)
7. Liberti, J.C., Rappapjrt, T.S.: Smart antennas for wireless communication IS-95 and third generation CDMA applications. Prentice Hall PTR, Upper Saddle River. N.J (1999)
8. Winters, J.: IEEE Pers. Commun. Magazine. **2**, 23 (1998)
9. Foschini, G.J., Gans M.J.: Wirel. Pers. Commun. **6**(4), 311 (1998)
10. Mozingo, P.A., Miller, P.A.: Adaptive antenna arrays. Moscow. Radio Commun (1986)
11. Aizenberg, G.Z.: VHF antennas. Moscow, « Nauka » (1957) (in Russian)
12. Nicolski, B.B., Nicolski, T.N.: Electrodynamics and propagation. Moscow, « Nauka » (1957) (in Russian)
13. Gomidze, N.Kh., Surmanidze, I.S., Makharadze, K., Abuladze, N.N., Kavtaradze, L.O.: Multi-component antennas in wireless connection. RSU Works. Series: "Natural Sciences and Medicine" vol. 13, pp. 225–229, Batumi (2008)
14. Hertz, H.: The forces of electrical oscillations treated according to maxwell's theory, Weidemann's Ann. 36, 1 (1889); reprinted in chap. 9 of H. Hertz, Electric Waves (Dover, New York, 1962). A translation by O. Lodge appeared in Nature 39, 402 (1889)
15. Jackson, J.D.: Classical electrodynamics, 3rd edn. Wiley, New York (1999)

16. McDonald K.T.: Radiation in the near zone of a hertzian dipole 22 April 2004
17. Feynman, R.P.: The Feynman Lectures on Physics. Vol. II: Mainly Electromagnetism and Matter (The New Millennium ed.), Basic Books, New York (2011) ISBN 978-0-465-02494-0
18. Bettini A.: A Course in Classical Physics, Vol. 4-Waves and Light. pp. 95, 103. Springer (2016) ISBN 978-3319483290

Photon-Coupled, Two-State Photoswitchable Protein-Based Multiple-Valued Logic

Balázs Rakos(✉)

Department of Automation and Applied Informatics, Budapest University of Technology and Economics, Budapest, Hungary
balazs.rakos@gmail.com

Abstract. Multi-valued logic can be applied in several areas, like robotics and artificial intelligence. It is utilized in various commercial applications, such as the StrataFlash, a NOR flash memory developed by Intel. Contemporary research focuses on the development of fast, nanometer-size, low power consuming electronic devices, therefore it is imperative to examine such concepts in the field of multiple-valued logic. We present a method by which simple, two-state, photon-coupled photoswitchable proteins can be utilized for multiple-valued computations. Its advantages are discussed, and an example, the ternary OR gate, using readily available fluorescent reversibly photoswitchable proteins is provided.

Keywords: Multiple-valued logic · Molecular electronics · Photoswitchable proteins · Logic circuits · Nanoelectronics

1 Introduction

In contrast with binary logic, which uses only two logic values (true, false), in the case of multiple-valued logic (MVL) at least three values are utilized [1, 2]. It retains the advantages of both binary logic, and analog signal processing [3], and requires less devices and connections in the circuitry. MVL is more efficient than Boolean logic, and it can be used in several applications [4]. Since contemporary research strives to develop nanometer-size, fast, and low power-consuming electronic devices, elementary building blocks with such properties, applicable for many-valued computations should be explored, as well.

Besides the commercially available CMOS VLSI technology, molecules have been considered as potential elements of MVL systems [5–7], as well. It has been suggested in [5] that with the aid of fluorescent photoionic devices [8] four-state MVL may be realized, where the input signal is a chemical species (e.g. H+), and the output is fluorescence. Tape porphyrins may also be used as molecular MVL memory units [7]. However, these concepts are not as practical as the presently used conventional semiconductor-based technology.

Previous studies suggest that proteins may be promising candidates for the realization of nanometer-size, fast, and low power-consuming elementary electronic devices due to their advantageous properties [9–13]. Many of these works apply photoswitchable proteins as building blocks of the proposed circuits.

A. R. Várkonyi-Kóczy (Ed.): INTER-ACADEMIA 2019, LNNS 101, pp. 129–134, 2020.
https://doi.org/10.1007/978-3-030-36841-8_13

When irradiated by appropriate photon pulses, reversibly photoswitchable proteins can be switched between different conformations (forms, states). Furthermore, fluorescent photoswitchable proteins can even emit photons with frequencies corresponding to their forms upon excitation by light.

In two recent papers [14, 15], we proposed protein-based architectures for MVL computations, one of them deals with photon-coupled, multiple-state (at least three), hypothetical photoswitchable proteins [14], the other one discusses field-effect protein-based circuits [15].

In this work, we examine the applicability of photon-coupled, simple, two-state photoswitchable proteins in MVL circuits. Their main advantage in contrast to multiple-state ones is their simplicity, and readily availability.

Section 2 of the present paper introduces the main concept of the two-state photoswitchable protein-based MVL, furthermore a ternary OR gate consisting of existing proteins is demonstrated. The results are discussed, and the work is concluded in Sects. 3 and 4 respectively.

2 Two-State Photoswitchable Protein-Based Multiple-Valued Logic

2.1 The Concept

Reversibly photoswitchable proteins can be switched between different forms by irradiating them with photons with appropriate frequencies. In the case of a two-state molecule, photons with frequency f_{12} switch it from $form_1$ to $form_2$, and light with frequency f_{21} switches it back to $form_1$. In [14] we showed that such proteins can potentially realize MVL if they possess more than two forms between which they can be switched by light with specific frequencies. Photoswitchable proteins with two forms are more simple, and readily available than those with multiple states. In the following we show through a simple example that such molecules can be used as basic elements of MVL circuits, as well.

Table 1 displays the switching properties of Dronpa and rsCherryRev, two artificially developed reversibly photoswitchable fluorescent proteins (data were taken from [16]).

Table 1. Switching properties of Dronpa and rsCherryRev.

Protein	λ_{ex}	λ_{em}	λ_{on}	λ_{off}
Dronpa	503 nm	517 nm	405 nm	488 nm
rsCherryRev	572 nm	608 nm	550 nm	450 nm

These proteins can emit radiation with a peak emission wavelength, λ_{em} when irradiated by light with peak excitation wavelength, λ_{ex}. The molecules can be reversibly switched between their fluorescent and non-fluorescent states by light pulses with λ_{on} and λ_{off} wavelengths (λ_{on} switches the protein to its fluorescent state, and λ_{off} switches it back to its non-fluorescent state).

By placing these two proteins next to each other, and treating this system as a single element of MVL circuits, we can switch it between three states, thereby realizing a ternary logic memory element in the following way. Let us assume that the three values of the ternary system are denoted by '−', '0', and '+', and let '−' correspond to the absence of radiation, '0' to radiation with $\lambda_O = 517$ nm, and '+' to radiation with $\lambda_+ = 608$ nm, furthermore let us assume that the Dronpa-rsCherryRev system is continuously irradiated by light with $\lambda_{ex1} = 503$ nm and $\lambda_{ex2} = 572$ nm wavelengths, and in the beginning both of the proteins are in their non-fluorescent state. If the system is irradiated by light with $\lambda_{on1} = 405$ nm wavelength, it switches Dronpa to its fluorescent state, and since it is continuously irradiated with the proper λ_{ex1}, it will emit λ_O, which corresponds to logic '0'. If the two-protein element is irradiated by light with $\lambda_{on2} = 550$ nm wavelength, it switches rsCherryRev to its fluorescent state, and since it is continuously irradiated with the proper λ_{ex2}, it will emit λ_+, which corresponds to logic '+'. When the system is irradiated by only λ_{ex1} and λ_{ex2}, it doesn't emit radiation, since both of the molecules remain in their non-fluorescent state, which corresponds to logic '−'. Since such two-protein elements can be integrated together with the aid of photon-coupling, the realization of MVL gates, and circuits are also possible.

2.2 An Example: The Ternary OR Gate

In the following, we will demonstrate through a simple example, the ternary OR gate, that the concept described in the previous section can be utilized for MVL computations. The truth table of the ternary OR gate is displayed in Table 2.

Table 2. Truth table of the ternary OR gate.

Input	−	0	+
−	−	0	+
0	0	0	+
+	+	+	+

The structural arrangement of the photoswitchable protein-based ternary OR gate is displayed in Fig. 1, where 2H5Q is the parent protein of rsCherryRev, and 2POX is the Dronpa molecule. The molecular structures of the proteins were plotted with VMD [17] utilizing their 3D molecular data obtained from the Protein Data Bank [18]. The two molecules, rsCherryRev and Dronpa, are placed next to each other, forming the logic gate. The distance between the molecules are not strict, since coupling is not involved.

Table 3 shows the operation of the gate. The rsCherryRev-Dronpa arrangement is continuously irradiated by radiation with $\lambda_{ex1} = 503$ nm and $\lambda_{ex2} = 572$ nm wavelengths, and initially both of the proteins are in their non-fluorescent states. If the gate isn't irradiated by additional radiation, which corresponds to logic '−' at the inputs, the molecules don't emit photons, which is equivalent with logic '−' at the output. However, when irradiation with $\lambda_{on1} = 405$ nm is present (logic '0' at the input), Dronpa switches to its fluorescent form, and since it is continuously excited with λ_{ex1}, it

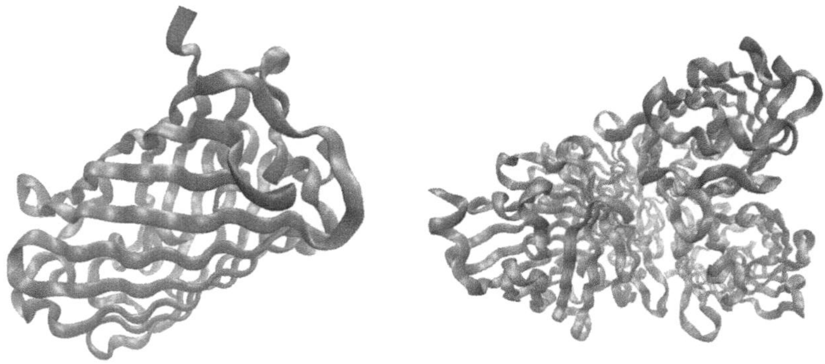

Fig. 1. Arrangement of the photoswitchable protein-based ternary OR gate.

will emit photons with λ_0 = 517 nm, which corresponds to logic '0'. If, on the other hand, the arrangement is irradiated by photons with λ_{on2} = 550 nm and λ_{off1} = 488 nm, subsequently (λ_{off1} is necessary, to switch Dronpa to its non-fluorescent form if λ_{on1} is present at the other input and switches it to its fluorescent state), which is equivalent with logic '+' at the input, it will emit radiation with λ_+ = 608 nm, which corresponds to logic '+' at the output. The structure can be reused after computation, if it is reset with a light pulse consisting of λ_{off1} = 488 nm, and λ_{off2} = 450 nm.

Table 3. Operation of the photoswitchable protein-based ternary OR gate (the system is continuously irradiated by λ_{ex1} and λ_{ex2}).

Input	No irradiation	λ_{on1}	λ_{on2}, λ_{off1}
No irradiation	No emission	λ_0	λ_+
λ_{on1}	λ_0	λ_0	λ_+
λ_{on2}, λ_{off1}	λ_+	λ_+	λ_+

3 Discussion

The main advantage of the concept outlined in this paper in contrast with the principle presented in [14] is the fact that MVL can be realized with simple, two-state photo-switchable proteins, there is no need for finding or developing more complicated molecules with more than two forms. Although the not-too-numerous, readily available photoswitchable proteins are quite slow (switching in milliseconds), because they were designed for other purposes, structural rearrangements in such molecules can occur even in picoseconds [11, 19]. In this way THz-frequency MVL circuits are possible with properly designed proteins. Since proteins are nanometer-size molecules, they can realize nanoscale MVL circuitry. Furthermore, due to the low power consumption and dissipation of the operation of the aforementioned devices, very energy-efficient architectures can be realized.

One additional, potential advantage of the proposed arrangement is that even the individual element of the architecture, which consists of more than one two-state protein, can interact within itself if its protein constituents influence each other via photon coupling. However, the exploration of this effect is not discussed in the present work.

Although we introduced our concept through ternary MVL operations, its extension to other MVL systems is trivial.

4 Conclusions

We showed that reversibly photoswitchable fluorescent proteins with two forms are potentially suitable for the realization of multiple-value logic architectures. One of their main advantages is their simplicity, and easier availability in contrast with photo-switchable proteins with more forms. We demonstrated that a system consisting of two, two-state photoswitchable proteins can be used as a ternary logic-based memory element. Furthermore, we presented a ternary OR gate realized using readily available molecules. The concept can be easily extended to other than ternary MVL systems. The author's opinion is that the presently available nanofabrication tools and procedures enable the realization of the proposed structures, enabling the experimental validation of the concept in the near future. One limitation of the suggested arrangements is speed, since presently available photoswitchable proteins switch rather slowly, therefore the development of fast-switching molecules via protein engineering is also a necessary task of the future.

References

1. Hayes, B.: Third base. Am. Scientist **89**, 490–494 (2001)
2. Yoeli, M., Rosenfeld, G.: Logical design of ternary switching circuits. IEEE Trans. Electron. Comput. **EC-14**(1), 19–29 (1965)
3. Current, K.W.: Current-mode CMOS multiple valued logic circuits. IEEE J. Solid-State Circuits **29**(2), 95–107 (1994)
4. Dubrova, E.: Multiple-Valued Logic Synthesis and Optimization. The Springer International Series in Engineering and Computer Science, vol. 654, pp. 89–114 (2002)
5. Prasanna de Silva, A., James, M.R., McKinney, B.O.F., Pears, D.A., Weir, S.M.: Molecular computational elements encode large populations of small objects. Nat. Mater. 5, 787–789 (2006)
6. de Silva, A.P., Uchiyama, S.: Molecular logic and computing. Nat. Nanotechnol. **2**, 399–410 (2007)
7. Li, E.Y., Marzari, N.: Conductance switching and many-valued logic in porphyrin assemblies. J. Phys. Chem. Lett. **4**(18), 3039–3044 (2013)
8. Callan, J.F., de Silva, A.P., Ferguson, J., Huxley, A.J.M., O'Brien, A.M.: Fluorescent photoionic devices with two receptors and two switching mechanisms: applications to pH sensors and implications for metal ion detection. Tetrahedron **60**, 11125–11131 (2004)

9. Rakos, B.: Simulation of Coulomb-coupled, protein-based logic. J. Autom. Mobile Robot. Intell. Syst. **3**(4), 46–48 (2009)
10. Rakos, B.: Coulomb-coupled, protein-based computing arrays. Adv. Mater. Res. **222**, 181–184 (2011)
11. Rakos, B.: Modeling of dipole-dipole-coupled, electric field-driven, protein-based computing architectures. Int. J. Circuit Theory Appl. **43**, 60–72 (2015)
12. Rakos, B.: Pulse-driven, photon-coupled, protein-based logic circuits. Adv. Intell. Syst. Comput. **519**, 123–127 (2016)
13. Rakos, B.: Photon-coupled, photoswitchable protein-based OR, NOR logic gates. Adv. Intell. Syst. Comput. **660**, 99–103 (2017)
14. Rakos, B.: Multiple-valued computing by photon-coupled, photoswitchable proteins. Lecture Notes in Networks and Systems, vol. 53, pp. 131–136 (2018)
15. Rakos, B.: Multiple-valued computing by dipole-dipole coupled proteins. Int. J. Circuit Theory Appl. **47**(8), 1357–1369 (2019)
16. Bourgeois, D., Adam, V.: Reversible photoswitching in fluorescent proteins: a mechanistic view. IUBMB Life **64**(6), 482–491 (2012)
17. Humphrey, W., Dalke, A., Schulten, K.: VMD-visual molecular dynamics. J. Molec. Graph. **14**(1), 33–38 (1996)
18. www.rcsb.org
19. Xu, D., Phillips, J.C., Schulten, K.: Protein response to external electric fields: relaxation, hysteresis, and echo. J. Phys. Chem. **100**, 12108–12121 (1996). https://doi.org/10.1021/jp960076a

Radiation Patterns of Double DNA-Like Helices as Elements of Metamaterials and Antenna Systems

Ivan Mikhalka(✉), Igor Semchenko, and Sergei Khakhomov

Francisk Skorina Gomel State University, Sovyetskaya Str. 104, 246019 Gomel, Belarus
mikhalkaivan@mail.ru, i.semchenko@gsu.by

Abstract. In this paper a double DNA-like helix as a promising element of metamaterials, metasurfaces and antenna systems for various frequency ranges is considered. The article demonstrates the possibility of creating the required radiation pattern of such a helix, excited by a plane electromagnetic wave by changing its angle of incidence. It is shown, that an incident wave at some angle can activate two wave modes in a helix, each of them propagates with its own phase velocity, and responsible for a certain type of radiation. This effect can be used in passive antenna devices and metamaterials, where the control of the direction of propagation of the reflected wave without changing the frequency of the excitation is required. In addition, the paper provides an overview of the main properties of cylindrical helical radiators, which in the future can be used in fabrication metamaterials and metasurfaces.

Keywords: DNA-like helix · Metamaterial · Helical antenna · Radiation pattern

1 Introduction

1.1 Presenting the Question

Last 15 years, researchers have had a constant interest in metamaterials and their particular case—metasurfaces, which can function in a wide frequency range, from microwaves to the visible light [1–6]. It is relevant to consider various artificial elements – microresonators, which can be used as unit cells of metamaterials and metasurfaces. Such promising microresonators include electrically conductive elements of a helical shape, which have previously proved successful in creating antenna systems. Before proceeding with the design of metamaterials or metasurfaces, it is necessary to study in detail the properties of its elements (unit cells) and optimize their parameters. The most important characteristics of the radiators include the radiation pattern and polarization of the radiated wave, which should be investigated for both active and passive modes of operation of the device. Not only widely used single-stranded cylindrical helices, but also double-stranded helices are of interest. Such double helices can be called DNA-like helices because of their geometric similarity with the DNA molecule [7–14].

A. R. Várkonyi-Kóczy (Ed.): INTER-ACADEMIA 2019, LNNS 101, pp. 135–143, 2020.
https://doi.org/10.1007/978-3-030-36841-8_14

Biophysics considers some complex molecules, including molecules with a helical structure, as resonance systems. Such approach can be applied to other complex molecules, in particular, those that have a great length. These molecules include some proteins, DNA, RNA, which have a helical structure, that is connected with the topic of this work. The resonant properties of these molecules have also their researchers.

1.2 Literature Review

Metamaterials and metasurfaces are usually considered as artificial structures whose properties are not found in objects that exist in nature. Such exotic properties of metasurfaces include, for example, complete transparency outside the resonant frequency band, when at resonant frequency complete absorption of incident radiation is realized. This property allows us to consider the metasurface as a "perfect" absorber [1]. Another extraordinary property of metasurfaces – the possibility to control the wave front of the reflected and transmitted waves and creation of anomalous reflection [2–6]. For designing such and other metasurfaces, it is necessary to study thoroughly and optimize the characteristics of individual constituent elements, for example, their radiation patterns.

In the past, conductors of helical form were widely used in antennas fabrication. The work of Yurtsev et al. [15] is devoted to questions of the theory and practice of helical antennas. It summarized significant theoretical and experimental material. The work of Balanis [16] is devoted to general questions of the antenna theory. In addition, this book contains the chapter about helical antennas.

In [17], using of helical antennas in smart implants is considered. In this work, the helical antenna was chosen mainly because of its compactness. In [18], the resonance mechanism in the DNA molecule is studied. The role of aromatic rings with generalized electrons in the resonance is indicated. In [19], a model of amino acid sequence as an electrical circuit with passive elements is presented.

2 Helical Antennas

2.1 General Characteristics

Among the various types of antennas, helical antennas are significant. Helical antennas are low- and mid-directional broadband antennas of elliptical and controlled polarization. They are used as independent antennas, as exciters of horn antennas, irradiators of mirror and lens antennas, elements of antenna arrays. Helical antennas are surface wave antennas. They can be cylindrical, equigonal (conical and flat) and irregular. In this paper, only cylindrical antennas are studied.

In multi-stranded helical antennas, the working frequency range is expanded due to the suppression of the nearest lower and higher types of waves in them, that distort the main radiation pattern. One-way winded helical antennas radiate a field with an elliptical, close to circular, polarization. The direction of rotation of the field vector corresponds to the direction of the helix winding. To obtain a linear and controlled polarization a helical antenna with a two-side (counter) winding is used.

2.2 Radiation Patterns of Helical Antennas

Helical antennas can have three types of radiation patterns: toroidal with a width $2\theta_{0.5} \approx (45\ldots90)^\circ$, unidirectional with width $2\theta_{0.5} \approx (25\ldots180)^\circ$ and conical with width $2\theta_{0.5} \approx (40\ldots60)^\circ$. Here θ – axial angle measured from the axis of the antenna. Figure 1 illustrates the typical types of helical antenna patterns. The polarization of radiation can be elliptical, almost circular, linear, controlled.

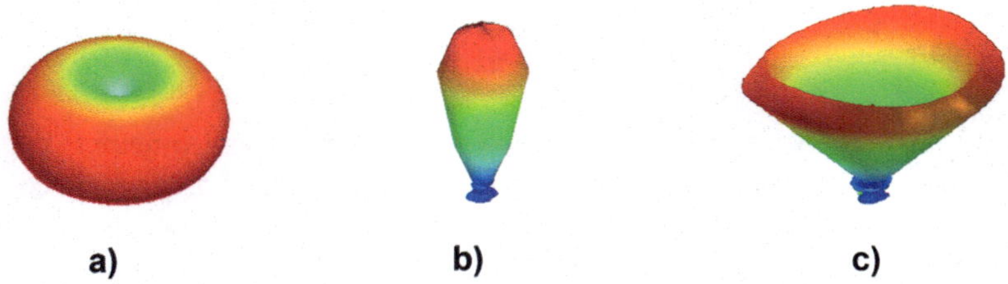

Fig. 1. Typical types of helical antenna patterns: (a) toroidal; (b) unidirectional; (c) conical

One or another type of radiation pattern in active antennas is obtained by changing the excitation frequency of the feeder line. In this paper, it will be demonstrated that it is possible to obtain various types of radiation patterns for passive radiators as a result of the excitation of a helix by an incident wave, by changing its angle of incidence.

Cylindrical helical antennas with constant pitch are not frequency-independent. Therefore, in the working frequency range, their radiation pattern changes more or less monotonously, in particular, with increasing frequency, the radiation pattern becomes narrower.

Different types of radiation patterns appear because various modes of electromagnetic wave, which propagate along the helix, can dominate depending on different types of helix excitation. So, the T_0 wave corresponds to the radiation in the direction, orthogonal to the axis of the helix, and the T_1 wave corresponds to the radiation in the axial direction.

Normal radiation mode

The normal mode of radiation is realized when the wavelength is much longer than the length of the turn of helix, $\lambda >> P$. In normal mode, the antenna radiates orthogonal to the axis of the helix, in the axial direction the radiation is zero, the radiation pattern has a toroidal appearance.

In the case of normal radiation, the currents along the branches of the helix are almost constant in phase and amplitude, it can practically be assumed that a quasistationary current flows along the helix. Components θ and φ of electric field, which was radiated in far zone, looks like [16]:

$$E_\theta = j\eta \frac{kI_0 h e^{-jkR}}{4\pi R} \sin\theta \tag{1}$$

$$E_{\phi} = \eta \frac{k^2 I_0 r^2 e^{-jkR}}{4R} \sin\theta, \qquad (2)$$

where h – pitch of the helix, r – radius of the helix, k – wave number, R – distance from helix to observation point, I_0 – current magnitude in the helix, j – imaginary unit, η – intrinsic impedance of the medium. The condition of circular polarization of the emitted wave is:

$$\frac{|E_{\theta}|}{|E_{\phi}|} = \frac{h}{\pi k r^2} = 1 \qquad (3)$$

When this ratio between the excitation wavelength and the geometrical parameters of the helix is fulfilled in the normal mode, circular polarization of the radiated wave in all directions is realized. It is possible to analyze how relation (3) depends on the pitch angle of the helix α. This angle is formed by a tangent to the spiral and a line perpendicular to the axis of the spiral. By using formulas

$$P \sin\alpha = h, P\cos\alpha = 2\pi r, k = \frac{2\pi}{\lambda}, \qquad (4)$$

we obtain trigonometric equation

$$(\frac{P}{2\lambda})^2 \cos^4\alpha + \cos^2\alpha - 1 = 0. \qquad (5)$$

For the considered case, where wavelength λ significantly exceeds the length of the turn P, solution of the Eq. (5) gives very small values for pitch angle α. For example, where $\lambda = 5P$, we obtain root of Eq. (5) $\alpha_5 = 5.68°$, and in case $\lambda = 10P$ equation gives $\alpha_{10} = 2.86°$. Fabrication helices with such low pitch angle is difficult. In practice, for helical antennas, in most cases the inequality $E_\theta >> E_\varphi$ is satisfied, so, these antennas radiate a linearly polarized wave in normal mode.

Axial radiation mode

When the wavelength is approximately equal to the length of the turn, $\lambda \approx P$, axial radiation mode for the helix can be realized. In this case, maximum of radiation is obtained along the axis of the helix. In addition, there are a number of side lobes. In axial mode, the radiation pattern is unidirectional.

In addition, in this mode, other remarkable properties are observed for the helical antenna: (1) the traveling wave mode is established along the turns of the helix, reflection from the edges of the helix is low; (2) impedance becomes purely active. The polarization of the wave in the axial mode of radiation is close to a circle, and polarization depends on number of turn: when number becomes larger, then polarization becomes closer to circular.

2.3 Calculation of Radiation Patterns of 20-Turns DNA-like Helix

Consider a double 20-turns DNA-like helix, which presented in Fig. 2. It is called DNA-like because the strands of the helix are mutually displaced along its axis. The double helix is excited by an incident plane-polarized wave. By changing the angle of incidence of the wave, it is possible to obtain any type of radiation pattern of the single-stranded helical antenna. During excitation at the different angles different current magnitudes and phases distribution occurs on the wires of the helix. Each distribution case has its own type of radiation pattern.

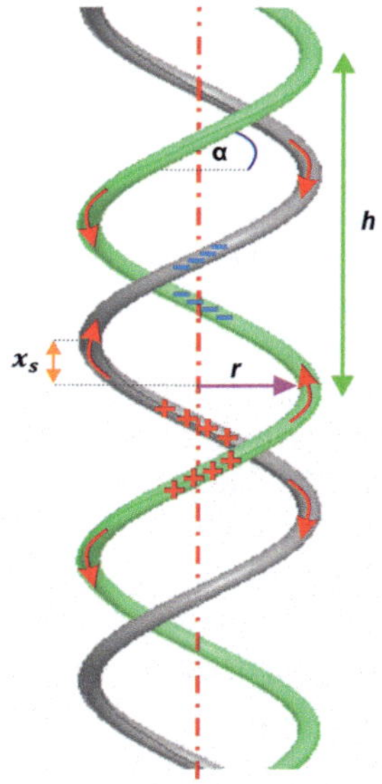

Fig. 2. Double DNA-like helix. In the figure marked: h – pitch of the helix, r – radius of the helix, x_s – mutual displacement of the wires, α – pitch angle. On the strands, currents and charges are shown, for the case, when currents in both strands have the same direction, relatively to the helix axis.

The main idea is that to change the shape of the pattern for an active helical antenna, excited by the feeder line, we need to change the frequency of the excitation. In the case of a passive double helix excited by the incident wave, it becomes possible to change only the angle of incidence of the wave, without changing its frequency. This is due to the fact that the vector of electric field can be decomposed into axial and orthogonal components (see Fig. 3), as a result, two different waves with different phase velocities will be activated in the double helix.

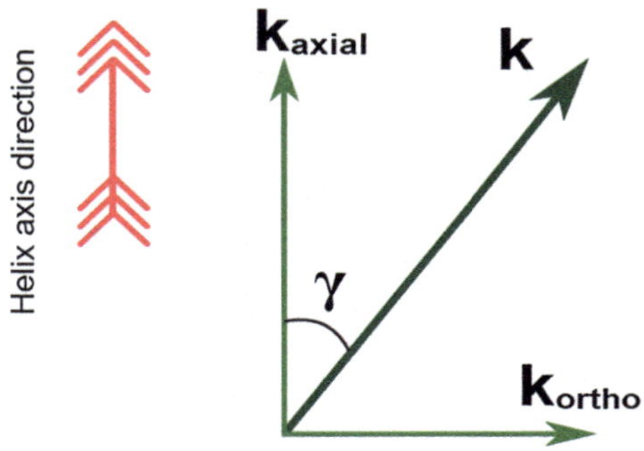

Fig. 3. Vector **k**, decomposed into axial and orthogonal components, γ – angle of incidence of wave, relatively to the axis of the helix.

Further, it will be demonstrated how by using an incident field with a wavelength, approximately equal to the length of the turn, $\lambda \approx P$, it is possible to obtain a conical radiation pattern, which in ordinary single-stranded helical antennas is realized at wavelengths smaller than the length of the turn, $\lambda < P$.

It can be seen from the Fig. 4, that is possible to obtain all radiation patterns which are possible for a single-stranded helix in different frequency ranges, by changing the angle of incidence of the wave on a double DNA-like helix without changing the frequency of the wave. Due to the high frequency of excitation at $\lambda \approx$ P toroidal patterns, corresponding to case $\gamma = 90°$, is very flattened.

In this case, the radiation, emitted by double DNA-like helix in orthogonal to the helix axis direction, is circularly polarized, as was shown in [7–11, 13]. This is significant difference from ordinary single-stranded helical antennas, which radiate at wavelength $\lambda \approx P$ circularly polarized wave only in axial direction.

Also, it should be noted that the radiation of a circularly polarized wave requires a strictly defined pitch angle of the helix. This pitch angle for helices has values $\alpha_{\text{double opt.}} \approx 25°$ for double DNA-like helix [7–11, 13] and $\alpha_{\text{single opt.}} \approx 15°$ for single-stranded helix, and, as we can see, significantly different for these two cases.

It should be noted that in all cases the helix forms a radiation pattern that is almost symmetrical in a plane, orthogonal to the axis of the helix. So, mainly, only axial angle, which is formed by wave vector of the incident wave and by axis of the helix, has significance.

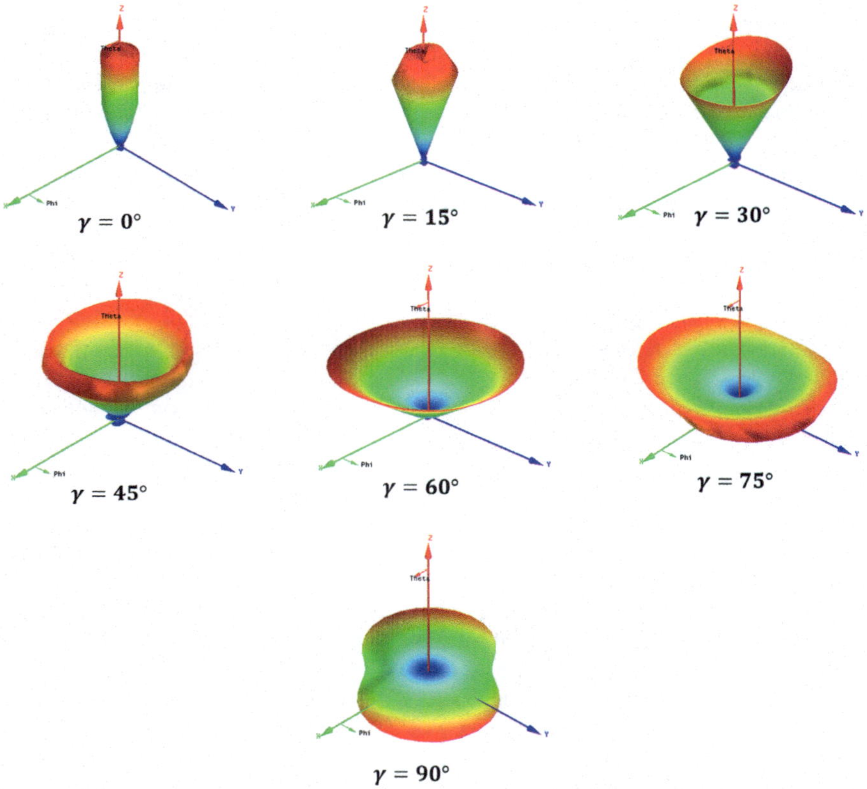

Fig. 4. Radiation patterns of double 20-turns DNA-like helix, at different angles of incidence γ relatively to the axis of the helix. Wavelength is approximately equal to the length of one turn, $\lambda \approx P$

3 Conclusion

In this paper, we compared the main characteristics of cylindrical helical radiators, the main modes of their operation, and the features that distinguish them from others. Traditional single-stranded helix and double DNA-like helix are considered. Main attention is paid to the radiation patterns, which are formed by these helices in active and passive modes. It is indicated that the helical antenna is capable of forming three types of diagrams: toroidal, unidirectional and conical. Using computer simulation, it has been demonstrated that all the above-described radiation patterns can be obtained for a double DNA-like helix in a passive mode at a single frequency of an electromagnetic wave. This is achieved by exciting such helix with an incident electromagnetic wave when the angle of incidence of this wave changes relatively to the axis of the helix.

This effect can be used in various devices, including those based on metasurfaces, where it is required to change the radiation directivity of the system without changing the frequency of excitation.

References

1. Asadchy, V., Faniayeu, I., Ra'di, Y., Khakhomov, S., Semchenko, I., Tretyakov, S.: Broadband reflectionless metasheets: frequency-selective transmission and perfect absorption. Phys. Rev. X, **5**(3), 031005(1–10) (2015)
2. Asadchy, V., Ra'di, Y., Vehmas, J., Tretyakov, S.: Functional metamirrors using bianisotropic elements. Phys. Rev. Lett. **114**(9), 095503, 1–5 (2015)
3. Asadchy, V., Albooyeh, M., Tretyakov, S.: Optical metamirror: all-dielectric frequency selective mirror with fully controllable reflection phase. J. Opt. Soc. Am. B **33**(2), A16–A20 (2016)
4. Asadchy, V., Albooyeh, M., Tcvetkova, S., D′iaz-Rubio, A., Ra'di, Y., Tretyakov, S.: Perfect control of reflection and refraction using spatially dispersive metasurfaces. Phys. Rev. B, **94**(7), 075142(1–14) (2016)
5. Asadchy, V., Wickberg, A., D′iaz-Rubio, A., Wegener, M.: Eliminating scattering loss in anomalously reflecting optical metasurfaces. ACS Photonics **4**(5), 1264–1270 (2017)
6. Quevedo-Teruel, O., Chen, H., Díaz-Rubio, A.: Roadmap on metasurfaces. J. Opt. **21**(7), 073002 (2019). https://doi.org/10.1088/2040-8986/ab161d
7. Semchenko, I., Khakhomov, S., Balmakov, A.: Polarization selectivity of electromagnetic radiation of deoxyribonucleic acid. J. Commun. Technol. Electron. **52**(9), 996–1001 (2007). https://doi.org/10.1134/S1064226907090070
8. Semchenko, I., Khakhomov, S., Balmakov, A.: Research of artificial DNA-Like structures in microwave: observation of polarization selectivity of reflection of waves. J. Electromagn. Waves Electron. Syst. **4**, 66–72 (2009)
9. Semchenko, I., Balmakov, A., Khakhomov, S.: Effect of maximum interaction of circularly polarized electromagnetic waves with the molecule of DNA. J. Autom. Mobile Robot. Intell. Syst. **3**(4), 207–209 (2009)
10. Semchenko, I., Khakhomov, S., Balmakov, A.: Polarization selectivity of interaction of DNA molecules with X-ray radiation. Biophysics **55**(2), 194–198 (2010). https://doi.org/10.1134/S0006350910020053
11. Semchenko, I., Khakhomov, S., Balmakov, A.: Polarization selectivity of artificial anisotropic structures based on DNA-like helices. Crystallogr. Rep. **55**(6), 921–926 (2010). https://doi.org/10.1134/S1063774510060040
12. Semchenko, I., Khakhomov, S., Samofalov, A., Songsong, Q.: The equilibrium state of bifilar helix as element of metamaterials. Japanese J. Appl. Phys. Conf. Proc. **4**, 011112-1–011112-6 (2016)
13. Semchenko, I., Khakhomov, S., Balmakov, A.: Electromagnetic model of DNA: observation of polarization selectivity of radiation. In: The 6th International Conference on Global Research and Education Inter-Academia 2007: proceedings, vol. 1, Hamamatsu, Japan: Shizuoka University, pp. 136–145 (2007)
14. Serdyukov, A., Semchenko, I., Tretyakov, S., Sihvola, A.: Electromagnetics of Bi-Anisotropic Materials - Theory and Application, vol. 11. Gordon and Breach Science Publishers, Amsterdam (2001)
15. Yurtsev, O., Runov, A., Kazarin, A.: The Helical Antennas. Sovietskoye radio, Moscow (1974)
16. Balanis, C.A.: Antenna Theory: Analysis and Design, 2nd ed. Wiley (1997)
17. Karnaushenko, D.M., Karnaushenko, D., Makarov, D., Schmidt, O.: Compact helical antenna for smart implant applications. NPG Asia Mater. (2015) https://doi.org/10.1038/am.2015.53

18. Savelyev, I., Zyryanova, N., Polesskaya, O., O'Mealy, C., Myakishev-Rempel, M.: On the DNA resonance code (2018). https://doi.org/10.13140/RG.2.2.13874.86720
19. Marshall, R.: Generalized macro level models of amino acid sequences using passive electrical circuits. In: Proceedings IWBBIO, pp. 608–622 (2014)

Thermally Annealed in Vacuum Undoped and Al-Doped ZnO Thin Films for Multifunctional Applications

Ion Lungu and Tamara Potlog(✉)

Physics Department and Engineering, Moldova State University, Chisinau, Moldova
tpotlog@gmail.com

Abstract. Undoped and Al-doped ZnO thin films have been prepared by spray pyrolysis in oxygen and argon atmospheres. The structural properties of ZnO thin films were investigated using atomic force microscopy (AFM) and X-ray diffraction (XRD). The optical properties were studied by UV-VIS spectroscopy and photoluminescence (PL) at room temperature. The electrical resistivity and Hall mobility were measured using the van der Pauw technique at room temperature. AFM studies show that the ambient atmosphere influences the roughness parameters of the ZnO surfaces. The XRD results revealed that undoped and Al-doped ZnO films are polycrystalline and developed [0002] preferred orientation. The best electrical parameters (conductivity, mobility carriers and carrier concentration) are obtained for 1.0 at % of Al-doped ZnO synthetized in Ar atmosphere. In all cases, the electrical parameters under Ar are higher than under O_2 atmosphere, unless they are not doped. Different applications of undoped and Al-doped ZnO thin films are discussed.

Keywords: Undoped and Al-doped ZnO · Spray pyrolysis · Surface properties · Optical and electrical properties · Photoluminescence

1 Introduction

For the design and realization of ZnO-based devices, one of the most significant issues is doping, which involves the heavy doping with trivalent elements (Al, Ga, In). Minami et al. [1] summarized the resistivity of impurity-doped binary compound TCO films reported among thirty years and found that the obtained minimum resistivity of impurity-doped ZnO films is still decreasing, while SnO_2 and In_2O_3 films have essentially remained unchanged [2, 3]. Such trend indicates the possible significant promotion of ZnO-based TCO films.

How to obtain thinner films with both excellent electrical and optical properties is one of the main research tasks. Therefore, in this paper, the systematic investigation of the effect of ambient atmosphere on the structural and electrical properties of undoped and Al-doping ZnO nanostructured thin films are investigated for its industrial applications.

A. R. Várkonyi-Kóczy (Ed.): INTER-ACADEMIA 2019, LNNS 101, pp. 144–158, 2020.
https://doi.org/10.1007/978-3-030-36841-8_15

2 Experimental Details

Undoped and Al-doped ZnO thin films were deposited on glass substrates by spray pyrolysis technique with schematic diagram illustrated in Fig. 1. The glass substrates were preventively degreased in toluene and isopropyl alcohol, corroded in methanol +5% Br for 10 min, dried in isopropyl alcohol vapor and then, were placed in the depositing chamber.

The synthesis parameters that are more relevant for spray pyrolysis are the concentration molarity of the precursor solution, the carrier gas flux rate, and the synthesis temperature. Purified argon and oxygen were used as the carrier gases.

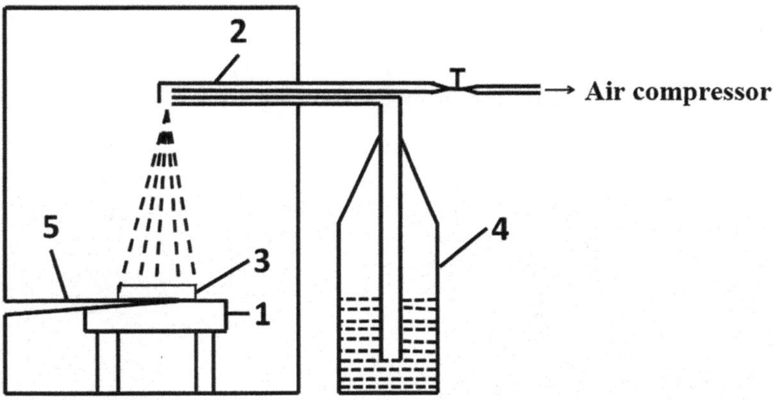

Fig. 1. Schematic diagram of spray pyrolysis setup: 1- substrate holder; 2 – glass tube; 3 – substrate; 4 – solution; 5 – thermocouple.

The initial solution was prepared by dissolving zinc acetate ($Zn(CH_3COO)_2 \cdot 2(H_2O)$) in methanol-water solution in the ratio 25:65 to obtain a concentration of 0.2 M. Zinc acetate shows higher solubility in methanol (higher dielectric constant).Water was added to the solvents to increase the dielectric constant and hence contribute to the dissolving of zinc acetate. In dissolving zinc acetate in methanol solution, each zinc ion is surrounded by six methanol molecules. In addition, a few drops of concentrated acetic acid were added to the starting solution. For doping of ZnO thin films, aluminum chloride was used. The solution was stirred for 40 min at 50 °C. The spray flow rate was fixed at a constant 2 ml/min. The nozzle-substrate distance was 30 cm.

The first set of undoped ZnO thin films were synthesized at different substrates temperatures onto glass substrates for investigation transmittance of the films. The second set was obtained at the atomic ratio of Al/Zn added in the starting solution fixed at 1, 2, 3 and 5. For the stabilization of physical properties of the ZnO thin films, both sets were subjected to a thermal vacuum annealing at 420 °C.

The surface morphology was investigated using a multimode commercial atomic force microscope (SOLVER Next – NT-MDT). Cone-shaped tips from monocrystalline silicon (tip radius ~10 nm) on cantilevers with a stiffness of about 17 N/m were used to perform the measurements. The AFM measurements were successfully carried out at two different scan sizes (10 × 10 um and 5 × 5 um) in tapping mode in which height (topography) and phase signals were acquired simultaneously. The crystal structures of ZnO thin films were measured by Ultima IV X-ray Diffracrometer (XRD) in a θ–2θ configuration with CuKα radiation (λ = 0.154 nm).

For the investigation of the optical properties of ZnO thin films a double-beam Jasco V-760 UV/VIS double spectrophotometer was used to record transmission spectra in the 300–1000 nm wavelength range. Energy resolution measurements do not exceed 2 meV. Photoluminescence measurements were recorded on an optical spectrometer based on an MDR-206 grating monochromator (LOMO, Saint Petersburg, Russia) using laser excitation λ = 337 nm.

3 Results and Discussion

3.1 Morphology and Structure of ZnO Thin Films

The AFM study of the samples allows us to determine the surface roughness. The roughness parameters of the surfaces, Root Mean Square (RMS) and arithmetic average Roughness (Ra), were estimated using image processing software (Nova Px) for topography. For example, the AFM images for ZnO undoped and non-annealed obtained in O_2 atmosphere with scan area 10 μm × 10 μm are presented in Fig. 2.

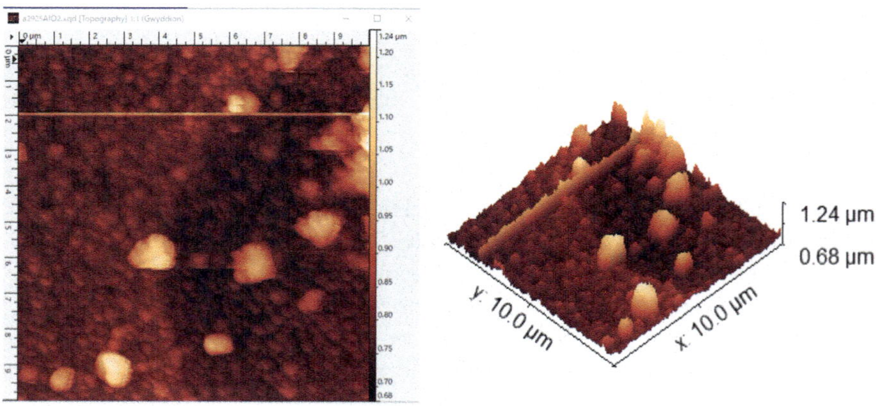

Fig. 2. AFM images with scan area 10 μm × 10 μm of undoped ZnO/Glass thin films deposited in O_2 atmosphere and annealed in vacuum at 420 °C.

The roughness parameters of the as-deposited ZnO surface obtained in O_2 atmosphere showed the highest values for RMS = 60.96 nm and Ra = 1.45 nm. The undoped ZnO surface deposited in Ar atmosphere showed the lowest values for RMS = 29, 48 nm and Ra = 0.45 nm in comparison with values for surfaces of thin films obtained in O_2 atmosphere. As one can see from Fig. 4, the lateral profile obtained for 2% Al–ZnO obtained in Ar atmosphere and annealed at 420 °C in vacuum makes clear the growth of the structures in both height and width. Similar behavior has been reported by Vasco et al. for ZnO deposited on InP (100) films grown by PLD technique [4]. A M Rosa et al. [5] have reported the similar growth for ZnO thin films deposited by RF magnetron sputtering. The changes in roughness are due to the growth evolution of the granular structures and the formation of mound-like structure (Fig. 3).

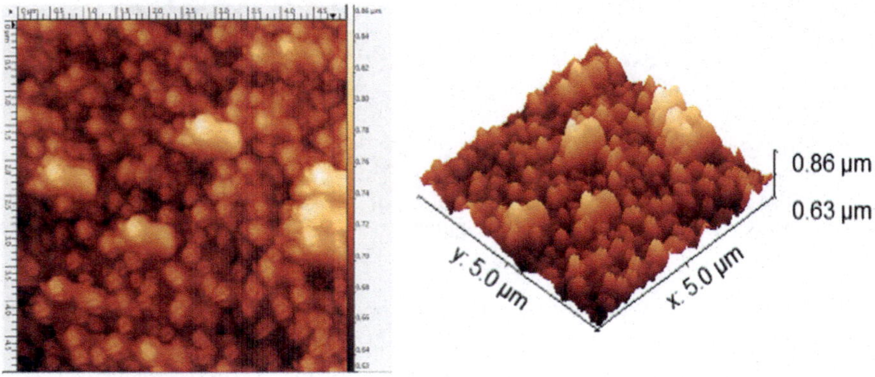

Fig. 3. AFM images with scan area 5 μm × 5 μm of undoped ZnO/Glass thin films deposited in Ar atmosphere and annealed in vacuum at 420 °C.

Fig. 4. AFM images with scan area 5 μm × 5 μm of 2% Al–ZnO/Glass thin films deposited in Ar atmosphere and annealed in vacuum at 420 °C.

Firstly, it was observed that the doping with Al increasing the RMS and Ra. For example, 2% Al–ZnO/Glass thin films deposited in Ar atmosphere and annealed in vacuum at 420 °C reached RMS = 30, 75 nm and Ra = 0.2 nm. These results indicate that the surface of undoped ZnO film is smoother than that of the Al-doped ZnO films. Secondly, smooth of the surface decreases as a result of increasing Al concentration. For ZnO thin films deposited in O_2 atmosphere and annealed in the same condition like

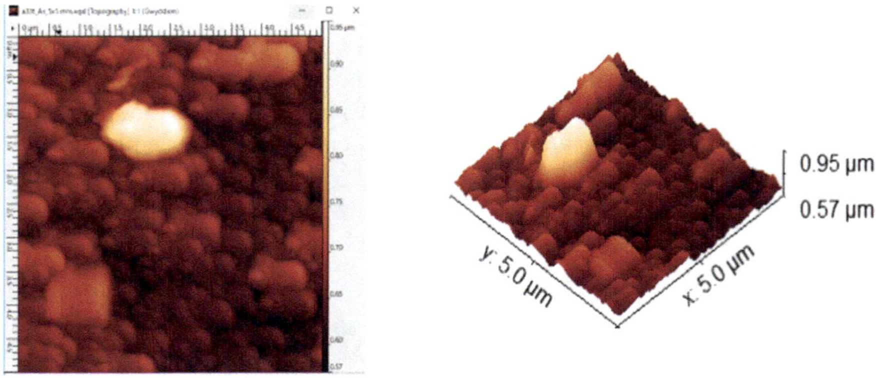

Fig. 5. AFM images with scan area 5 µm × 5 µm of 2% Al–ZnO/Glass thin films deposited in Ar atmosphere and annealed in vacuum at 420 °C.

the ones obtained in Ar atmosphere (Fig. 5) the RMS = 43.64 nm and Ra = 0.5016 nm decrease.

Figures 6 (a) and (b) show the XRD spectra of the thermal annealed at 420 °C undoped ZnO and Al-doped thin films in O_2 and Ar atmospheres.

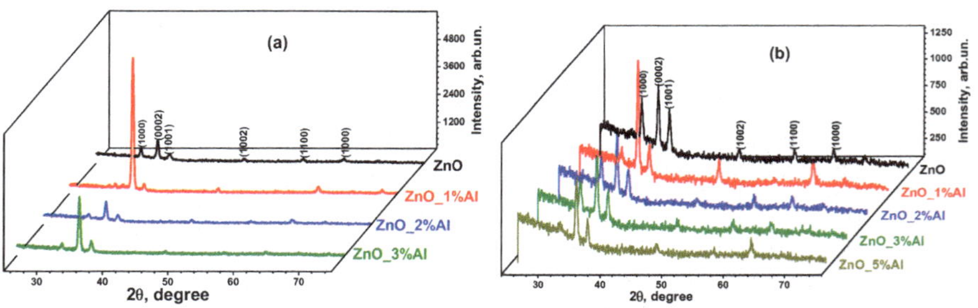

Fig. 6. The XRD patterns of the vacuum annealed undoped and Al-doped ZnO thin films in O_2 (a) and Ar (b) atmospheres.

The XRD results show that the all ZnO thin films have polycrystalline structure with the (0002), (1000) and (1001) reflections corresponding to the ZnO hexagonal wurtzite crystal structure with c-axis perpendicular to the substrate surface [6]. As can be seen from Fig. 6 (a, b), the film doped with 1% Al resulted in the highest (0002) orientation for both types of atmospheres. Regardless of the concentration of the Al, the (0002) reflection dominates in both atmospheres. For example, when the films are obtained in the Ar atmosphere, structural aspects of the (0002) plane remain unaltered (Fig. 7 (b), while for obtained in O_2 atmosphere this diffraction plan splites (see Fig. 7 (a)).

In the case of ZnO obtained in O_2 atmosphere, two XRD peaks arise at 32.99° and 34.48°. Comparison of these diffraction angles values with values for other phases

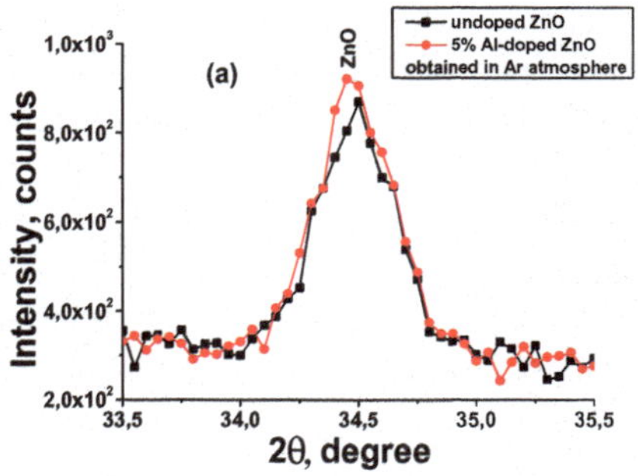

Fig. 7. The XRD peak position for undoped and Al dopant concentration of the vacuum annealed ZnO thin films in O_2 (a) and Ar (b) atmospheres.

allow us to conclude that in the case of ZnO deposited in O_2 atmosphere the $2\theta = 32.99^o$ belongs to the zinc hydroxide ($Zn(OH)_2$) phase. $Zn(OH)_2$ is an amphoteric compound, which normally crystallizes in wűlfingite ε-$Zn(OH)_2$ with orthorhombic structure (space group $P2_12_12_1$) under ambient conditions. [7–11]. The ICSD #50447 data on $Zn(OH)_2$ confirms our assumption of forming a new phase ε-$Zn(OH)_2$. Probable, when ZnO obtained in oxygen atmosphere, O_2 molecules are adsorbed on the surface of ZnO, while synthetized in Ar, O_2 molecules are desorbed from the surface of ZnO thin films. A small deviation in the (0002) peak position was found when doped with Al at different concentrations, indicating that some residual stress inside the film may exist.

Jeong et al. [12] observed an increase in the position of the (0002) peak with increased doping and attributed this shift to residual stress in the film caused by the difference in ionic sizes between Zn^{2+} (72 pm) and Al^{3+} (53 pm). In our study, this slightly increase in the (0002) position peak is observed for ZnO thin films obtained in O_2 atmosphere.

The crystallite size and lattice strain determined from the (0002) diffraction peak for the ZnO samples obtained in O_2 atmosphere as a function of Al concentration are given in Table 1. According to the Table 1, for a preferential growth along (0002) plane, the crystallite size changes from 29.26 nm for undoped film to 30.82 nm for 1% Al-doped ZnO. In addition, the smallest lattice strain is revealed for the 1% Al-doped ZnO.

Table 1. Structural properties of the undoped ZnO and Al-doped ZnO thin films obtained in O_2 atmosphere.

O_2 atm.	Nr.	2θ (deg)	d (ang)	FWHM (rad)	D (ang)	$\varepsilon \times 10^{-3}$
Vacuum anneal. Al-doped ZnO thin films	Undoped	34.46	2.6008	0.00551	292.6	4.4
	1%	34.48	2.6003	0.0052	308.2	4.2
	2%	34.49	2.5996	0.00553	291.7	4.5
	3%	34.48	2.6004	0.00541	298.3	4.4

For the ZnO layer obtained in Ar atmosphere in the same condition, according to the Fig. 6 (b), the amorphous phase is also presented. The (1000) and (1001) reflections are more pronounced in comparison with those of ZnO obtained in the O_2 atmosphere. The structural parameters of the undoped ZnO and Al-doped ZnO thin films obtained in Ar atmosphere with different Al concentrations are shown in Table 2.

Therefore, the values of grain sizes for all ZnO films obtained in both gas atmospheres decrease with the increasing of the Al concentration, and indicate a small difference between the values. The values of 2θ for the diffraction peaks are very close to that of the standard ZnO crystal [13].

Table 2. Structural properties of the undoped ZnO and Al-doped ZnO thin films obtained in Ar atmosphere.

Ar atm.	Nr.	2θ (deg)	d (ang)	FWHM (rad)	D (ang)	ε (lattice strain) × 10^{-3}
Vacuum anneal. Al-doped ZnO thin films	As dep.	34.45	2,5997	0.00565	285.6	4.55
	1%	34.44	2.5994	0.00525	307.3	4.23
	2%	34.44	2.5986	0.00529	304.8	4.26
	3%	34.44	2.5991	0.00545	296.1	4.39
	5%	34.44	2.6001	0.00571	282.6	4.6

3.2 Optical Properties

The transmission spectra of undoped ZnO thin films deposited on glass substrates at 400 °C in the O_2 atmosphere, from a different volume of spray solution, are shown in Fig. 8 (a). For all sprayed volumes the transmittance curves have the same behaviour in the (200–1200) nm wavelength range and reached 80–85% values. A transmission spectrum (50 ml) has a shape with characteristic interference maxima.

The optical transmission spectra of undoped ZnO thin films obtained within the temperature range of the glass substrate (250…450)°C in the O_2 atmosphere at a constant volume of 50 ml are shown in Fig. 8 (b). As one can see, the optical transmission has (80–85)% values within the wavelength range (300…1000) nm.

For the thin films obtained at different substrate temperatures the same amount of the solution was sprayed. It was established that the optical transmittance changes slightly and was practically the same with the exception of the thin films grown at temperatures of 400 °C and 450 °C, whose crystalline structure is probable more qualitative.

In order to appreciate the optical parameters of ZnO thin films obtained in both atmospheres were measured reflectance (R) and transmission spectra (T) at normal incidence of light on the sample surface.

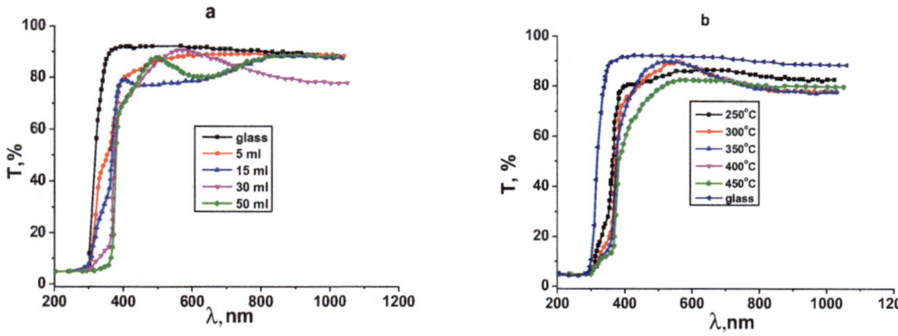

Fig. 8. The transmission spectra of the undoped ZnO thin films obtained at Tsub = 400 °C at different volume of spray solution (a) and different substrate temperatures (b).

From the transmission and reflection spectra of the ZnO thin films with 300 nm thickness obtained in O_2 and Ar atmospheres were estimated absorption coefficient in the interval 200–900 nm wavelength range. The absorption coefficient, α, was calculated from expression [14].

$$\alpha = \frac{1}{d} ln \frac{(1 - R)^2}{T} \tag{1}$$

where d is the film thickness, and R and T represent the reflection and transmission spectra, respectively.

In order to reveal the nature of the band gap of the ZnO films, α and hv values were used to fit-in with the following equation $(\alpha hv) = A \cdot (hv - Eg)^n$. Values of n for allowed direct and indirect, and forbidden direct and indirect transitions are ½, 2, 3/2 and 3, respectively. We can obtain a set of values of (n, E_g) from the plots of $\alpha^{1/n}$ versus hv for above n values. The $(\alpha hv)^2$ versus (hv) plots for all samples are presented in Fig. 9 (a, b).

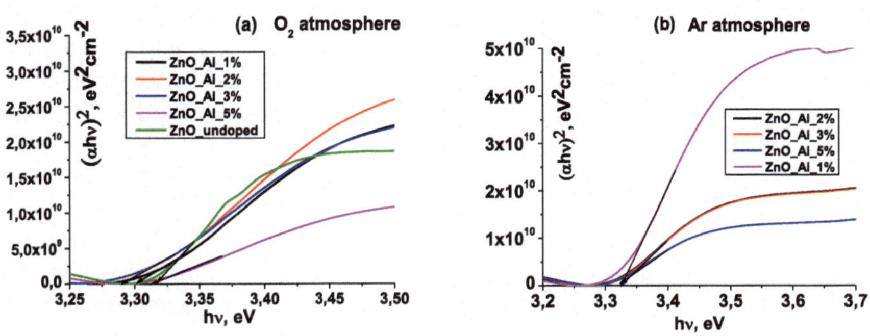

Fig. 9. Dependencies of $(\alpha hv)^2$ vs. hv of the vacuum annealed undoped and Al-doped ZnO thin films in O_2 (a) and Ar (b) atmospheres.

It is evident that curves for all samples have straight-line portions in the low energy region, which confirms the direct band gap nature. The values of band gap are presented in Table 3.

Table 3. Band gap values of undoped and Al-doped ZnO thin films vacuum thermal annealed.

Samples	$E_{g,}$ eV	d ± 0, 0 2 nm
ZnO_Al_2%_Ar	3,32	302
ZnO_Al_3%_Ar	3,32	300
ZnO_Al_5%_Ar	3,31	298
ZnO_Al_1%_Ar	3,32	298
ZnO_Al_1%_O_2	3,31	301
ZnO_Al_2%_O_2	3,30	304
ZnO_Al_3%_O_2	3,29	303.5
ZnO_Al_5%_O_2	3,31	298
ZnO_O_2 undoped	3,32	298

As one can see, for the films grown in O_2 atmosphere, the bandgap values decrease with increasing Al concentration. A decrease in the bandgap values from 3.32 eV for undoped to 3.29 eV for 3% Al-doped ZnO thin films obtained in O_2 atmosphere was depicted. It may be due to the presence of zinc hydroxide forms such as $Zn(OH)_2$. The band gap values decrease are due to the removal of zinc hydroxide and the decrease of defect levels which is a more common phenomena in chemically deposited thin films. In addition, the decrease in the values of the band gap energies can be attributed to the increase in grain sizes, also. Another tendency of behavior is observed for ZnO films grown in Ar atmosphere. The value of band gap does not change with the increasing the Al concentration in ZnO thin films.

So, for ZnO thin films obtained in O_2 atmosphere the value of E_g is changing slightly in the range of 3.29–3.32 eV due to the presence of zinc hydroxide. The band gap values decrease is due to the removal of zinc hydroxide and the decrease of defect levels. For ZnO films grown in Ar atmosphere, the value of band gap does not change with the increasing of the Al concentration in ZnO thin films. The lower values of E_g for synthetized thin films in both atmospheres in comparison with in ZnO bulk are a result of defects in the nanostructured thin films.

3.3 Electrical Properties

Figure 10 shows the Arrhenius plot (ln σ vs $1/T$) for the vacuum annealed 1% Al-doped ZnO thin films prepared in O_2 and Ar atmospheres in the range 300 K to 77 K. The same behavior of plots (ln σ vs $1/T$) shows all undoped and Al-doped ZnO thin films regardless of the gas atmosphere in which they were obtained.

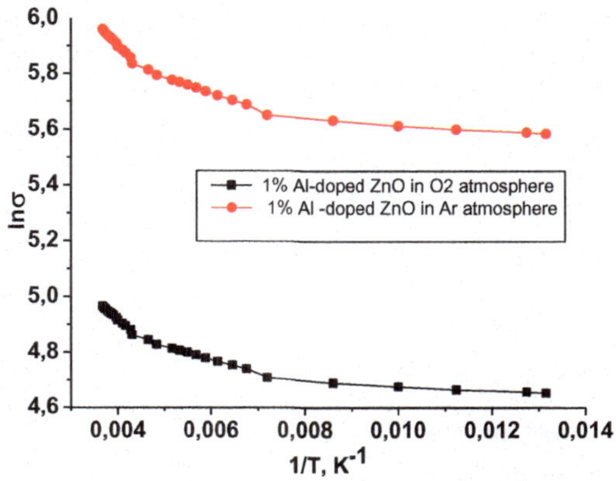

Fig. 10. The electrical conductivity in function of the measurement temperatures for 1% Al-doped ZnO obtained in O_2 and Ar atmospheres.

From the plot ln σ vs *1/T*, it is observed that the conductivity of the ZnO samples increases with increasing temperature. From the straight-line fits, it has been observed three linear regions indicating that the conduction mechanism is thermally activated. The activation energy (ΔE_a), for the 1% Al-doped ZnO film synthetized in Ar atmosphere for the 1st, 2nd and 3rd regions equal with 14.4 meV, 4.9 meV and 1.035 meV, respectively, while for 1% Al-doped synthetized in O_2 for the same regions equal with 12.17 meV, 4.23 meV and 0.85 meV, respectively. According to the mentioned results, the 1.0 at.% Al-doped ZnO thin films have the best structural, optical and electrical properties that can be adapted for several applications such as window layers in solar cells.

The best Al-doped conductor with a maximum conductivity 1316 $(\Omega \cdot cm)^{-1}$ and carrier concentration of 1.25×10^{22} cm^{-3} was obtained for 1.0 at.% of Al-doped ZnO thin film synthetized in Ar atmosphere. Therefore, spray pyrolysis is an efficient method to get Al-doped ZnO thin films with high quality and low cost.

4 ZnO Applications

4.1 Undoped ZnO Nanostructured Thin Films as Antireflective Coating

There are several papers proposing ZnO as AR coating for silicon and CIGS solar cells to improve the efficiency [15–18] but, to best of our knowledge, no one yet reported use of ZnO as antireflective (AR) coating for InP/CdS solar cell.

The heterojunction InP/CdS solar cells were fabricated by deposition of InP by Halide Vapor Phase Epitaxy (HVPE) on InP substrates with carrier concentration 10^{18} cm^{-3} and orientation [100] at T_{sub} = 680 °C in hydrogen atmosphere [19]. The CdS thin films were deposited by close space sublimation method at $_{Tsub}$ = 650 °C in hydrogen atmosphere, also.

Figure 11 (a) show the schematic of the InP/CdS solar cell model proposed in this study. As contact ohmic was used Ag + 5%Zn for InP, and In for CdS thin film, both deposited by thermal evaporation. Fortunately, refractive index of ZnO is very close to the ideal value required for antireflection (AR) coating of InP surface.

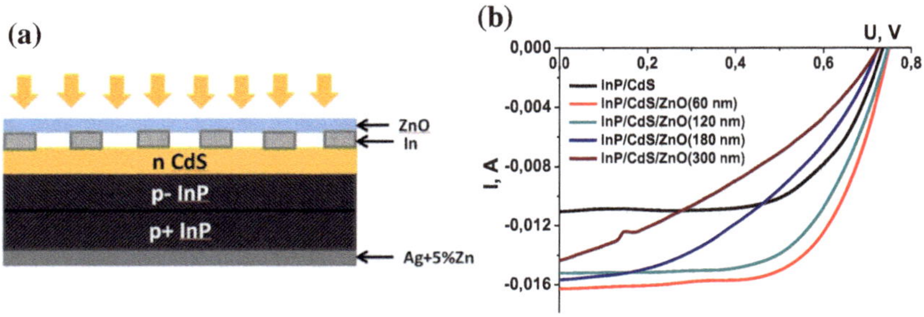

Fig. 11. The schematic of the InP/CdS solar cell with antireflective ZnO thin film (a) and current-voltage characteristics of the *n*CdS/*p*InP device in dependence on the thickness of the ZnO AR layer, *d*, nm: 1–0; 2–60; 3–120; 4–180; 5–300 (b).

In our proposed model, front region of the solar cell, which is directly exposed to solar radiation, is composed of n-type ZnO/CdS layers. Therefore, transparency of the ZnO layer for visible is very important. The nanometric ZnO anti-reflective layers deposited on nCdS-pInP structure frontal surface have been sprayed at 250 °C for not damage the grid of the generated charge carrier's collection.

The impact of ZnO thickness on open circuit voltage (U_{oc}), short circuit current density (J_{sc}), fill factor (FF), and power conversion efficiency (η) of the ZnO/CdS/p^{-}InP/p^{+}InP solar cell are investigated. The photovoltaic parameters dependence on ZnO antireflective layer thickness of *n*CdS/*p*InP structure is given in Table 4. Thicknesses of (60…100) nm are the most favorable for increasing the photovoltaic cell efficiency (Fig. 11 (b)).

Table 4. Photovoltaic parameters of the *n*CdS/*p*InP device in dependence on the AR ZnO layer thickness.

Solar cell	d_{AR}, nm	J_{sc}, mA/cm^2	U_{oc}, V	FF, %	η, %	R_{ser}, Ω
*n*CdS/*p*InP	0	18.33	0.73	64.75	8.56	5.2
	60	27.21	0.74	63.35	12.74	3.67
	120	25.36	0.74	61.51	11.54	4.33
	180	26.11	0.72	42.83	8.05	5.57
	300	23.96	0.72	34.7	5.98	11.88

Thus, for these thicknesses of the antireflective layer, the photovoltaic cell efficiency increases by approximately 3%. At the thickness of antireflection layers higher than 100 nm, the short circuit current decreases along with the decrease of the fill factor

FF. The decrease in short circuit current density and the fill factor is largely due to the increase in the structure series resistance. The increase in series resistance of the photovoltaic structure is related to the keeping of the photovoltaic structure at the temperature of 250 °C repeatedly.

4.2 Al-Doped ZnO Nanostructured Thin Films as UV Detector

ZnO-UV detector with planar electrodes was fabricated by the vacuum evaporation of gold (Au). Au was used as the contact metal because of its lower electron-negativity [20].

The measured dark and illuminated I-U characteristics for the vacuum annealed 1% Al-doped ZnO films obtained in Ar atmosphere are shown in Fig. 12 (a). The wavelength of 365 nm was used for the illuminated I-U measurements. The dark current at 5 V bias voltages was 816 μA. The photocurrent upon 365 nm UV light illumination is 44 mA at 5 V biases, which is obviously higher than the dark current. The linear I-U relations under both forward and reverse bias exhibit ohmic metal-semiconductor contacts. The detector operates in the photoconductive mod.

Figure 12 (b) shows the PL spectra of the undoped 1% Al-doped ZnO films deposited in Ar atmosphere on glass substrate. Laser excitation wavelength (λ_{ext} = 337 nm) was used to excite the electrons from the material valence band. It was established that the photoluminescence curves measured at 300 K in the region of 1.5…3.5 eV consist of three emission bands and an ultraviolet luminescence intensive broad band situated at 2.94 eV with a shoulder.

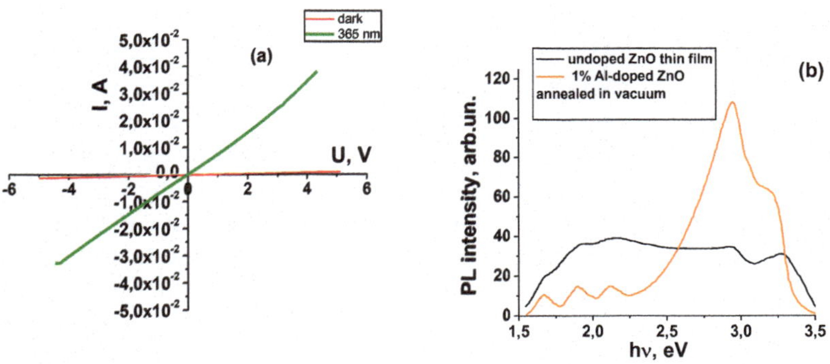

Fig. 12. The dark and illuminated I-U characteristics (a) and the photoluminescence spectra at room temperature (b) for the vacuum annealed 1% Al-doped ZnO films.

The deconvolution procedure of broadband show the superposition of the three PL bands centered at 3.18 eV, 2.94 eV and 2.51 eV. These PL bands can be assigned to the near-band edge (NBE) and defect-related emissions.

Except the intensive broadband, the PL spectrum of the vacuum annealed 1% Al-ZnO thin film deposited in Ar atmosphere still contains three emission bands located at 1.67 eV, 1.89 eV and 2.12 eV. These PL bands are observed at energies significantly lower than the band gap. Since the absorption coefficient is lower in this spectral range, according to the [21], the PL is not strongly reabsorbed and the interference fringes are observed. These emission bands could be probably due to the oxygen vacancies in the ZnO, which could result from the deposition conditions, i.e., ZnO obtained in Ar and annealed in vacuum atmospheres of limited oxygen supply.

5 Conclusions

Nanostructured undoped and Al-doped ZnO thin films were deposited by spray pyrolysis in O_2 and Ar atmospheres at 450 °C substrate temperature. The key results are outlined as follows:

1. The AFM study showed the surfaces of the ZnO films roughen more in O_2 atmosphere.
2. XRD analysis revealed that the ZnO thin films have polycrystalline structure with the [0002] preferred orientation. XRD analysis of Al-doped ZnO deposited in O_2 atmosphere revealed the zinc hydroxide ($Zn(OH)_2$) phase. The crystallite size determined from the (0002) reflection as a function of Al concentration changed from from 29.26 nm for undoped film to 30.82 nm for 1% Al-doped ZnO obtained in Ar atmosphere.
3. The highest transmittance value reached (80–85)% for all nanostructured ZnO thin films, regardless of the nature of the ambient atmosphere.
4. The optical studies indicated the presence of direct band transition across a bandgap of all ZnO thin films. The band gap values changed slightly for Al-doped ZnO thin films obtained in O_2 with Ar.
5. For ZnO thin films obtained in O_2 atmosphere the value of E_g is changing in the range of 3.29–3.32 eV due to the presence of zinc hydroxide.
6. The position of the most intensive PL ultraviolet band is related to electronic transition between defect level and the valence band edge of 1% Al-doped ZnO thin films.
7. The best Al-doped conductor with a maximum conductivity 1316 $(\Omega\cdot cm)^{-1}$ and carrier concentration of 1.25×10^{22} cm^{-3} was obtained for 1.0 at.% of Al-doped ZnO thin film synthetized in Ar atmosphere.

The desired structural and optical properties of undoped and Al-doped ZnO thin films make its promising material for industrial applications, especially in optoelectronics.

Acknowledgements. The authors thank the Ministry of Education, Culture and Research of Republic of Moldova for funding the grant 15.817.02.39A. In addition, a part of this research is supported by Research Center for Biomedical Engineering through Cooperative Project, RIE, Japan. Authors also thank Vasile Botnariuc, Moldova State University for supplying InP/CdS solar cells and Mr. Miyake Taku, Graduate School of Integrated Science and Technology, Shizuoka University, for taking AFM images.

References

1. Tadatsugu, M., Hidehito, N., Shinzo, T.: Highly conductive and transparent aluminum doped zinc oxide thin films prepared by RF magnetron sputtering. Japanese J Appl Phys **23** (5), 2 (1984)
2. Manifacier, J.C., Szepessy, L., Bresse, J.F., Perotin, M., Stuck, R.: In_2O_3: (Sn) and SnO_2: (F) films - application to solar energy conversion part II—Electrical and optical properties. Mater. Res. Bull. **14**(2), 163–175 (1979)
3. Zhilova, O.V., Pankov, S.Yu., Sitnikov, A.V., Kalinin, Yu.E., Volochaev, M.N., Makagonov, V.A.: Structure and electrophysical properties of thin-film SnO_2–In_2O_3 heterostructure. J. Mater. Sci.: Mater. Electron. **30**(13), 11859–1186 (2019)
4. Vasco, E., Rubio-Zuazo, J., Va′zquez, L., Prieto, C., Zaldo, C.: Submicron structure and acoustic properties of ZnO films deposited on (100) InP by pulsed laser deposition. J. Vac. Sci. Technol. B **19**(1), 224–229 (2001)
5. Rosa, A.M., da Silva, E.P., Amorim, E., Chave, M., Catto, A.C., Lisboa-Filho, P.N., Bortoleto, J. R.: Growth evolution of ZnO thin films deposited by RF magnetron sputtering. J. Phys.: Conf. Ser. 370 012020 (2012)
6. Fang, Z., Wang, Y., Peng, X., Liu, X., Zhen, C.: Structural and optical properties of ZnO films grown on the AAO templates. Mater. Lett. **57**(26–27), 4187–4190 (2003)
7. Mingsong, W., Lingxia, J.: Eui Jung Kim and Sung Hong Hahn,: Electronic structure and optical properties of Zn(OH)2: LDA + U calculations and intense yellow luminescence. RSC Adv. **5**, 87496–87503 (2015)
8. Qu, X.-R., Jia, D.-C.: Synthesis of octahedral ZnO mesoscale superstructures via thermal decomposing octahedral zinc hydroxide precursors. J. Cryst. Growth **311**, 1223–1228 (2009)
9. Shaporev, A.S., Ivanov, V.K., Baranchikov, A.E., Polezhaeva, O.S., Tretyakov, Y.D.: Russ. J. Inorg. Chem. **52**, 1811–1816 (2007)
10. Mukhopadhyay, S., Das, P.P., Maity, S., Ghosh, P., Devi, P.S.: Solution grown ZnO rods: Synthesis, characterization and defect mediated photocatalytic activity. Appl. Catal. B **165**, 128–138 (2015)
11. Uekawa, N., Iahii, S., Kojima, T., Kakegawa, K.: Synthesis of ZnO sols by low-temperature heating of ethylene glycol solution and control of their photoluminescence with addition of glucose. Mater. Lett. **61**, 1729–1734 (2007)
12. Jeong, S.H., Park, B.N., Lee, S.B., Boo, J.H.: Metal-doped ZnO thin films, synthesis and characterisations. Surf. Coat. Technol. **201**, 5318–5322 (2007)
13. Norton, D.P., Heo, Y.W., Ivill, M.P., et al.: ZnO: growth, doping & processing. Mater. Today **7**(6), 34–40 (2004)
14. Illiberi, A., Kniknie, B., van Deelen, J. et al.: Industrial high-rate ($\sim$ 14 nm/s) deposition of low resistive and transparent ZnOx:Al flms on glass. Solar Energy Mater. Solar Cells **95**(7), 1955–1959 (2011)
15. Belghit, K., Subhan, M.A, Rulhe, U., Duchemin, S., Bougnot: Sprayed ZnO thin Films as optical window in $CuInSe_2$ based solar cells. J. Eur. Photovolt. Sol. Energy Conf. (1991)
16. Godavarti, U., Mote, V.D., Dasari, M.: Role of cobalt doping on the electrical conductivity of ZnO nanoparticles. J. Asian Ceram. Soc. **5**(4), 391–39 (2017)
17. Chant, E., Wongratanaphisan, D., Gardchareon, A., et al.: Effect of ZnO double layer as anti-reflection coating layer in ZnO Dye-Sensitized. Solar Cells Energy Proced. **79**(November), 879–884 (2015)
18. Simaşchevici, A., Şerban, D., Bruc, L., Coval, A., Gorceac, L., Monaico, E., Usatîi, Iu.: Int. Sci. J. Altern. Energy Ecology ISJAEEET2, 3451–54 (2006)

19. Potlog, T., Botnariuc, V., Gorceac, L., Spalatu, N., Maticiuc, N., Raievschi, S.: The caracterization of the CdS-based solar cell heterojunctions. In: The Proceedings of the International Semiconductor Conference, 11–13 October, Sinaia, Romania, Vol. 01, pp. 105–108 (2010)
20. Mead, C.A.: Surface barriers on ZnSe and ZnO. Phys. Lett. **18**, 218 (1965)
21. Larsen, J.K. et al.: Interference effects in photoluminescence spectra of Cu_2ZnSnS_4 and Cu $(In,Ga)Se_2$ thin films. J. Appl. Phys. **118**, 035307-1- 035307-9 (2015)

Schottky Diode Detectors with Low Leakage Current at High Operating Voltage

V. M. Sklyarchuk[1], V. A. Gnatyuk[2,3](✉), V. G. Pylypko[1], and T. Aoki[3]

[1] Yuriy Fedkovych Chernivtsi National University, Kotsubynsky Str. 2, Chernivtsi 58012, Ukraine
[2] V.E. Lashkaryov Institute of Semiconductor Physics, National Academy of Sciences of Ukraine, Prospekt Nauky 41, Kiev 03028, Ukraine
gnatyuk@ua.fm
[3] Research Institute of Electronics, Shizuoka University, 3-5-1 Johoku, Naka-ku, Hamamatsu 432-8011, Japan

Abstract. For the first time, $Hg_3In_2Te_6$ (MIT) based Schottky diode photodetectors with the lowest reverse dark currents at high bias voltages were created. Both the Schottky rectifying and near Ohmic contacts were obtained by thermal vacuum deposition of Cr onto the MIT crystal surfaces pre-treated with Ar-ion bombardment at different regimes. The crystal surface morphology was monitored by AFM. Cr/MIT/Cr photodiodes were sensitive in the range of 0.6–1.8 μm and operated at increased bias voltage up to 300 V with low current density <150 and <20 μA/cm^2 at 1 V at room temperature. *I–V* characteristics of the Cr/MIT/Cr diodes were investigated and showed high rectification ratio up to 10^3 at 1 V. A noticeable increase of the monochromatic current photosensitivity of Cr/MIT/Cr photodetectors was observed with increasing bias voltage and this parameter was weakly temperature dependent at voltages >10 V.

Keywords: $Hg_3In_2Te_6$ crystals · Schottky diode · *I–V* characteristic · Photosensitivity · IR photodetector

1 Introduction

Widespread use of infrared (IR) detectors and related devices in many various fields causes the search for optimal radiation-sensitive materials and development of a variety of techniques to create sensors with high efficiency and photosensitivity in required spectral ranges, enhanced degradation resistance and stable parameters [1]. It is know that various chalcogenide compounds are suitable semiconductors for fabrication of device structures sensitive in different IR regions [1–10]. One of the practically important IR spectral regions is the vicinity of 1.55 μm that corresponds to the minimum optical transmission losses and, hence the transparency window of quartz and silica glasses used in optical fibers. $Hg_3In_2Te_6$ (MIT) compound is the most attractive semiconductor for uncooled IR detectors operating in this spectral range [4–10].

A. R. Várkonyi-Kóczy (Ed.): INTER-ACADEMIA 2019, LNNS 101, pp. 159–167, 2020.
https://doi.org/10.1007/978-3-030-36841-8_16

An important advantage of MIT-based photodetectors and other devices, which operate in the near IR range, over other semiconductors is the ability to function reliably under the action of various external influences and factors including conditions of an enhanced level of ionization radiation [4–10]. This is due to the fact that MIT single crystals have tremendous radiation resistance thanks to a large number of stoichiometric cation vacancies [4–6]. However, this structure feature of the semiconductor is the reason of difficulties to form barrier structures for photodetectors. It limits the ability of employing MIT to create, for example, rectifier diodes which can operate at elevated reverse bias voltages to achieve higher detection efficiency [4–10].

We have shown that the appropriate characteristics of the Cr/*n*-MIT Schottky contact can be achieved by optimizing the corresponding surface processing of MIT crystals before the formation of a rectifying contact [6, 10]. It is known that the surface state density significantly affects the bending of energy bands in a semiconductor [11, 12]. As our study has shown, if a density of surface states at the surface of *n*-type MIT crystals is high enough, the semiconductor work function does not necessarily exceed the metal work function to create a Schottky contact. By varying the technological conditions of semiconductor surface treatment, it is possible to modify the band bending at the crystal surface and, hence to change the potential barrier height [12]. At large band bending, rectifying properties of the contact with employing almost any metal can be obtained. At slight band bending, a metal with a large work function can form a contact with *n*-MIT, which is close electrical properties to Ohmic one [10].

Thus, the band bending at the semiconductor surface and, hence the potential barrier height can be changed by applying appropriate treatments of the opposite surfaces of MIT crystals before metal electrode deposition. This has allowed us to obtain both a high rectifying Schottky contact and Ohmic one even using the same metal (Cr) for electrodes [10]. The electrical and photoelectric properties of the developed Cr/MIT/Cr Schottky diode photodetectors have been studied and analyzed.

2 Formation of MIT-Based Photodiodes

For the fabrication of photodiodes, n-type MIT single crystals, grown by the modified Bridgman method, were used [6, 7]. The wafers with an area of 5×5 mm^2 and thickness of 0.5–0.7 mm were sliced from the MIT ingot by the string cutting technique. The electron concentration $n = (4\text{–}8) \times 10^{14}$ cm^{-3}, mobility $\mu_n = 200\text{–}300$ cm^2/(V · s) and resistivity $\rho = 2\text{–}4 \times 10^3$ Ω · cm were obtained from the Hall effect and conductivity measurements of MIT crystals at room temperature. The techniques of formation of both the Ohmic and Schottky contacts included preliminary chemical etching of MIT samples in a Br-methanol solution during 80–120 s and washing them in methanol and deionized water. Electrical contacts were created by vacuum evaporation of Cr on the opposite crystal faces. Before deposition of the metal onto the preliminary heated (~100 °C) MIT substrates, the crystal surfaces were treated by Ar-ions with energy of 200–500 eV at different regimes (ion beam density and processing duration). Using Cr as a metal for the electrical contacts ensured good adhesion of the metal film to the MIT crystal surface at relatively moderate metal evaporation temperature without a significant elevation of the substrate temperature (only up to ~100 °C). Utilization of Cr for both the electrodes in the diodes also provided a certain unification of the technology under the Schottky and Ohmic contact formation. Moreover, the use of Cr increased radiation resistivity of the contacts and stability of diode parameters [13].

The fabricated Cr/MIT/Cr Schottky diodes were examined by electrical and spectral measurements. The *I–V* characteristics were measured in dark condition at room temperature ($T = 293$ K). Forward current flowed when the rectifying contact was biased positively with respect to the Ohmic one. Since Cr/MIT/Cr detectors operate in reverse bias mode, it is important that the fabricated diode structures have quite low reverse dark currents even at higher bias voltages. This make it possible increasing bias voltage to extend the depletion region up to the whole thickness of the MIT crystals and thus, to achieve better collection of photogenerated charge carriers. Spectral dependencies of the monochromatic current photosensitivity of the Cr/MIT/Cr Schottky diode detectors were measured at reverse bias with two voltage values: $V = 1$ V and $V = 25$ V, and at different temperatures: $T = 283$ K and $T = 343$ K.

3 Electrical Properties of Cr/MIT/Cr Schottky Diodes

Figure 1(a) shows the *I–V* characteristics of the Cr/MIT/Cr Schottky diode detectors measured at room temperature. It should be emphasized that the Cr/MIT/Cr diodes demonstrated very high rectification ratio of 500–700 at applied bias $V = 1$ V. The best value of $\sim 10^3$, observed in the experiment, was higher than rectification ratios of the MIT diode structures formed using an interfacial oxide layer grown chemically [4] or formed by oxygen plasma [8].

Room temperature reverse dark current densities of the Cr/MIT/Cr Schottky diodes with the rectifying contact area $S = 4$ mm^2 were very low: 10–16 μA/cm^2, 50–75 μA/cm^2 and 100–150 μA/cm^2 at reverse bias voltages of 1 V, 120 V and 300 V, respectively, i.e. lower than that in the corresponding published results [4–9]. For our

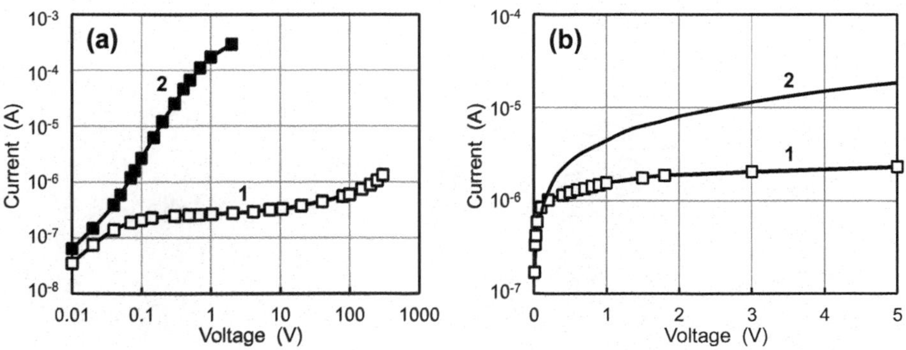

Fig. 1. *I–V* characteristics of the Cr/MIT/Cr Schottky diode at reverse (curve 1) and forward (curve 2) bias voltages (a); comparison of the initial region of the reverse *I–V* characteristic of the Cr/MIT/Cr Schottky diode (curve 1) with the corresponding results (curve 2) from [4] for the same sizes of electrical contact areas and $T = 293$ K (b).

Cr/MIT/Cr diodes, reverse dark current density decreased with decreasing the electrode (rectifying contact) area and it was ~40 μA/cm^2 at V = 300 V for S = 2–3 mm^2.

In [4, 5, 8], the I-V characteristics of the MIT-based diodes with an interfacial oxide layer are presented but only for low bias voltages (up to V = 5–10 V). A comparison of the reverse current of the fabricated Cr/MIT/Cr Schottky diode (curve 1) in the range V = 0–5 V with the corresponding results reported in [4] (curve 2) is shown in Fig. 1 (b). In the initial part (up to 10 V) of the reverse I-V characteristic of the Cr/MIT/Cr diodes, a weaker increase in dark current with increasing bias voltage was observed (Fig. 1(a), curve 1) than that for the MIT-based diodes described in [4]. A square-root voltage dependency of the current was observed in the range V = 20–120 V and then a linear one at higher voltages V = 130–300 V. The mechanisms of dark current in the Cr/MIT/Cr Schottky diodes have been partly discussed in our recent publication [10] and will be analyzed in future studies.

4 AFM Morphology Study of MIT Crystals Subjected to Surface Processing

Low dark currents at high reverse bias voltages in the Cr/MIT/Cr diodes were not only due to high quality of initial MIT single crystals, but also by applying efficient surface processing (Ar-ion bombardment) prior to formation of both the Schottky and Ohmic contacts. The processing time, voltage and current density of argon plasma were varied and selected for optimal treatment of the MIT surface. Such optimization of the MIT crystals resulted in transformation of the surface morphology.

The AFM micrographs of the MIT surfaces after non-optimized (a) and optimized (b) argon plasma treatment show differences in the surface morphology (Fig. 2). The surface, intended for creation of the Schottky contact, is shown in Fig. 2(b). The MIT crystal surface consist of small (10–50 nm) pyramid-shaped nanograins which can be combined into larger (100–300 nm) grains. The density of nanograins is higher and they are evenly distributed over the MIT crystal surface after optimized treatment with argon plasma (Fig. 2(b)). It was such surface processing that created the corresponding

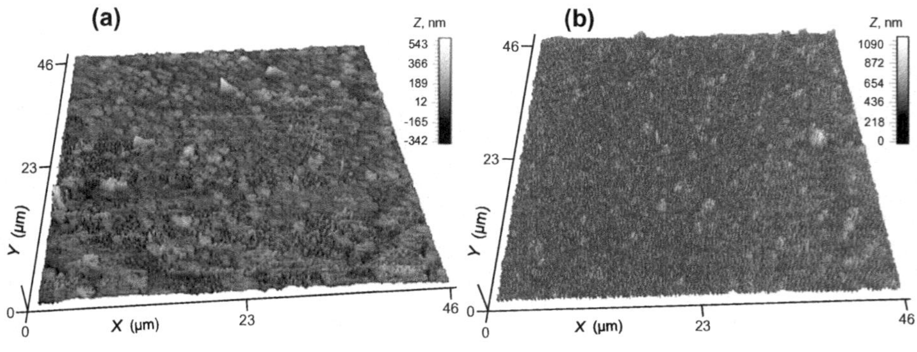

Fig. 2. AFM photomicrographs of the MIT crystal after non-optimized (a) and optimized (b) surface processing with Ar-ion bombardment performed for the rectifying contact formation.

surface states to form a highly rectifying contact and create the Cr/MIT/Cr Schottky diodes with low leakage currents even at high reverse bias voltages.

5 Photoelectric Properties of Cr/MIT/Cr Schottky Diodes

Figure 3 shows the voltage dependencies of the monochromatic current photosensitivity at two temperatures: $T = 283$ K and $T = 343$ K. It is noteworthy that the photocurrent increases significantly with increasing bias voltage in about 35 times (curve 1) and 70 (curve 2) times at $T = 283$ K and $T = 343$ K, respectively (Fig. 3). In addition, the temperature stability of the monochromatic current photosensitivity is significantly improved at applies bias voltages higher than 10 V.

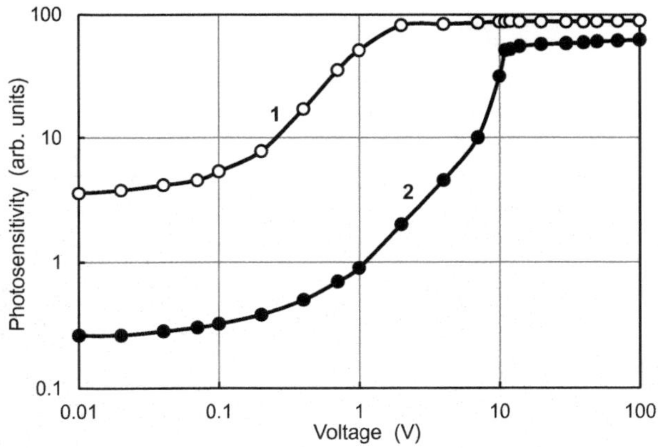

Fig. 3. Voltage dependencies of the monochromatic current photosensitivity of the Cr/MIT/Cr Schottky diode for $\lambda = 1.55$ μm at $T = 283$ K (curve 1) and $T = 343$ K (curve 2).

Figure 4 shows the photosensitivity spectra of the fabricated Cr/MIT/Cr Schottky diode detectors biased with $V = 1$ V at different temperatures: $T = 283$ K (curves 1) and $T = 343$ K (curves 2). For clarity, the spectra are plotted in linear (a) and semi-logarithmic (b) scales (Fig. 4). The detectors demonstrated a relatively high photoelectric sensitivity in the spectral range $\lambda = 0.6$–1.8 μm. The spectrum profile is the similar as demonstrated in our previous study [10].

As seen from Fig. 4, the monochromatic current sensitivity significantly decreased when the Cr/MIT/Cr detector was heated (curves 2). However, the temperature stability of the detectors can be significantly improved if we increase reverse bias voltage. The spectra of the monochromatic current photosensitivity at different temperatures ($T = 283$ K (curves 1) and $T = 343$ K (curves 2)), taken at applied bias voltage $V = 25$ V, are shown in Fig. 5. For visibility, the spectra are also plotted in linear (a) and semi-logarithmic (b) scales (Fig. 5). The temperature stability of the monochromatic current photosensitivity is noticeably improved at elevated bias voltage (Fig. 5). Most likely, the surface recombination velocity significantly decreases with

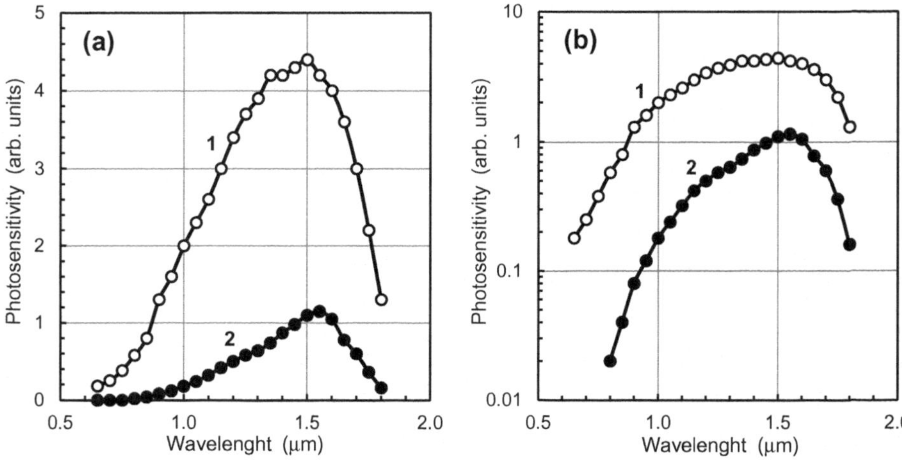

Fig. 4. Spectral dependencies of the monochromatic current photosensitivity of the Cr/MIT/Cr Schottky diode in linear (a) and semi-logarithmic (b) scales at reverse bias voltage V =1 V, at T = 283 K (curves 1) and T = 343 K (curves 2).

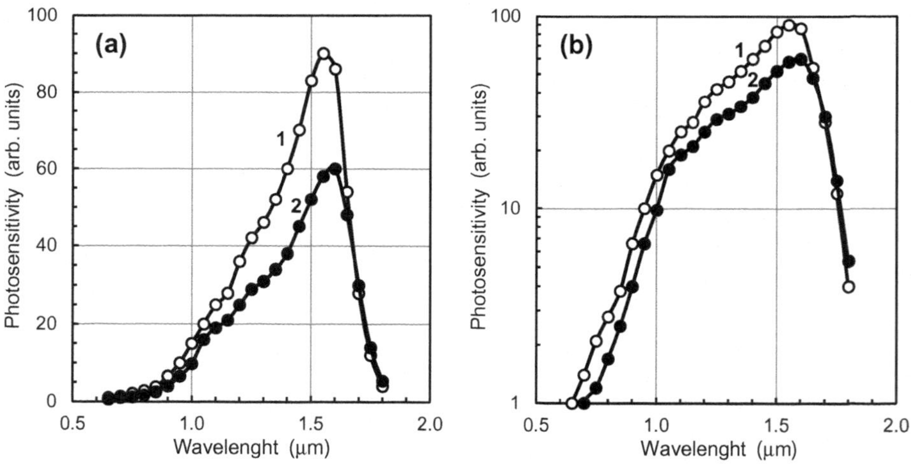

Fig. 5. Spectral dependencies of the monochromatic current photosensitivity of the Cr/MIT/Cr Schottky diode in linear (a) and semi-logarithmic (b) scales at reverse bias voltage V = 25 V, at T = 283 K (curves 1) and T = 343 K (curves 2).

increasing voltage that slows the decrease in the photocurrent with increasing temperature of the detector, i.e. leads to an increase the temperature stability of the light sensitivity. The obtained result is quite important from a practical point of view, when such a photodiode is used as a temperature stable sensor.

Attention is drawn to a particularly weak temperature dependency of the monochromatic current photosensitivity in the wavelength range of 1.6–1.8 μm

(Fig. 5). As seen from Fig. 6, the spectral dependencies of the monochromatic current photosensitivities, measured at temperatures $T = 283$ K (curve 1) and $T = 343$ K (curve 2), intersect at $\lambda \approx 1.66$ μm. This means that the monochromatic current photosensitivity is actually temperature independent at $\lambda \approx 1.66$ μm. It should be noted that for this wavelength the light sensitivity signal is ~35% of the maximum (Fig. 5(a), curve 2) that is quite acceptable for practical use.

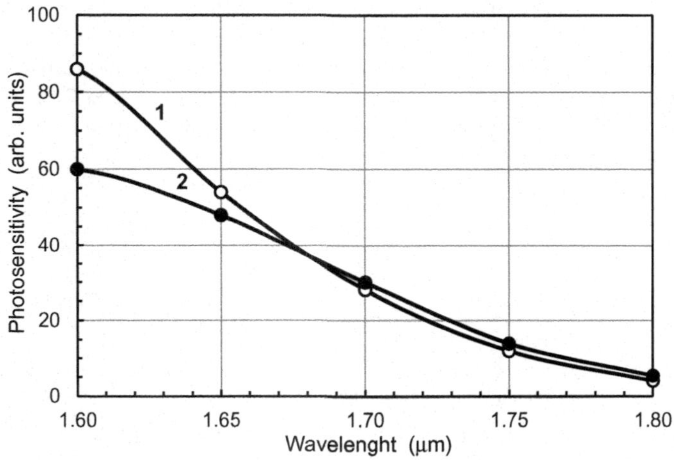

Fig. 6. Wavelength dependencies of the monochromatic current photosensitivity of the Cr/MIT/Cr Schottky diode in the range $\lambda = 1.6$–1.8 μm at $V = 25$ V, and at $T = 283$ K (curve 1) and $T = 343$ K (curve 2).

The monochromatic current sensitivity reached ~1–1.2 A/W at the spectrum maximum ($\lambda = 1.55$–1.60 μm) at reverse bias $V = 0.5$–1.0 V. This is a great advantage of the fabricated Cr/MIT/Cr photodiodes because that spectral range ($\lambda \approx 1.55$ μm) corresponds to the optical transparency window of common glasses used in fiber optics. Therefore, the developed Cr/MIT/Cr Schottky diodes are very promising for application as efficient detectors for fiber optic communication lines.

6 Conclusions

The most important result of our study is the development and fabrication of MIT-based photodiodes with low leakage current which can operate at high bias voltages under conditions of enhanced radiation including X/γ-rays. The obtained Cr/MIT/Cr Schottky diode detectors have quite low reverse dark currents which are comparable (even better for the best Cr/MIT/Cr diodes) than the currents of the photodetectors based on Ge_xSi_{1-x}/Si multilayer superlattices [14]. The reverse dark current density, achieved for those detectors was 20 μA/cm^2 at 1 V, while for the best Cr/MIT/Cr diodes, it was 10–16 μA/cm^2 at the same bias voltage and room temperature. It is important to note, when created Cr/MIT/Cr diodes are used as photosensitive sensors, a significant increase in the photocurrent is observed and the temperature stability of the photodiodes noticeably increases with an increase in applied reverse bias voltage.

Acknowledgement. This research was partly supported by the following research projects: "Development of Cd(Zn)Te-based X/gamma-ray detectors with high resolution for security and diagnostics instruments" (The 2019 Cooperative Research at Research Center of Biomedical Engineering, Japan, grant number 2022), "Development of perovskite single crystal X/gamma-ray detectors for environmental radioactive contamination monitoring" (The Short-term Recruitment Program of Foreign Experts in Anhui (APFEP, 2019, China), "Exploring novel perovskite single-crystal based gamma-ray detector for trace environmental radioactivity monitoring" (Program of the Joint Ukraine—The People's Republic of China R&D Projects for the period of 2019–2020 adopted by the Ministry of Science and Technology of the People's Republic of China, grant number CU03-15).

References

1. Rogalski, A.: History of infrared detectors. Opto-Electron. Rev. **20**, 279–308 (2012)
2. Gnatyuk, V.A., Gorodnychenko, O.S., Mozol', P.O., Ponedilok, A.V., Vlasenko, O.I.: Photoelectric properties of polycrystalline $Cd_xHg_{1-x}Te$ films grown on alternative substrates. Semicond. Sci. Techn. **13**(11), 1304–1308 (1998)
3. Gnatyuk, V.A., Mozol', P.O., Prokopenko, I.V., Smirnov, O.B., Vlasenko, O.I.: Laser-induced changes in the photoelectric properties of $Mn_xHg_{1-x}Te$ crystals. Inf. Phys. Tech. **42** (2), 69–75 (2001)
4. Malik, A., Vieira, M., Femandes, M., Macarico, F., Grushka, Z.M.: Near-infrared photodetectors based on HgInTe-semiconductor compound. Proc. SPIE **3629**, 433–442 (1999)
5. Grushka, Z.M., Gorley, P.N., Grushka, O.G., Horley, P.P., Radevych, Y.I., Zhuo Z.: Mercury indium telluride—a new promising material for photonic structures and devices. Proc. SPIE **6029**, 60291A-1–9 (2006)
6. Maslyanchuk, O.L., Kosyachenko, L.A., German, I.I., Rarenko, I.M., Gnatyuk, V.A., Aoki, T.: Electrical and optical properties of $Hg_3In_2Te_6$ single crystals. Phys. Status Solidi C **6**(5), 1154–1157 (2009)
7. Kosyachenko, L.A., Kabanova, I.S., Sklyarchuk, V.M., Sklyarchuk, O.F., Rarenko, I.M.: $Hg_3In_2Te_6$-based photodiodes for fiber optic communication. Phys. Status Solidi A **206**(2), 351–355 (2009)
8. Zhang, L., Zhang, X.L., Sun, W.G., Lu, Z.X.: Photoelectron characteristics of HgInTe detector. Proc. SPIE **8193**, 81932X-1–7 (2011)
9. Rarenko, I., Korbutyak, D., Koshkin, V., Danilchenko, B., Kosyachenko, L., Fochuk, P., Sklyarchuk, V., Zakharuk, Z., Dremlyuzhenko, S., Rarenko, A., Nikonyuk, Ye., Klad'ko, V., Demchyna, L., Budzulyak, S., Vakhnyak, N., Medvid', A., Dauksta, E.: Materials and photosensor devices with high radiation stability. Adv. Mater. Res. **1117**, 107–113 (2015)
10. Sklyarchuk, V., Gnatyuk, V., Aoki, T.: $Hg_3In_2Te_6$-based radiation and temperature stable photodetectors. Proc. SPIE **10766,** 107660R-1–7 (2018)
11. Sze, S.M., Ng, K.K.: Physics of semiconductor devices, 3rd edn. Wiley, Hoboken, New Jersey (2007)
12. Mönch, W.: Semiconductor surfaces and interfaces. Springer, Berlin, Heidelberg, New York (1993)

13. Kochubey, V.A., Atuchin, V.V., Pokrovsky, L.D., Soldatenkov, I.S., Troitskaia, I.B., Kozhukhov, A.S., Kruchinin, V.N.: Structure, microrelief and optical properties of chromium films deposited by sublimation in vacuum. Lett. Mater. **3**(4), 326–329 (2013)
14. Yakimov, A.I., Dvurechenskii, A.V., Nikiforov, A.I., Chaikovskii, S.V., Tiis S.A.: Ge/Si photodiodes with embedded arrays of Ge quantum dots for the near infrared (1.3–1.5 μm) region. Semiconductors **37**(11), 1345–1349 (2003)

Machine Intelligence and Computer Science

Augmented Smart Refrigerator—An Intelligent Space Application

B. Tusor, Š. Gubo(✉), T. Kmeť, and J. T. Tóth

Department of Mathematics and Informatics, J. Selye University, Bratislavská Str. 3322, P.O. BOX 54, 945 01 Komárno, Slovakia
{tusorb, gubos, kmett, tothj}@ujs.sk

Abstract. Recently, with the advancement in computation technology, ubiquitous computing paradigms like Intelligent Spaces are not only gaining popularity but are also slowly getting into the price range of average households. However, while many everyday devices and services can be accessed and afforded by middle class families, smart refrigerators are still too expensive, even though they can be very useful to aid the economics and budgeting of the household. In this paper, an affordable smart refrigerator framework is proposed that can be implemented by using cheap, easily accessible devices to augment older, regular refrigerator models, integrating the core functionalities that many expensive models have, for a much lower cost.

Keywords: Household economics · Intelligent space · Smart fridge · Raspberry pi · Single board computers · Assisted daily life · Augmented fridge

1 Introduction

Nowadays it is not rare for a household to have a smart television, environmental (temperature, lighting etc.) control or even virtual assistants (e.g. Amazon Alexa, Microsoft Cortana, Google Assistant [1] etc.). Such products have recently become available for the average consumers. However, smart refrigerators, which can be a very useful part of home economics and budgeting are still out of reach as the existing models in the market are rather expensive.

A smart refrigerator, in theory, can account for its contents, issue warnings if the expiration date of a product is near or has already passed, give the users advices on what meal can be prepared given the available consumables and known recipes (and thus promoting a healthier diet [2–4]), send an alarm if the fridge door is ajar, help compiling a shopping list, etc. Thus, it is basically a consumables management and decision support system.

Besides investing a large sum of money, another way to acquire a smart fridge is creating one by augmenting an older model with the necessary devices that can implement most (if not all) of the functions that an expensive model can provide. Many research recent proposals have made this approach the focus of their work, using the ubiquitous computing paradigm [5–7], using microcontrollers or different Raspberry Pi single board computers [8] to implement a certain level of intelligence into classic refrigerator models.

A. R. Várkonyi-Kóczy (Ed.): INTER-ACADEMIA 2019, LNNS 101, pp. 171–178, 2020.
https://doi.org/10.1007/978-3-030-36841-8_17

The Intelligent Space (iSpace) framework has been proposed by [9]. The base idea is that any space can be turned into an iSpace by installing one or more distributed intelligent networked devices (DINDs) into it, preferably in a non-invasive way, in order to implement new services or expand existing ones, e.g. requesting information or a given service from the system. With single board computers readily available, nowadays this can be implemented at a relatively low cost.

In this paper, an affordable, Intelligent Space-based smart refrigerator solution is proposed that can be built with cheap, easily accessible devices (single board computers and sensors) in order to augment older, 'not-smart' refrigerators to integrate the core functionalities that expensive models have, for a fraction of its price. While other researches tend to primarily focus on dietary (and thus, health) management and food waste reduction, this research considers the economic side of the topic, aiding its users to make well-informed decisions when shopping for groceries.

The rest of the paper is as follows. Section 2 describes the general architecture of the system, while Sect. 3 sums up the functions that the system is designed to realize. In Sect. 4 the hardware implementation is summed up. The final Section concludes the paper and outlines future work and development directions.

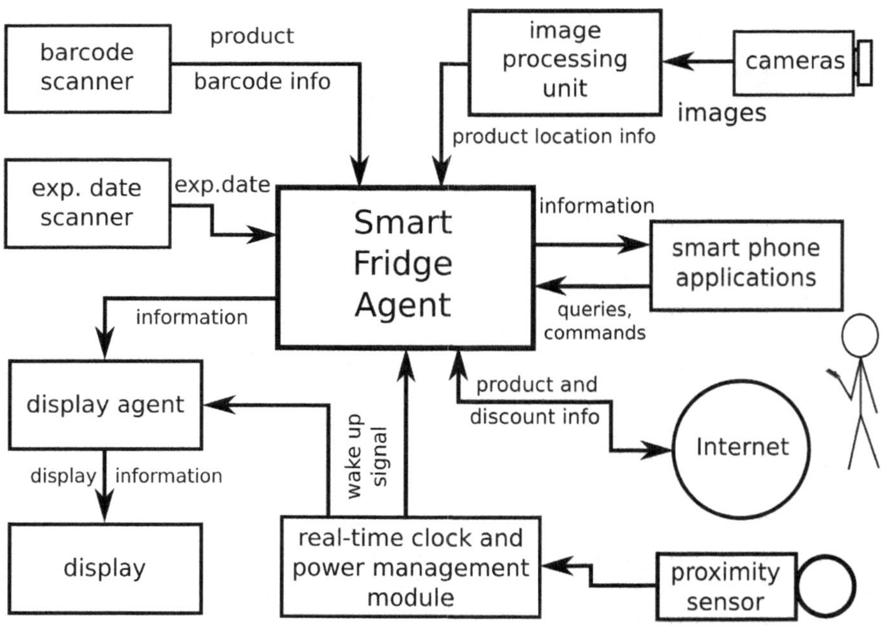

Fig. 1. The architecture of the augmented smart fridge system.

2 The Architecture of the Augmented Smart Fridge

The architecture of the proposed augmented smart fridge system is illustrated in Fig. 1. It mainly consists of an *image processing unit* (IPU) that processes camera information, a *display agent* (DA) that handles a display, and a dedicated agent (*Smart Fridge Agent*, SFA) that implements the central intelligence.

The system works as follows. While idle, the SFA, IPU and DA units are in sleep mode. They get waken up periodically (once or twice a day) by a real-time cock and power management module in order to do routine functions; or when the proximity of a user is detected using a proximity sensor.

If the user intends to add a new item, signals it to the system (by using a graphical user interface (GUI) on either a touch screen connected to the SFA, or the menu of a smart phone application), then scans the barcode with the barcode scanner, followed by the expiration date using another scanner (or the barcode scanner module, if one is available). The SFA checks the barcode on the Internet and retrieves information regarding it (name, type, nutritional values, etc.), then creates a profile for it and saves it to its database. Meanwhile, it communicates with the user through either the display agent, or a smart phone application. If visual sensors are available (waterproof IP cameras that take pictures of the inside of the fridge), then it can also get an approximate location of the object (Fig. 2), which makes it possible to monitor its presence: if an item was observed to occur many times, the system can presume that it is something that the user enjoys having (and thus, marks it as a *frequent item*), so if its absence is detected, then the system will add it to the shopping list. Furthermore, with access to the mailing lists of (trusted and secure) stores online, the system can check if the item is available at a discounted price at either of the stores and direct the users toward them.

Fig. 2. An illustration for the item tracking function of the image processing unit.

3 Functions and Data Structures

The system is designed to realize the following functions:

- **Wake up on human proximity**. Since it is not necessary to have the system run constantly, power can be saved and the longevity of its parts can be ensured by putting the system into sleep mode after all necessary operations are done, until a user approaches the fridge. In that case, a proximity sensor should send a signal to the real-time clock and power management module that is attached to the SFA.
- **Warn users if the fridge door is ajar**. In order to be efficient, the fridge door should be closed tightly by default. However, the door might be left accidentally ajar, which can be overlooked by the user. Fortunately, it can be detected with suitable methods (e.g. magnetic sensors pasted on the side of the door) and issue a warning to the user about the situation.
- **Warn users if the expiration dates of the contained items are close or have already passed**. To reduce the chance of spoilage, the user should be notified if any items are close to its expiration date. Furthermore, the estimated expiration date of food items put into containers should also be calculated (typically a couple of days) and treated accordingly.
- **Register new item if the user puts it into the fridge**. When the user intends to put a new item into the fridge, they can use a suitable GUI to register it in the system and read the barcode of it using the barcode scanner or a dedicated camera. The SFA then attempts to look up the product belonging to the barcode in given databases on the Internet. If this is not available, then the user is able to add the data considering the item (brand/product name, type (e.g. milk)) and have the expiration date scanned.
- **Register it if item had been taken out of the fridge, warn if necessary**. There are occasions when items are taken out and not consumed fully, but the user had forgotten to put them back, potentially causing them to go to waste. Thus, the system should register that an item was taken out and ask the user if they intend to put it back. If this does not happen, the system should warn the user about it.
- **Keep tab on frequently seen items**. The system should learn what items the user usually have in the fridge (e.g. milk), and automatically adds it to the shopping list of the user. Both item types and brand products should be considered: e.g. the user might have milk in general or have a favorite brand of milk that occurs more often than the other brands.
- **Parse promotional emails and collect information about discounts online**. The system maintains a database, storing the shops where the frequently seen items can be purchased for a discounted price, and take it into consideration when putting the shopping list together.

Figure 3 depicts the databases that the proposed system uses. Information about items in general are stored in the General database: the name, type, regularity, nutritional data for each product, as well as the lowest known price at the moment and the store that offers it. In the figure, data is marked with rounded boxes, while boxes with sharp corners denote indexing structures that enhance data processing speed. For example, using the name of a given product can quickly lead to the data of the given

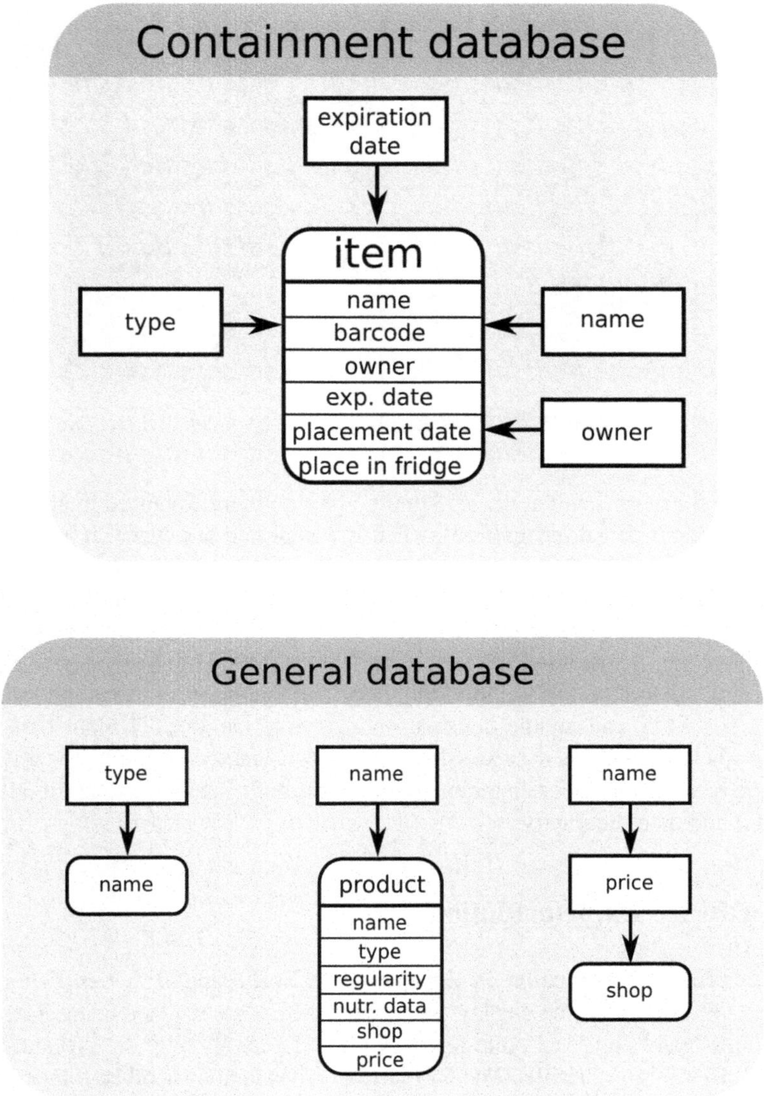

Fig. 3. The indexed data structures used to store what items are in the fridge (top) and the general information about products (bottom).

product (instead of having to search through a whole list to find the appropriate one). The product names are also indexed by type, so any product can of a given type (e.g. meat) can be quickly found. Furthermore, for each regular item, shops are indexed by the price they offer each given item: knowing the name of the item, the lowest price can be found very fast along with the store it is offered by.

The items registered in the fridge are stored in the Containment database, storing the name, barcode and expiration date of the item, as well as who put it there (the user

Fig. 4. Raspberry Pi Zero W (left) and Raspberry Pi 3B+ models (with cooling fans, right).

can identify him- or herself at the registration; or with more advanced image processing capabilities, using facial recognition), when it was placed and where it is in the fridge. It is indexed by product name, type, owner name and most importantly, by the expiration date of the items. The latter makes it possibly to quickly determine if any item is close to be spoiled.

For indexing, Sequential Fuzzy Indexing Tables (SFIT, [10]) are used that realize a sequence of 2D lookup tables, providing very fast operation in exchange for memory storage. The SFIT can handle constant and variable length [11] input data as well. Among others, it had been successfully used in a dietary assistant system [12, 13] where the SFIT is applied as an indexed storage for both nutritional data (of 34 different nutrients) and as a dictionary.

4 Hardware Implementation

In the current implementation in the Intelligent Space and Robotics Laboratory of J. Selye university, both the display agent and the SFA are realized using Raspberry Pi (RPI) single board computers (the former with RPI Zero W (Fig. 4., left) and the latter with RPI 3B+ (Fig. 4., right)). The RPI has the advantage over other competing SBCs that there are numerous extensions and sensors cheaply and readily available in the market that are compatible with it (or even designed specifically for it). Many of those sensors (proximity, temperature, magnetic door sensors) are significant for the project, along with other extensions like uninterruptable power supply units (to ensure undisturbed operation even in cases of power outage) and the real-time clock and power management module (that wakes the system up from sleep mode, because the RPI is not built with such function by default). To implement the display agent, the RPI Zero W module is connected to an e-ink screen that can retain information for a very small amount of power.

Since single board computers have limited computational capabilities, the image processing tasks (such as object detection) are outsourced to a specific hardware, instead of executing them on the SFA. In the proposed project, a Tensor Processing Unit (TPU, [14]) and a dedicated developer board (Coral Edge Dev Board) is used to quickly recognize objects in the camera images.

5 Conclusion

In this paper, an affordable smart refrigerator augmentation system is proposed in the framework of the Intelligent Space. While many other researches and development in this topic focus on the nutritional aspect, this research focuses on household economics, implementing a decision supporting system for more efficient and better-informed shopping.

The system uses single board computers as distributed intelligent networked devices, a Raspberry Pi Zero W to handle output devices such as screens, a Coral Edge Dev Board to process input information acquired from image data from the cameras in the fridge, and an RPI 3B+ to store and process commands received from the users and information gathered through the internet from relevant and trusted sources (the vendors of given consumable products). The information about the products are stored in lists that are enhanced with indexing to make searches and queries faster.

In the next phase of the implementation, further important aspects will be taken into accounts, such as privacy and data security, as the system is connected to the Internet. Another challenge is making the computer vision-based product identification better. Ideally, the ambient intelligence should be able to see and recognize the item after it had been placed, but as most refrigerators in family homes are jam-packed with products, it is virtually impossible to place cameras in any way that provides a perfect field of view. A way to work around this problem is instead of looking at the inside of the fridge, the system could look at the hand of the user as the item is put in or taken out, as it will be less obstructed from view.

Furthermore, the personal taste of the user should also be considered, so the system could recommend other products of that can be cheaply acquired. For example, if the user keeps using one brand of flavored milk, but another brand is on sale, then the system could offer it the next time it puts together a shopping list. The user on the other hand should be able to refuse the recommendation, in which case the system should tune the taste model of the user accordingly.

Acknowledgement. The publication was prepared with the financial support of Pallas Athéné Domus Educationis Foundation, project number: PADE-0117-5.

References

1. López, G., Quesada, L., Guerrero, L.A.: Alexa vs. Siri vs. Cortana vs. Google Assistant: a comparison of speech-based natural user interfaces. In: Nunes I. (eds.) Advances in Human Factors and Systems Interaction. AHFE 2017. Advances in Intelligent Systems and Computing, vol 592, pp. 241–250. Springer, Cham (2018)
2. Shadangi, V., Jain, N.: Medical internet refrigerator. In: International Conference on Control, Instrumentation, Communicational Technologies, pp. 363–366 (2015)

3. Pescosolido, L., Berta, R., Scalise, L., Revel, G.M., De Gloria, A., Orlandi, G.: An IoT-inspired cloud-based web service architecture for e-health applications. In: IEEE International Smart Cities Conference, pp. 1–4 (2016)
4. Luo, S., Jin, J.S., Li, J.: A Smart fridge with an ability to enhance health and enable better nutrition. In: Int. J. Multimed. Ubiquitous Eng. 69–76 (2009)
5. Rouillard, J.: The pervasive fridge. A smart computer system against uneaten food loss. In: 7th International Conference on Systems, pp. 135–140 (2012)
6. Wu., H.-H., Chuang, Y.-T.: Low-cost smart refrigerator. In: 1st International Conference on Edge Computing, pp. 228–231 (2017)
7. Edward, M., Karyono, K., Meidia, H.: Smart fridge design using NodeMCU and home server based on Raspberry Pi 3. In: 4th International Conference on New Media Studies, pp. 1–4 (2017)
8. Gay, W.W.: Raspberry Pi Hardware Reference (2014). https://doi.org/10.1007/978-1-4842-0799-4
9. Hashimoto, H.: Intelligent space: interaction and intelligence. In: Artificial Life and Robotics, vol. 7, no. 3, pp. 79–85 (2003)
10. Várkonyi-Kóczy, A.R., Tusor, B., Tóth, J.T.: A multi-attribute classification method to solve the problem of dimensionality. In: Advances in Intelligent Systems and Computing, vol. 519, pp. 403–410. Springer, Heidelberg (2017)
11. Tusor, B., Várkonyi-Kóczy, A.R., Tóth, J.T.: A fuzzy data structure for variable length data and missing value classification. In: Recent Advances in Technology Research and Education. Advances in Intelligent Systems and Computing, vol. 660, pp. 297–304. Springer, Heidelberg (2017)
12. Tusor, B., Simon-Nagy, G., Várkonyi-Kóczy, A.R., Tóth, J.T.: Personalized dietary assistant-an intelligent space application. In: 21st IEEE International Conference on Intelligent Engineering Systems (INES 2017), pp. 27–32 (2017)
13. Tusor, B., Várkonyi-Kóczy, A.R., Bukor, J.: An iSpace-based dietary advisor. In: 2018 IEEE International Symposium on Medical Measurements and Applications (MeMeA), pp. 1–6 (2018)
14. Cass, S.: Taking AI to the edge: Google's TPU now comes in a maker-friendly package. IEEE Spectr. **56**(5), 16–17 (2019)

Domain Reduction Techniques for Sequential Fuzzy Indexing Tables – A Case Study

B. Tusor(✉), J. Bukor, L. Végh, and O. Takáč

Department of Mathematics and Informatics, J. Selye University, Bratislavská Str. 3322, 945 01 Komárno, Slovakia
{tusorb,bukorj,veghl,takaco}@ujs.sk

Abstract. In recent years, single-board computers have been gaining popularity because they make it possible to use low cost, low energy consumption devices to solve complex tasks that would be much harder to solve with microcontrollers. However, these devices have much more limited capabilities both computation and memory-wise compared to traditional computers, which in turn limits the options available to use them for classification problems. One of the available options is using Lookup Table-based classifiers that require minimal computation, although in return they require more memory space. Sequential Fuzzy Indexing Tables are improved versions of Lookup Tables that require less memory, but for large problem spaces their storage cost is still very high. This is due to the size of its structure, which can be reduced with suitable domain conversion techniques. In this paper, multiple options are investigated, analyzed and compared in order to solve this problem.

Keywords: Data hashing · Text processing · Machine learning · Domain reduction

1 Introduction

Recently, single-board computers (SBCs), such as Raspberry Pi models [1], have been gaining popularity due to their low cost, low energy consumption rates and ease of use. However, they have much more limited capabilities both computation and memory-wise compared to traditional computers, which in turn limits the options available for tasks that require a significant amount of processing or memory, such as machine learning. Many classic machine learning methods can be considered, but most of them require modern hardware solutions like parallelization to work sufficiently fast.

One method that was designed for the least amount of necessary computational steps is based on indexing tables or Lookup Tables (LUT, [2]). LUT-based classifiers have been used to solve various classification problems, mainly in image processing or network routing. They store pre-processed function values in arrays, so during runtime they can be easily accessed. The disadvantage is that the whole problem space is needed to be stored in the memory.

In previous works, the authors have developed a classification method called Sequential Fuzzy Indexing Tables [3] that expands on the base idea of indexing tables by dividing the problem space into a sequence of lower dimension parts that preserve

A. R. Várkonyi-Kóczy (Ed.): INTER-ACADEMIA 2019, LNNS 101, pp. 179–190, 2020.
https://doi.org/10.1007/978-3-030-36841-8_18

all known value combinations in 2D arrays. The classifier had been shown [4] to be able to handle constant and variable length inputs (e.g. words) and is capable of a fast operation in exchange for a considerable amount of memory that mainly depends on the domain size of the attributes in the given problems. Since it uses Lookup Tables that are addressed directly using the input attribute values, such values are needed to be of positive integer format ($x \in \mathbb{N}^+$). Other kinds of input values (such as real numbers and character strings) are needed to be converted into positive integer formats.

There are many techniques to map a value in a given domain onto another value in a more restricted domain. The goal is to reduce the domain for each attribute if necessary, without sacrificing much information (by assigning the same value in the new domain to two different values of the old domain). A small (arbitrary) level of lenience can be allowed: values that are very close to each other can be mapped into the same value, but none that are farther from each other in the respective domain than a given threshold.

In this paper, multiple mapping function options are investigated and analyzed that can be used to improve Sequential Fuzzy Indexing Table structures or any other Lookup Table-based classifiers. Two main approaches are examined: the case where the mapping is done through run-time calculations (using linear functions, hash functions [5], artificial neural networks [6]) and the application of specific structures that store pre-calculated values (LUT principle) in certain ways (such as Key-Value Lists, LUTs and even small SFITs).

The rest of the paper is as follows. In Sect. 2 a brief summary of SFITs are given. In Sect. 3 the two approaches mentioned above are examined: Subsect. 3.1 analyzes run-time calculations used for data conversion and Subsect. 3.2 investigates structure-based methods. The approaches are compared and analyzed in Sect. 4. Finally, conclusions are drawn.

2 Sequential Fuzzy Indexing Tables

Sequential Fuzzy Indexing Tables are Lookup Table-based classifiers that only store relevant parts of the problem space (i.e. parts that contain interesting data), through a sequence of 2D LUTs (one LUT constituting one layer).

The architecture of the SFIT classifier is as follows. The first layer L_0 evaluates the first attribute value (x_0), then each consequent layer L_i adds another attribute value x_i. Each layer L_i contains an index matrix Λ^{L_i}, storing index markers. Each index marker η_i in Λ^{L_i} indicates which row is considered in $\Lambda^{L_{i+1}}$. Thus:

$$\eta_i = \begin{cases} \Lambda^{L_i}_{x_i} & \text{if } i = 0 \\ \Lambda^{L_i}_{\eta_{i-1}, x_i} & \text{if } i \in [1, N-1] \end{cases} \tag{1}$$

where N is the number of attributes (i.e. the length of the input sample). This is calculated for $\forall i \in [1, N-1]$, and the output (the label assigned to the input samples) is given in the last layer.

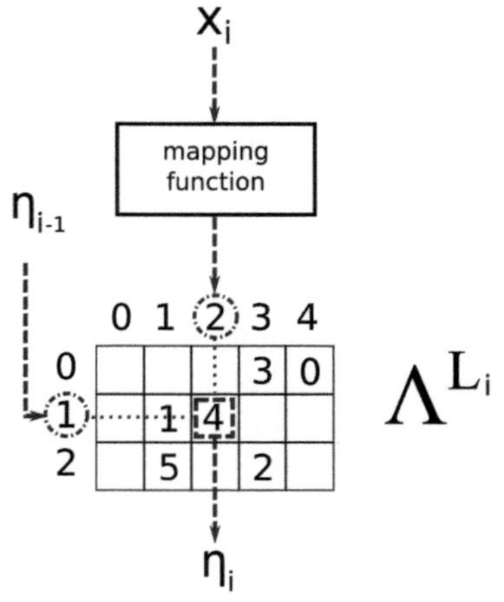

Fig. 1. The role of the mapping function in the SFIT architecture.

Figure 1 depicts a layer (L_i). Input value x_i is mapped into a suitable domain, then it is combined with the index marker from the previous layer (η_{i-1}) in order to get η_i. In the example of the image, x_i is mapped to value 2, so $\eta_i = \Lambda^{L_i}_{\eta_{i-1},x_i} = \Lambda^{L_i}_{1,2} = 4$.

Since only positive integer values can be used directly as array coordinates, if the input values are not formatted such then they are needed to be converted into the positive integer domain by a mapping function. This raises a need for a suitable mapping function for data with non-integer values.

3 Domain Reduction Approaches

The goal is to transform the domain of the given attributes to one that is easier to handle by the given data structure (e.g. associative arrays) or classifier. In case of SFITs, domains with integer values and a clearly defined range is needed, one that does not lead to an unfulfillable memory requirement.

Two main approaches are considered in this paper:

- **Run-time calculations**, in which the data values are converted directly: linear mapping functions, hash functions [4], artificial neural networks [5].
- **Structures** that store pre-calculated values (LUT principle) that can be quickly accessed: Key-Value Lists, LUTs and SFITs.

3.1 Run-Time Calculations

There are many methods that can be used to gain the target value from a mathematical combination of the input attribute values, from using a simple linear mapping functions to more complex machine learning methods, like Artificial Neural Networks (ANNS).

Linear Mapping Functions. The simplest way to convert a real value x to a bounded integer value $\tilde{x}$ can be done with a linear mapping function:

$$\tilde{x} = \lfloor a \cdot x \rfloor + b \tag{2}$$

where a is the scaling factor and b is an appropriate bias that ensures that the lower bound of $\tilde{x}$ is at zero. The advantage of this method is that it is computationally inexpensive, as it only requires an addition, a multiplication and one rounding steps, which can be done in linear time. Parameters a and b can be set manually or calculated from the given data set, by considering the lowest (x_L) and highest values (x_H) of the given attribute. Furthermore, if each input value is needed to get a unique output value, then the smallest difference of each attribute value is needed to be considered and the scaling coefficient set accordingly. Generally, the lower it is set, the more different input values will be mapped to the same output, lowering the size of the output domain but also potentially losing information. Furthermore, the mapping can be done for multiple attribute values (of numeric type) at the same time, e.g.:

$$\tilde{x} = \left\lfloor \sum_{i=1}^{N} (a_i \cdot x_i) \right\rfloor + b \tag{3}$$

where a_i is the scaling factor for the respective input x_i, for N attributes.

Hash functions are functions that map data of any arbitrary size onto data of a given, constant size. Linear mapping functions thus can be considered hash functions as well, though the latter is often more complex in implementation. They have been widely used in cryptography, to convert data into given strings of hash values. They are usually designed and optimized to generate unique hash values, as fast as possible. Their downside when used for domain reduction is that the function is needed to be carefully (manually) made in order to avoid collusion (mapping two different input values onto the same output value) as much as possible.

Neural Networks. Artificial Neural Networks (ANNs, [6]) are multilayered structures in which each basic element (neuron) calculates the linear combination of its inputs with the appropriate weight parameters:

$$y_j = \left\lfloor \sum_{i=1}^{m} (w_{ij} \cdot x_i) \right\rfloor + w_{0j} \tag{4}$$

where w_{ij} is the weight parameter belonging to input x_i, and w_{ij} is the bias parameter and m is the number of inputs for the given neuron.

The weight parameter values are set either through analytical calculations [7] or training algorithms (e.g. back propagation [8]). The classic variants of ANNs are notoriously slow to train, so modern implementations generally use parallel computing to accelerate the process.

Recurrent Neural Networks (RNNs, [9]) are neural networks with a feedback loop that takes the output of the network from the previous timestep as input as well, giving them the ability to learn the context of the input data as well. They are usually used on a stream of input or time series. Furthermore, they can be very advantageously used for processing variable length inputs [10]. They are trained by an extended version of the back-propagation algorithm (Back-Propagation Through Time, BGTT [10]). Their disadvantage is that the vanishing and exploding gradient problem [12] (which is caused by the feedback loop) makes their successful training significantly harder than that of ANNs. This problem is typically solved by setting thresholds (e.g. LSTM-RNN [13]).

3.2 Domain Reduction Using Structures

Key-Value Lists. Key-Value Lists are simple lists that consist of pairs of data points: a *value* and a *key* by which the data is ordered. It can be utilized for data conversion, by using input data x (that is needed to be converted) as key and the target data $\tilde{x}$ (that x is needed to be converted into) as value. Thus, the problem is reduced to finding the item on the list where the key matches the input data, and simply output the value of the item. The complexity of this is linear with the length of the list if linear search is used, and logarithmic if binary search is used. For the latter, the list is needed to be ordered at any given moment.

The main disadvantage of using KV-lists as data converter is that all possible key values are needed to be known and be had on the list, which makes it less useful to convert floating point values. However, it can easily map words or categorical values onto integer numbers, variable length inputs are not an issue.

Lookup Tables. Lookup Tables or hash tables have been widely used for various tasks like database indexing, error checking, network caching and even classification [14].

The time to retrieve an element given its index number is independent on the size of the array and the position of the element in the array.

In case of hash tables, the position of each element in the array is calculated using a suitable hash function on the input values. For example, for character strings the ascii codes are used to calculate the position (a positive integer number) of each word, and thus the position itself will be the output that is assigned to the given word.

Sequential Fuzzy Indexing Tables. SFITs can also be used for data conversion and domain reduction, thus applied in tandem: one SFIT as a dictionary that converts a word into an index value, and the other that uses that index value. The method had been already been successfully used in a dietary assistant system [15, 16] where the SFIT is applied as an indexed storage for nutritional data and as a dictionary.

4 Comparative Analysis

4.1 Complexity Analysis

Table 1 compares the different domain reduction and conversion algorithm considered in this paper. Their memory requirements, training and run-time speeds are taken into account, as well as their implementation complexity and their ability to handle variable length data (e.g. words).

In terms of memory need, linear and hash functions are with the lowest requirement as only the coefficients are needed to be stored. They are also fast to train (setting up the coefficient values, though for more complex hashing some manual tuning of the formula is sometimes needed), and very fast to apply run-time. Their complexity is also generally low (as fast operation is one of the main design principles for hash functions).

Classic ANNs have to store the weight parameters of the neurons, which in turn depends on the number of used neurons (m). They are on the end notoriously slow to train ($O(m^2)$), and not very fast ($O(m)$) to run if parallel computing is not available. They are also quite complex to implement, although many models are readily available to download from the Internet. Recurrent Neural Networks and other deep NNs have a higher memory need, and due to their complexity increase they usually require parallel computing to work efficiently.

Lists only contain the relevant data so their memory need scale linearly with the amount of the data. Training a simple, linear Key-Value list is very fast as new values are simply added to the end of the list. On the other hand, keeping the list ordered requires a logarithmic search before each insertion. Still, this is easy to implement, and makes using the structure much faster ($O(\log_2 P)$, considering a P-long list).

Of the considered methods, LUTs are the easiest to implement as they only consist of a low-dimensional array. Their training and run-time speed is very fast ($O(P)$ and $O(1)$, respectively) as well, but in turn their usage is very limited due to their large memory requirement (since a large part of the problem space is needed to be kept in the memory).

Finally, SFITs are a bit more complicated to implement than LUTs, their training and operation are both a bit slower ($O(NP)$ and $O(N)$, respectively) but their memory requirement is also often significantly lower (especially for data with lower domain ranges), and thus they can be used for higher dimension problems.

Technically, most methods (even classic ANNs) can be used for variable length data by padding the inputs with zeros, though it usually does not result in any better performance than using methods that can handle variable length data by default [4]. However, LUTs are for the most part not suitable for variable length input data due to their hard restriction on dimensionality. Hash functions can be used for variable length data, but it requires complicated formulas and significantly more memory than hashing constant length data.

Table 1. Comparative analysis of various domain reduction algorithms.

Method		Memory requirement	Training speed	Run-time speed	Implementation complexity	Variable length data?
Linear and hash functions		Very low	Fast	Very fast	Very low	Yes, but not efficient
ANNs		Medium	Slow	Medium	Medium	Yes, but not efficient
RNNs		High	Very slow	Medium	High	Yes
Key-Value Lists	Lin.	Low	Very fast	Slow	Very low	Yes
	Log.	Low	Fast	Fast	Low	Yes
LUTs		Very high	Very fast	Very fast	Very low	No
SFITs		High	Fast	Fast	Medium	Yes

4.2 Practical Analysis

In order to illustrate the differences between most of the methods discussed above, three sets of experiments were conducted on an average laptop PC (Asus ROG GL503VM, Intel® Core™ i7-7700 CPU @ 2.80 GHz, 16 GB RAM, Gigabyte® GeForce™ GTX 1060 graphic card with 6 GB RAM), using Matlab R2018b. This configuration was chosen instead of directly using a Single Board Computer in order to more easily measure the memory requirement of each method without risking damaging the SBC.

In the first experiment, the case of mapping floating point values to a positive integer domain is investigated. Linear mapping functions and Key-Value Lists are considered, because out of the discussed methods they are the most suited to handle single, floating point data (neural networks for single inputs are basically linear mapping functions, while SFITs and LUTs require integer inputs). The data is taken from the Dynamic gas mixtures data sets (from the UCI Machine Learning Repository [17]), in which a gas sensor array is exposed to the mixture of two dynamic gas mixtures at varying concentrations. The data of the second channel of the array (mostly negative floating-point values), is taken to be mapped to a positive integer domain. The data is cleaned (the missing values are erased), yielding 1675550 values.

Table 2 summarizes the results. For linear mapping, (2) is used. The training of the mapping function is done by manually setting the scaling coefficient (a), then finding the bias coefficient (b, by taking the smallest value of the scaled value array). Thus gaining $a = 100$, $b = -2079$, which takes ~1.538 ms to calculate.

The application of the method is simply using (2) on the data, which takes ~0.247 ms on average. The method uses ~128 KB of memory.

The training of Key-Value Lists first requires filtering the data for redundancies (as it is pointless and wasteful to store one value more than once), which results in 1340621 data values (20% reduction in size), then ordered (thus creating the list of the Keys). An array of incremented values (from 1 to 1340621) is created as the list of Values. Its memory requirement is measured to be 13116 KB (with the *profile report* function of Matlab) and takes 68 ms on average. Executing the method involves finding the given key in the list of Keys, which takes ~2.9154 ms on average (with the *find* function of Matlab).

These results show that for single floating-point values linear mapping functions are indeed the most beneficial both in terms of memory requirement and speed.

Table 2. Comparative analysis of linear mapping and Key-Value lists on the 2nd sensory data vector of the Dynamic gas mixtures data set.

Method	Memory requirement	Training speed	Run-time speed
Linear mapping function	128 KB	1.538 ms	0.247 ms
Key-value list	13116 KB	68.9 ms	2.9154 ms

In the second set of experiments, short strings of characters (i.e. words with fixed lengths) are converted into numerical ($\mathbb{N}^+$) values. 1015 three letters long English words are used (e.g. act, zoo etc.). The data is processed with KV Lists (similarly to that of the previous experiment), SFITs, LUTs and ANNs (for which the implementation of [18] is used, which is a simple network with backpropagation training). For KV Lists, the list of character strings can be directly used as keys, but for the latter three methods the normalized ASCII codes of each letter are used (since 'a' = 97, a bias value of -96 is added to each value so they can fit into the range of [1, 27], e.g. "a b o u t" → "97 98 115 117 116" →"1 2 15 21 20", thus reducing the output domain).

Table 3 summarizes the results. The KV List takes about 71 KB of memory and only ~5.884 ms to set up. The ANN main structure (weights and biases) take up about 16 KB (though with additional structures, its operation requires ~36 MB). However, it also takes more than ~10 min to train (1500 iterations with the backpropagation algorithm using 100 and 150 neurons in the hidden layers). The SFIT structure takes ~17.2 MB of memory and ~26.05 ms to train on average, while the LUT takes ~8 MB of memory but 266.9 ms to train (since a larger structure (an array with the size of 27^3) is needed to be maintained). In terms of run-time speed, predictably the LUT is the fastest (~0.124 ms), followed by the KV List (0.226 ms). The ANN takes ~0.585 ms to operate while the SFIT takes ~2.973 ms on average.

In the third set of experiments, strings of characters with different lengths are converted into numerical ($\mathbb{N}^+$) values. Four, five and six letters long English words are used (e.g. able, above, advice etc.), 500 of each. The indexing (the integer number assigned to them as output) is done by length, then alphabetical order (so the four-long words are set from A to Z (1…500), then the five-long words from A to Z (501…1000)

Table 3. Comparative analysis of Artificial Neural Networks, Sequential Fuzzy Indexing Tables, Lookup Tables and Key-Value Lists on 3-long English words.

Method	Memory requirement	Training speed	Run-time speed
ANN	16 KB	634312 ms	0.585 ms
SFIT	17.2 MB	26.05 ms	2.973 ms
LUT	~8 MB	266.9 ms	0.124 ms
Key-value list	71 KB	5.884 ms	0.226 ms

Table 4. Comparative analysis of Artificial Neural Networks, Sequential Fuzzy Indexing Tables and Key-Value Lists on 4, 5 and 6 letters long English words.

Method	Memory requirement	Training speed	Run-time speed
ANN	16 KB	952546 ms	33.947 ms
SFIT	17.2 MB	105.93 ms	0.6794 ms
Key-value list	71 KB	5.884 ms	0.226 ms

and finally the six-long words (1001…1500)). KV lists, ANNs and SFITs are investigated (because the problem is too large for LUTs).

Table 4 summarizes the results. The ANNs have the smallest memory need (~16 KB with 150 and 50 neurons in the hidden layers; though they still need 36 MB to operate), followed by the KV Lists (~71 KB), then the SFITs with ~17.2 MB. The ANN is again the slowest to train with ~15.87 min. The SFITs are significantly faster with 105.93 ms, and the KV Lists with 5.884 ms on average. The latter method is the fastest as well with 0.226 ms, followed by the SFIT (0.6794 ms) and the ANN with 33.947 ms on average.

The experiments clearly show that linear mapping functions are the most beneficial to use for data with single, floating point values, both in terms of memory requirement, training and operation speed. If the data consists of a few (1 to 3) low dimensional attributes (e.g. short words, color triplets (HSV, RBG) etc.), then LUTs are recommended. Figure 2 illustrates the scaling of the memory requirement in function of input attribute dimensions for a 3D LUT (uniform sizes are considered, starting from 26^3 and ending with 1100^3, at which more than 10 GB of memory is needed to store the structure). The memory requirement exceeds the available 1 GB of RAM that RPI 3B + models have, for LUT structures that are larger than 500^3.

If the data consists of more than 4 attributes and it can be presumed that all possible value combinations (that can occur run-time) are known, then KV Lists are the most beneficial. On the other hand, the downside of KV Lists, compared to ANNs and SFITs, is that the latter two have a generalization ability: if a value (e.g. 1.3) is known, then the system would recognize not yet seen, but very similar values (e.g. 1.31) as well and assign the output values accordingly. Thus, if it cannot be presumed that all possible values are known, then SFIT is recommended.

The only two advantages that the ANN has over the SFIT are the lower memory need, and that it can work directly with floating point values. Their main downside, aside from their long training phase, is that it even though they are considered universal approximators, i.e. being able to learn any functions with the right number of neurons, finding the correct setup can be very complicated. In the 2nd experiment the network could only learn 1009 of the 1015 input values (which counts as high precision in general neural network operation, but in the current problem the precision rate is needed to be 100%). In the third experiment the precision is only 1477 out of 1500. With more fine tuning it is possible to reach 100% precision, but if there are alternative solutions (linear mapping + SFIT) that can provide the same precision faster and with less effort, then it is worth using them over ANNs in domain reduction problems.

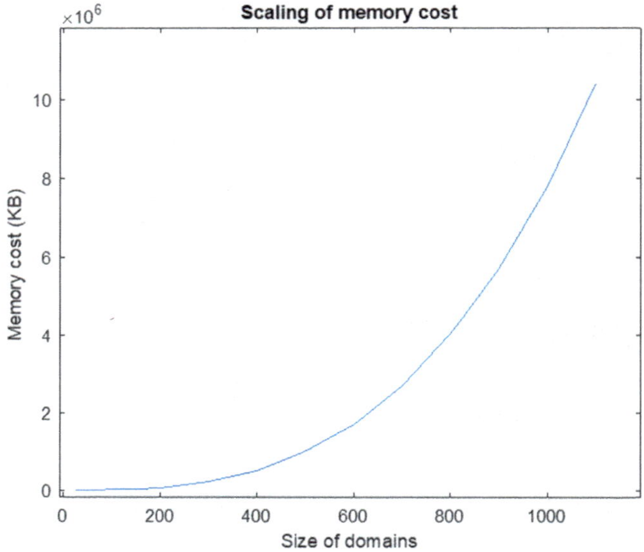

Fig. 2. The scaling of the memory requirement of 3D Lookup Tables in function of the sizes of the input domains (considering the same size for all three domains, e.g. $D_1 = D_2 = D_3 = 300$).

5 Conclusions

In this paper, multiple mapping functions are investigated and compared in order to reduce the size of the structure of Sequential Fuzzy Indexing Table classifiers, so they can be effectively used in single board computers as well. Linear functions, hash functions, Key-Value Lists, Lookup Tables, SFITs, Artificial Neural Networks and Recurrent Neural Networks are considered and analyzed regarding their computational complexity and additional memory requirements.

Considering the results of the comparison, for floating point data to integer conversion, linear mapping functions are the most beneficial due to their simplicity and fast operation speed. If more than one data value is needed to be converted into an integer one, but their number is still low (e.g. 2 or 3), then Lookup Tables can be used efficiently (with linear mapping functions if they are not in integer format).

If there are more than 3 attributes, and their original domains are relatively small (e.g. a letter in English can be considered an integer number between 1 and 27) but not all possible input values are known (there can be previously unseen input values during its operation), then SFITs can be used to convert them into a single integer value. Otherwise, Key-Value Lists are recommended.

Unfortunately, since parallelization plays a big role in reducing the training and operation time for most modern ANNs (e.g. RNNs), they are less likely to be used efficiently on SBCs like the Raspberry Pi 3 or older models (unless they are outfitted with additional devices, like a Tensor Processing Unit (TPU [19])). On the other hand, newer models bring more possibilities, so in future work this investigation will be expanded to a wider variety of newer SBC models.

Acknowledgement. The publication was prepared with the financial support of Pallas Athéné Domus Educationis Foundation, project number: PADE-0117-5.

References

1. Gay, W.W.: Raspberry Pi Hardware Reference (2014). https://doi.org/10.1007/978-1-4842-0799-4
2. Campbell-Kelly, M., Croarken, M., Robson, E.: The History of Mathematical Tables from Sumer to Spreadsheets. NY, USA (2003)
3. Tusor, B., Várkonyi-Kóczy, A.R., Tóth, J.T.: Active problem workspace reduction with a fast fuzzy classifier for real-time applications. In: IEEE International Conference on Systems, Man, and Cybernetics, pp. 4423–4428, Budapest, Hungary (2016)
4. Várkonyi-Kóczy, A.R., Tusor, B., Tóth, J.T.: Robust variable length data classification with extended sequential fuzzy indexing tables. In: 2017 IEEE International Instrumentation and Measurement Technology Conference (I2MTC), pp. 1881–1886, Torino, Italy (2017)
5. Konheim, A.G.: Hashing in Computer Science: fifty Years of Slicing and Dicing, p. 393. Wiley-Intersciencem (2010)
6. Minsky, M., Papert, S.: Perceptrons: an Introduction to Computational Geometry. MIT Press (1969)
7. Tapson, J., van Schaik, A.: Learning the pseudoinverse solution to network weights. Neural Netw. **45**, 94–100 (2013)
8. Rumelhart, D.E., Hinton, G.E., Williams, R.J.: Learning representations by back-propagating errors. Nature **323**(6088), 533–536 (1986)
9. Kamwa, I., Grondin, R., Sood, V.K., Gagnon, C., Nguyen, V.T., Mereb, J.: Recurrent neural networks for phasor detection and adaptive identification in power system control and protectio. IEEE Trans. Instr. Meas. **45**(2), 657–664 (1996)
10. Mikolov, T., Karafiát, M., Burget, L., Černocký, J., Khudanpur, S.: Recurrent neural network based language model. In: INTERSPEECH-2010, pp. 1045–1048 (2010)
11. Mozer, M.C., Chauvin, Y., Rumelhart, D.: A Focused Backpropagation Algorithm for Temporal Pattern Recognition, pp. 137–169. Lawrence Erlbaum Associates, Hillsdale, NJ (1995)
12. Bengio, Y., Simard, P., Frasconi, P.: Learning long-term dependencies with gradient descent is difficult. IEEE Trans. Neural Netw. **5**(2), 157–166 (1994)
13. Hochreiter, S., Schmidhuber, J.: Long short-term memory. Neural Comput. **9**(8), 1735–1780 (1997)

14. Zarit, B.D., Super, B.J., Quek, F.K.H.: Comparison of five color models in skin pixel classification. In: Proceedings International Workshop on Recognition, Analysis, and Tracking of Faces and Gestures in Real-Time Systems, Corfu, Greece, pp. 58–63 (1999)
15. Tusor, B., Simon-Nagy, G., Várkonyi-Kóczy, A.R., Tóth, J.T.: Personalized dietary assistant-an intelligent space application. In: 21st IEEE International Conference on Intelligent Engineering Systems (INES 2017), pp. 27–32 (2017)
16. Tusor, B., Várkonyi-Kóczy, A.R., Bukor, J.: An ISpace-based dietary advisor. In: 2018 IEEE International Symposium on Medical Measurements and Applications (MeMeA), pp. 1–6 (2018)
17. Fonollosa, J., Sheik, S., Huerta R., Marco S.: Reservoir computing compensates slow response of chemosensor arrays exposed to fast varying gas concentrations in continuous monitoring. Sens. Actuators B: Chem **215**, 618–629 (2015)
18. Tshitoyan, V.: Simple Neural Network. https://www.github.com/vtshitoyan/simpleNN, GitHub. Accessed 10 August 2019
19. Cass, S.: Taking AI to the edge: Google's TPU now comes in a maker-friendly package. IEEE Spectr. **56**(5), 16–17 (2019)

Building Energy Information: Demand and Consumption Prediction with Machine Learning Models for Sustainable and Smart Cities

Sina Ardabili[1], Amir Mosavi[2,3](✉), and Annamária R. Várkonyi-Kóczy[4]

[1] Institute of Advanced Studies Koszeg, Koszeg, Hungary
[2] Kalman Kando Faculty of Electrical Engineering, Obuda University, Budapest, Hungary
amir.mosavi@kvk.uni-obuda.hu
[3] School of the Built Environment, Oxford Brookes University, Oxford OX3 0BP, UK
[4] Department of Mathematics and Informatics, J. Selye University, Komarno, Slovakia

Abstract. Building energy consumption plays an essential role in urban sustainability. The prediction of the energy demand is also of particular importance for developing smart cities and urban planning. Machine learning has recently contributed to the advancement of methods and technologies to predict demand and consumption for building energy systems. This paper presents a state of the art of machine learning models and evaluates the performance of these models. Through a systematic review and a comprehensive taxonomy, the advances of machine learning are carefully investigated and promising models are introduced.

Keywords: Machine learning · Smart cities · IoT · Deep learning · Big data · Soft computing · Sustainable urban development · Building energy · Energy demand · And consumption · Sustainable cities

Nomenclatures

Generalized boosted regression	GBR
Deep learning	DL
Artificial neural network	ANN
Extreme learning machine	ELM
Machine learning	ML
Support vector machine	SVM
Wavelet neural networks	WNN
Support vector regression	SVR
Genetic algorithm	GA
Multi layered perceptron	MLP
Long short-term memory	LSTM

A. R. Várkonyi-Kóczy (Ed.): INTER-ACADEMIA 2019, LNNS 101, pp. 191–201, 2020.
https://doi.org/10.1007/978-3-030-36841-8_19

Decision tree	DT
Response surface methodology	RSM
Back propagation neural network	BPNN
Centroid mean	CM
Adaptive neuro fuzzy inference system	ANFIS
Analytic network process	ANP
Radial basis function	RBF
Feed-forward neural networks	FFNN
Particle swarm optimization	PSO
Random forest	RF
Non-random two-liquid	NRTL
Recurrent neural network	RNN
Partial least squares	PLS
Discriminant analysis	DA
Principal component analysis	PCA
Linear discriminant analysis	LDA
Autoregressive integrated moving average	ARIMA
Least-squares	LS
Sparse Bayesian	SB
Multi criteria decision making	MCDM
Genetic programming	GP
Multi linear regression	MLR
Step-wise Weight Assessment Ratio Analysis	SWARA
Multi Objective Optimization by Ratio Analysis	MOORA
Nonlinear autoregressive exogenous	NARX

1 Introduction

The energy is one of the essential aspects of smart cities [1]. The sustainability factor of urban development is a direct function of energy production and consumption of every city [2]. The energy consumption of buildings is responsible for a significant amount of energy used in cities [3]. From this perspective, the prediction of demand and consumption is essential in the development of smart cities of the future [4].

Machine learning (ML) has recently well contributed to advancing the accurate and reliable prediction models [5–9]. Literature includes the various application domains and the ML learning methods used in these areas [10–26]. ML methods have shown to outperform most of the conventional modeling techniques considering performance, accuracy, robustness, speed, and computation costs [27–38]. A number of surveys have investigated the usage of ML methods in various application domains [39–43]. However, there is a gap in review studies considering the building energy information for demand and consumption prediction in the era of smart cities, despite the increasing popularity of ML methods. Figure 1 shows the exponential increase in using ML models in this realm within the past decade. Consequently, the contribution of this paper is to investigate the application of novel ML models in shaping the future of smart and sustainable cities in terms of energy.

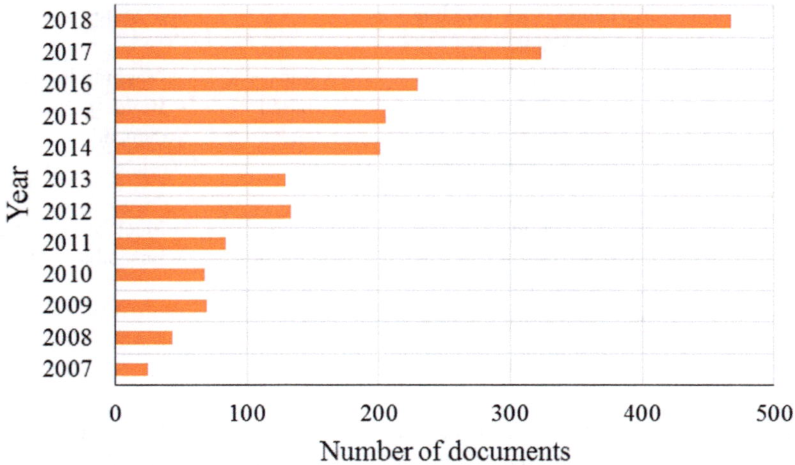

Fig. 1 Rapid increase of using ML models in various scientist domains (Source: web of science)

2 Building Energy Demand Prediction

Prediction of demand in building energy sector is essential for planning and managing energy systems. Table 1 presents top six studied developed by ML methods in building energy demand prediction.

Table 1. The notable ML methods in building energy demand prediction

References	Contribution	ML method	Keywords
[44]	To employ machine learning for Quantifying the effect of landscape composition and configuration on urban land surface temperatures	Generalized boosted regression (GBR)	– Land surface temperature – Machine learning
[45]	To present a comprehensive review about the application of machine learning as a solution in smart buildings	ANN, DL, SVM, GA and SVR	– Internet of things – Machine learning
[46]	To compare different forecasting models for estimating the natural gas demand	empirical models, RNN and LR	– Machine learning – Deep learning

(*continued*)

Table 1. (*continued*)

References	Contribution	ML method	Keywords
[47]	To employ different machine learning methods for Building Performance Simulation	ANN, LSTM	– Deep learning – Machine learning
[48]	To present a comprehensive state of the art of machine leaning methods for the prediction of building energy demand	ANN and SVM based machine learning methods	– Machine leaning – Building energy demand
[49]	To present a hybrid ensemble method to increase the accuracy of load demand estimation of PV for building energy sector	Single and hybrid machine learning methods	– Machine learning – Ensemble method

Osborne and Alvares-Sanches [44] developed an innovative approach in the presence of machine learning technique (GBR) for Quantifying the effect of landscape composition and configuration on urban land surface temperatures. Based on findings of the study, GBR could successfully predict land surface temperatures with a high correlation coefficient (0.956) using 102,935 data.

Djenouri et al. [45] presented a comprehensive state of the art about the application of ML methods in buildings. ML has been employed as solution for occupants and energy or devices. ML can be used as a multi-disciplinary solution for building purposes but in general the type and the size of the building are main factors for considering the effectiveness of ML method. But the use of ML methods can be more successful in energy demand purposes in buildings by developing innovative approaches. Hribar et al. [46] developed a study for evaluating the forecasting capabilities of different methods including empirical, deep learning and LR models for the prediction of natural gas demand in the presence of daily and hourly datasets. All the methods have been employed in their single form. Evaluations have been performed by the use of MAE and MAPE factors. Based on results deep learning method have the best performance (with MAE 1.06 and 18.3 for hourly and daily datasets, respectively) compared with that of the other techniques.

Singaravel et al. [47] employed ANN while comparing with single, two and three layer LSTM method for the estimation of building sustainability. Machine learning methods can play an important role in reducing the processing time and increasing the sustainability by increasing the model accuracy. Comparing the accuracy values of models for the prediction have been performed by employing determination coefficient values. Based on results all of methods could successfully cope with the prediction task but two layered LSTM method have the best performance compared with others. Ahmad et al. [48] provided a comprehensive state of the art of the machine learning based prediction models for the estimation of building energy demand sector. In general methods have been divided into two main categories including ANN and SVM based machine learning methods as the most frequently used methods in this field of science. This paper also indicated the importance of machine learning methods in the sustainability of buildings energy demand. SVM based methods provided a higher accuracy compared with that of the ANN based methods.

Reza et al. [49] developed a novel hybrid ensemble method including neural ensemble, Bayesian model and wavelet transform method for the prediction of PV performance in the building energy demand sector. This method has been developed by comparing different single and hybrid machine learning techniques in the term of the normalized root mean square error. This study wants to emphasize on the importance of the hybrid methods over the single methods. The proposed hybrid method could successfully estimate the demand forecasting factors and increased the accuracy of the model significantly.

Table 2 present a brief comparison about the accuracy, reliability and sustainability of methods developed for forecasting the energy demand in building sector. Accuracy factor has been generated from the performance factors related to the training step and reliability has been generated from the performance factors related to the testing step. But, sustainability was a little difference and has been generated by comparing reliability, accuracy, processing time and other factors which have been considered by results of the reviewed articles.

Table 2. The comparison results of methods for energy demand in building sector

Method	Application	Accuracy	Reliability	Sustainability	References
GBR	Regression	++	+	+++	[44]
ANN	Classification regression	+	+	+	[45]
SVM	Classification regression	++	++	+	[45]
DL	Classification regression	+++	+++	+++	[45]
Hybrid ML	Classification regression	+++	++	++	[45]
RNN	Regression	+++	+++	+++	[46]
LR	Regression	+	+	+	[46]
LSTM	Simulation	+++	++	++	[47]
ANN	Simulation	++	+	+	[47]
ANN-based	Regression	+	+	+	[48]
SVM-based	Regression	++	++	++	[48]
BPNN	Regression	–	–	–	[49]
ARIMA	Regression	+	+	+	[49]
RBF-PSO	Regression	+	+	+	[49]
FNN-PSO	Regression	++	+	+	[49]
WT-BPNN	Regression	+	–	–	[49]
WT-ARIMA	Regression	+	+	+	[49]
WT-FNN-PSO	Regression	++	++	++	[49]

3 Building Energy Consumption Perdition

Building energy consumption is important as much as the importance of building energy demand. Prediction of energy consumption in building energy sector can be one of the main steps for reaching the sustainable buildings and is essential for planning and managing of energy systems. Table 3 presents top six studied developed by ML methods in building energy consumption prediction.

Table 3. The notable ML methods in building energy consumption prediction

References	Contribution	ML method	Keywords
[50]	To present a robust artificial neural network to explore complex building energy consumption data which have been generated from the simulation-Based Multi-Objective Optimization model	ANN	– Energy consumption – Machine learning
[51]	To develop an accurate machine learning method for energy prediction in buildings using data generated from internet of things technology	MLP, LR, RF, SVM and GBM	– Internet of things – Machine learning
[52]	To develop a long short term memory (LSTM) network to predict the energy consumers' behaviour based on their recent energy consumptions	LSTM	– Machine learning; – Smart grid
[53]	To develop an innovative hybrid deep learning method for the prediction of energy consumption in buildings	Hybrid LSTM-GA	– Deep learning – Machine learning
[54]	To develop a comprehensive survey about different machine learning methods developed for the prediction of energy consumption in buildings	Single, hybrid and ensemble machine learning methods	– Data mining – Machine learning
[55]	In order to develop machine learning methods for the prediction of energy load in building sectors.	SVM and NARX-RNN	– Machine learning – Deep learning

Sharif and Hammad [50] developed a robust ANN method to explore complex building energy consumption data which have been generated from the simulation-Based Multi-Objective Optimization model. In fact, this study focuses on developing an accurate prediction method for the energy consumption of buildings. Evaluating of results indicated that the developed ANN method benefits less time consuming as well as high accuracy, which increases the sustainability of the developed method.

Chammas et al. [51] developed a study for the prediction of the energy consumption in buildings using data generated from the IoT technology embedded in buildings. The proposed method is a prediction model based on MLP while comparing with LR, SVM, GBM, and random forest. Methods have been compared in terms of determination coefficient, MAPE, and RMSE. Dataset for the training process was separated into three categories (no light, no date, and weather only) for finding the effective variables on the modelling process. Based on the results, eliminating lights data have an essential effect on increasing the accuracy of the target model. The developed MLP model has a higher determination coefficient and a lower RMSE and MAPE compared with that for other methods.

Fenza et al. [52] developed a LSTM method for the prediction of consumers' behaviors in the term of energy consumption. Time series data have been employed in order to develop the target network. Results have been evaluated using RMSE factor. Based on the results, the proposed method has successfully cop with the task as well as providing the required sustainability for the prediction phase. Almalaq and Zhang [53] developed an innovative prediction model for the estimation of energy consumption in buildings using LSTM and optimizing its parameters by GA methodology to take an evolutionary DL method. The evaluation phase for this study has been performed by the use of datasets related to residential and commercial buildings. Results indicated that the hybrid methods which take an evolutionary DL method present an accurate and sustainable method for the prediction of energy consumption in buildings over the DL methods.

Chou and Tran [54] developed a comprehensive survey for studying different machine learning techniques developed for the prediction of energy consumption in building sectors. Methods have been categorized into three main categories, including single, hybrid, and ensemble machine learning methods. Methods have been compared in terms of performance factors and sustainability index. Results indicated that in the case of using single and ensemble methods, ANN-based methods have the best prediction performance, but in the case of using hybrid methods, SVM based methods could present the best performance. In general, hybrid methods is the proposed method from the viewpoint of accuracy and sustainability.

Koschwitz et al. [55] developed predictive models in order to estimate the building energy load. The target models include the RBF based SVM and Nonlinear Autoregressive Exogenous Recurrent Neural Networks which have been developed by the historical data from residential buildings in Germany. Based on the results, NARX-RNN provided higher performance and sustainability in comparison with those for the SVM method. Table 4 present a brief comparison of the accuracy, reliability, and sustainability of methods developed for forecasting the energy consumption in the building sector.

Table 4. The comparison results of methods for energy consumption in the building sector

Method	Application	Accuracy	Reliability	Sustainability	Reference
ANN	Regression	++	++	++	[50]
MLP	Regression	++	++	++	[51]
LR	Regression	–	–	–	[51]
SVM	Regression	+	+	+	[51]
GBM	Regression	+	+	+	[51]
RF	Regression	+	+	+	[51]
LSTM	Regression	+++	+++	+++	[52]
LSTM-GA	Regression	+++	+++	+++	[53]
Single-ANN based	Regression	++	++	++	[54]
Hybrid-ANN based	Regression	+++	++	++	[54]
Ensemble-ANN based	Regression	++	++	++	[54]
Single-SVM based	Regression	++	++	++	[54]
Hybrid-SVM based	Regression	+++	+++	+++	[54]
Ensemble-SVM based	Regression	+++	++	++	[54]
SVM	Regression	++	+	+	[55]
NARX-RNN	Regression	+++	+++	+++	[55]

4 Conclusion

This paper concludes that the usage of machine learning in building energy information applications is growing at a higher rate than we have seen during the last decade. The ensemble and hybrid models have emerged and continue to advance for higher accuracy and better performance. Deep learning models also will bring a tremendous amount of intelligence for better prediction models. This trend has been also reported in other research domains, e.g. biofuels, agriculture, hydrology, and production [56–64].

Acknowledgments. This publication has been supported by the Project: "Support of research and development activities of the J. Selye University in the field of Digital Slovakia and creative industry" of the Research & Innovation Operational Programme (ITMS code: NFP313010T504) co-funded by the European Regional Development Fund.

References

1. Ahmad, T., et al.: Supervised based machine learning models for short, medium and long-term energy prediction in distinct building environment. Energy **158**, 17–32 (2018)
2. Jiang, Z., Lin, R., Yang, F.: A hybrid machine learning model for electricity consumer categorization using smart meter data. Energies **11**(9) (2018)
3. Kim, S.H., et al.: Deep learning based on multi-decomposition for short-term load forecasting. Energies **11**(12) (2018)
4. Laib, O., Khadir, M.T., Mihaylova, L.: Toward efficient energy systems based on natural gas consumption prediction with LSTM recurrent neural networks. Energy **177**, 530–542 (2019)

5. Li, Z., et al.: An ANN-based fast building energy consumption prediction method for complex architectural form at the early design stage. In: Building Simulation (2019)
6. Liu, L., Ran, W.: Research on supply chain partner selection method based on BP neural network. Neural Comput. Appl. (2019)
7. Protić, M., Fathurrahman, F., Raos, M.: Modelling energy consumption of the republic of Serbia using linear regression and artificial neural network technique. Tehnicki Vjesnik **26** (1), 135–141 (2019)
8. Rahman, A., Smith, A.D.: Predicting heating demand and sizing a stratified thermal storage tank using deep learning algorithms. Appl. Energy **228**, 108–121 (2018)
9. Seyedzadeh, S., et al.: Machine learning for estimation of building energy consumption and performance: a review. Visualization Eng. **6**(1) (2018)
10. Mosavi, A., Ozturk, P., Chau, K.W.: Flood prediction using machine learning models: literature review. Water (Switzerland) **10**(11) (2018)
11. Mosavi, A., Rabczuk, T.: Learning and intelligent optimization for material design innovation. In: Kvasov, D.E., et al. (eds.), pp. 358–363. Springer (2017)
12. Mosavi, A., Rabczuk, T., Varkonyi-Koczy, A.R.: Reviewing the novel machine lea0072ning tools for materials design. Luca, D., Sirghi, L., Costin, C., (eds.), pp. 50–58. Springer (2018)
13. Mosavi, A., et al.: State of the art of machine learning models in energy systems, a systematic review. Energies **12**(7) (2019)
14. Mosavi, A., et al.: Prediction of multi-inputs bubble column reactor using a novel hybrid model of computational fluid dynamics and machine learning. Eng. Appl. Comput. Fluid Mech. **13**(1), 482–492 (2019)
15. Mosavi, A., Varkonyi-Koczy, A.R.: Integration of machine learning and optimization for robot learning. Jablonski, R., Szewczyk, R., (eds.), pp. 349–355. Springer (2017)
16. Nosratabadi, S., et al.: Sustainable business models: a review. Sustainability (Switzerland) **11**(6) (2019)
17. Qasem, S.N., et al.: Estimating daily dew point temperature using machine learning algorithms. Water (Switzerland) **11**(3) (2019)
18. Rezakazemi, M., Mosavi, A., Shirazian, S.: ANFIS pattern for molecular membranes separation optimization. J. Mol. Liq. **274**, 470–476 (2019)
19. Riahi-Madvar, H., et al.: Comparative analysis of soft computing techniques RBF, MLP, and ANFIS with MLR and MNLR for predicting grade-control scour hole geometry. Eng. Appl. Comput. Fluid Mech. **13**(1), 529–550 (2019)
20. Shabani, S., Samadianfard, S., Taghi Sattari, M., Shamshirband, S., Mosavi, A., Kmet, T., Várkonyi-Kóczy, A.R.: Modeling daily pan evaporation in humid cli-mates Using Gaussian Process Regression (2019). arXiv:2019070351. https://doi.org/10.20944/preprints201907.0351.v1
21. Shamshirband, S., Hadipoor, S., Baghban, M., Mosavi, A., Bukor, A., Annamaria, J., Varkonyi-Koczy, R.: Developing an ANFIS-PSO model to predict mercury emissions in Combustion Flue Gases (2019). arXiv:2019070165. https://doi.org/10.20944/preprints201907.0165.v1
22. Shamshirband, S., et al.: Ensemble models with uncertainty analysis for multi-day ahead forecasting of chlorophyll a concentration in coastal waters. Eng. Appl. Comput. Fluid Mech **13**(1), 91–101 (2019)
23. Shamshirband, S., Mosavi, A., Rabczuk, T.: Particle swarm optimization model to predict scour depth around bridge pier (2019). arXiv:1906.08863
24. Taherei Ghazvinei, P., et al.: Sugarcane growth prediction based on meteorological parameters using extreme learning machine and artificial neural network. Eng. Appl. Comput. Fluid Mech **12**(1), 738–749 (2018)

25. Torabi, M., et al.: A Hybrid clustering and classification technique for forecasting short-term energy consumption. Environ. Prog. Sustain. Energy **38**(1), 66–76 (2019)
26. Torabi, M., et al.: A hybrid machine learning approach for daily prediction of solar radiation. In: Lecture Notes in Networks and Systems, pp. 266–274. Springer (2019)
27. Aram, F., et al.: Design and validation of a computational program for analysing mental maps: aram mental map analyzer. Sustainability (Switzerland) **11**(14) (2019)
28. Asadi, E., et al.: Groundwater Quality Assessment For Drinking And Agricultural Purposes In Tabriz Aquifer, Iran (2019)
29. Asghar, M.Z., Subhan, F., Imran, M., Kundi, F.M., Shamshirband, S., Mosavi, A., Csiba, P., Várkonyi-Kóczy, A.R.: Performance evaluation of supervised machine learning techniques for efficient detection of emotions from online content (2019). arXiv:2019080019. https://doi.org/10.20944/preprints201908.0019.v1)
30. Bemani, A., Baghban, A., Shamshirband, S., Mosavi, A., Csiba, P., Várkonyi-Kóczy, A.R.: Applying ANN, ANFIS, and LSSVM models for estimation of Acid Sol-vent solubility in Supercritical CO_2 (2019). arXiv:2019060055 https://doi.org/10.20944/preprints201906.0055.v2
31. Choubin, B., et al.: Snow avalanche hazard prediction using machine learning methods. J. Hydrol. **577** (2019)
32. Choubin, B., et al.: An ensemble prediction of flood susceptibility using multivariate discriminant analysis, classification and regression trees, and support vector machines. Sci. Total Environ. **651**, 2087–2096 (2019)
33. Dehghani, M., et al.: Prediction of hydropower generation using Grey wolf optimization adaptive neuro-fuzzy inference system. Energies **12**(2) (2019)
34. Dineva, A., et al.: Review of soft computing models in design and control of rotating electrical machines. Energies **12**(6) (2019)
35. Dineva, A., et al.: Multi-label classification for fault diagnosis of rotating electrical machines (2019)
36. Farzaneh-Gord, M., et al.: Numerical simulation of pressure pulsation effects of a snubber in a CNG station for increasing measurement accuracy. Eng. Appl. Comput. Fluid Mech **13**(1), 642–663 (2019)
37. Ghalandari, M., et al.: Investigation of submerged structures' flexibility on sloshing frequency using a boundary element method and finite element analysis. Eng. Appl. Comput. Fluid Mech **13**(1), 519–528 (2019)
38. Ghalandari, M., et al.: Flutter speed estimation using presented differential quadrature method formulation. Eng. Appl. Comput. Fluid Mech **13**(1), 804–810 (2019)
39. Karballaeezadeh, N., et al.: Prediction of remaining service life of pavement using an optimized support vector machine (case study of Semnan-Firuzkuh road). Eng. Appl. Comput. Fluid Mech **13**(1), 188–198 (2019)
40. Menad, N.A., et al.: Modeling temperature dependency of oil-water relative permeability in thermal enhanced oil recovery processes using group method of data handling and gene expression programming. Eng. Appl. Comput. Fluid Mech **13**(1), 724–743 (2019)
41. Mohammadzadeh, S., et al.: Prediction of compression index of fine-grained soils using a gene expression programming model. Infrastructures **4**(2), 26 (2019)
42. Mosavi, A., Edalatifar, M.: A hybrid neuro-fuzzy algorithm for prediction of reference evapotranspiration. In: Lecture Notes in Networks and Systems, pp. 235–243. Springer (2019)
43. Mosavi, A., Lopez, A., Varkonyi-Koczy, A.R.: Industrial applications of big data: state of the art survey. Luca, D., Sirghi, L., Costin, C., (eds.), pp. 225–232. Springer (2018)
44. Osborne, P.E., Alvares-Sanches, T.: Quantifying how landscape composition and configuration affect urban land surface temperatures using machine learning and neutral landscapes. Comput. Environ. Urban Syst. **76**, 80–90 (2019)

45. Djenouri, D., et al.: Machine learning for smart building applications: review and taxonomy. ACM Comput. Surv. **52**(2) (2019)
46. Hribar, R., et al.: A comparison of models for forecasting the residential natural gas demand of an urban area. Energy, 511–522 (2019)
47. Singaravel, S., Suykens, J., Geyer, P.: Deep-learning neural-network architectures and methods: using component-based models in building-design energy prediction. Adv. Eng. Inform. **38**, 81–90 (2018)
48. Ahmad, T., et al.: A comprehensive overview on the data driven and large scale based approaches for forecasting of building energy demand: a review. Energy Build. **165**, 301–320 (2018)
49. Raza, M.Q., Nadarajah, M., Ekanayake, C.: Demand forecast of PV integrated bioclimatic buildings using ensemble framework. Appl. Energy **208**, 1626–1638 (2017)
50. Sharif, S.A., Hammad, A.: Developing surrogate ANN for selecting near-optimal building energy renovation methods considering energy consumption, LCC and LCA. J. Build. Eng. **25** (2019)
51. Chammas, M., Makhoul, A., Demerjian, J.: An efficient data model for energy prediction using wireless sensors. Comput. Electr. Eng. **76**, 249–257 (2019)
52. Fenza, G., Gallo, M., Loia, V.: Drift-aware methodology for anomaly detection in smart grid. IEEE Access **7**, 9645–9657 (2019)
53. Almalaq, A., Zhang, J.J.: Evolutionary deep learning-based energy consumption prediction for buildings. IEEE Access **7**, 1520–1531 (2019)
54. Chou, J.S., Tran, D.S.: Forecasting energy consumption time series using machine learning techniques based on usage patterns of residential householders. Energy, 709–726 (2018)
55. Koschwitz, D., Frisch, J., van Treeck, C.: Data-driven heating and cooling load predictions for non-residential buildings based on support vector machine regression and NARX recurrent neural network: a comparative study on district scale. Energy **165**, 134–142 (2018)
56. Ardabili, S., Mosavi, A., Mahmoudi, Mesri Gundoshmian, T, Nosratabadi, S., Var-konyi-Koczy, A.: Modelling temperature variation of mushroom growing hall us-ing artificial neural networks (2019)
57. Mesri Gundoshmian, T., Ardabili, S., Mosavi, A., Varkonyi-Koczy, A.: Prediction of combine harvester performance using hybrid machine learning modeling and re-sponse surface methodology (2019)
58. Ardabili, S., Mosavi, A., Varkonyi-Koczy, A.: Systematic review of deep learning and machine learning models in biofuels research (2019)
59. Ardabili, S., Mosavi, A., Varkonyi-Koczy, A.: Advances in machine learning model-ing reviewing hybrid and ensemble methods (2019)
60. Ardabili, S., Mosavi, A., Varkonyi-Koczy, A.: Building energy information: demand and consumption prediction with machine learning models for sustainable and smart cities (2019)
61. Ardabili, S., Mosavi, A., Dehghani, M., Varkonyi-Koczy, A.: Deep learning and machine learning in hydrological processes climate change and earth systems a systematic review (2019)
62. Mohammadzadeh D., Karballaeezadeh, N., Mohemmi, M., Mosavi, A., Varkonyi-Koczy A.: Urban train soil-structure interaction modeling and analysis (2019)
63. Mosavi, A., Ardabili, S., Varkonyi-Koczy, A.: List of deep learning models (2019)
64. Nosratabadi, S., Mosavi, A., Keivani, R., Ardabili, S., Aram, F.: State of the art sur-vey of deep learning and machine learning models for smart cities and urban sustainability (2019)

List of Deep Learning Models

Amir Mosavi[1,2](✉), Sina Ardabili[3], and Annamária R. Várkonyi-Kóczy[4]

[1] Institue of Automation, Kalman Kando Faculty of Electrical Engineering, Obuda University, Budapest, Hungary
amir.mosavi@kvk.uni-obuda.hu
[2] School of the Built Environment, Oxford Brookes University, Oxford OX3 0BP, UK
[3] Institute of Advanced Studies Koszeg, Koszeg, Hungary
[4] Department of Mathematics and Informatics, J. Selye University, Komarno, Slovakia

Abstract. Deep learning (DL) algorithms have recently emerged from machine learning and soft computing techniques. Since then, several deep learning (DL) algorithms have been recently introduced to scientific communities and are applied in various application domains. Today the usage of DL has become essential due to their intelligence, efficient learning, accuracy and robustness in model building. However, in the scientific literature, a comprehensive list of DL algorithms has not been introduced yet. This paper provides a list of the most popular DL algorithms, along with their applications domains.

Keywords: Deep learning · Machine learning · Convolutional neural networks (CNN) · Recurrent neural networks (RNN) · Denoising autoencoder (DAE) · Deep belief networks (DBNs) · Long short-term memory (LSTM)

1 Introduction

There has been an enormous evolution in system modeling and intelligence after introducing the early models for deep learning [1–8]. Deep learning methods very fast emerged and expanded applications in various scientific and engineering domains. Health informatics, energy, urban informatics, safety, security, hydrological systems modeling, economic, bioinformatics, and computational mechanics have been among the early application domains of deep learning. State of the art surveys on the data-driven methods and machine learning algorithms, e.g., [9–26], indicates that deep learning, along with the ensemble and hybrid machine learning methods are the future of data science. Further comparative studies, e.g., [26–42], report that deep learning models and hybrid machine learning models often outperform conventional machine learning models. Figure 1 represents the rapid rise in the applications of various deep learning methods during the past five years.

Deep learning methods are fast evolving for higher performance. Literature includes adequate review papers on the progressing algorithms in particular application domains, e.g., renewable energy forecasting, cardiovascular image analysis,

A. R. Várkonyi-Kóczy (Ed.): INTER-ACADEMIA 2019, LNNS 101, pp. 202–214, 2020.
https://doi.org/10.1007/978-3-030-36841-8_20

super-resolution imaging, radiology, 3D sensed data classification, 3D sensed data classification, multimedia analytics, sentiment classification, text detection, transportation systems, activity recognition in radar, hyperspectral, medical ultrasound analysis, image cytometry, and apache spark [43–59]. However, a simplified list of deep learning methods has not been communicated so far. Thus, there is a gap in research in introducing the deep learning methods and summarize the methods and application in a brief, yet communicative paper. Consequently, this paper aims at providing a comprehensive list of the most popular deep learning methods and their notable applications. In every section, one deep learning method is introduced and the notable applications related to that method are listed. The description of each deep learning method and the function of each building block is explained.

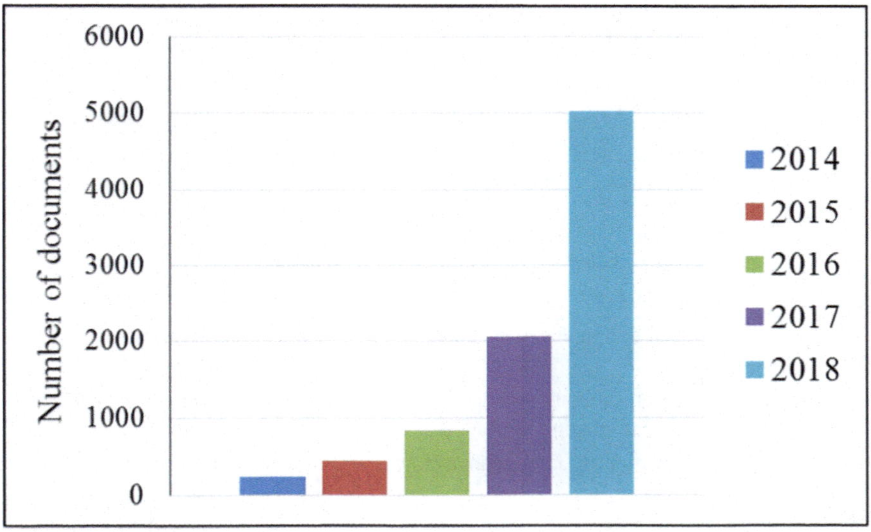

Fig. 1. The rapid increase of using DL models in various application domains (source: web of science)

2 Deep Learning Methods

Convolutional neural network (CNN) Recurrent neural network (RNN), Denoising autoencoder (DAE), deep belief networks (DBNs), Long Short-Term Memory (LSTM) are the most popular deep learning methods have been widely used. In this section, the description of each method is described along with the notable applications.

2.1 Convolutional Neural Network (CNN)

CNN is one of the most known architectures of DL techniques. This technique is generally employed for image processing applications. CNN contains three types of layers with different convolutional, pooling, and fully connected layers (Fig. 1). In each

CNN, there are two stages for the training process, the feed-forward stage, and the back-propagation stage. The most common CNN architectures are ZFNet [60], GoogLeNet [61], VGGNet [62], AlexNet [63], ResNet [64] (Table 1).

Although CNN is primarily known for image processing applications, the literature includes other application domains, e.g., energy, computational mechanics, electronics systems, remote sensing, etc.

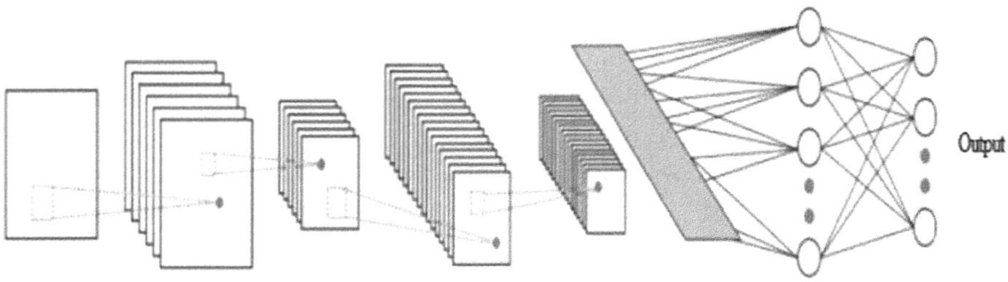

Fig. 2. CNN architecture

Table 1. The CNN notable applications

References	Application	Journal
Kong et al. [65]	Condition monitoring of wind turbines	Renewable Energy
Lossau et al. [66]	Motion estimation and correction of medical imaging	Computerized Medical Imaging and Graphics
Bhatnagar et al. [67]	Prediction of aerodynamic flow	Computational Mechanics
Nevavuori et al. [68]	Crop yield prediction	Computers and Electronics in Agriculture
Ajami et al. [69]	Advanced image processing	Remote Sensing

2.2 Recurrent Neural Networks (RNN)

RNN is designed to recognize sequences and patterns such as speech, handwriting, text, and such applications. RNN benefits cyclic connections in the structure which employ recurrent computations to sequentially process the input data [70]. RNN is basically a standard neural network that has been extended across time by having edges which feed into the next time step instead of into the next layer in the same time step. Each of the previous inputs data are kept in a state vector in hidden units, and these state vectors is utilized to compute the outputs. Figure 2 shows the architecture of RNN (Table 2).

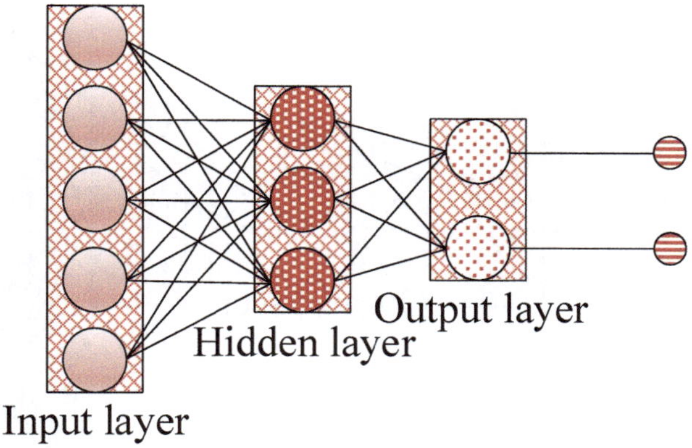

Fig. 3. RNN architecture

Table 2. Notable RNN applications

References	Application	Journal
Zhu et al. [71]	Wind speed prediction	Energy Conversion and Management
Pan et al. [72]	Tropical cyclone intensity prediction	Electronics Letters
Bisharad et al. [73]	Music genre recognition	Expert Systems
Zhong et al. [74]	Ship Trajectory Restoration	Journal of Navigation
Jarrah et al. [75]	Stock price trends predict	Advanced Computer Science and Applications

RNN is relatively newer deep learning method. This is why the application domains are still young and plenty of rooms remains for research and exploration. The energy, hydrological prediction, expert systems, navigation, and economics are the current applications reported in the literature.

2.3 Denoising AutoEncoder (DAE)

DAE has been extended from AE as asymmetrical neural network for learning features from noisy datasets. DAE consists of three main layers, including input, encoding, and decoding layers [76]. DAE is able to be aggregated for taking high-level features. Stacked Denoising AutoEncoder (SDAE), as an unsupervised algorithm, is generated by the DEA method, which can be employed for nonlinear dimensionality reduction. This method is a type of feed-forward neural network and employs a deep architecture with multiple hidden layers and a pre-training strategy [77, 78]. Figure 3 presents the architecture of DEA methodology (Table 3).

DEA is slowly starting to be known among researchers as an efficient DL algorithm. DEA has already been used in various application domains with promising

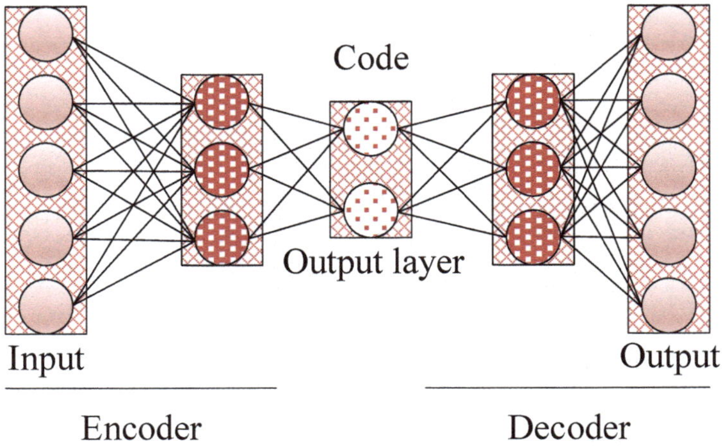

Fig. 4. DEA architecture

Table 3. The notable DEA applications

References	Application	Journal
Chen et al. [79]	Improving the cyber-physical systems	Journal on Wireless Communications
Liu et al. [80]	Electric load forecasting	Energies
Nicolai et al. [81]	Laser-based scan registration	IEEE Robotics and Automation
Yue, et al. [82]	Collaborative Filtering	Computer Science and Technology
Roy et a. [83]	Noisy image classification	Journal of Information and Communication Technology
Tan et al. [84]	Robust Speaker Verification	IEEE Transactions on Audio Speech

results. The energy forecasting, cybersecurity, banking, fraud detection, image classification, and speaker verification are among the current popular applications of DEA.

2.4 The Deep Belief Networks (DBNs)

DBNs are employed for high dimensional manifolds learning of data. This method contains multiple layers, including connections between the layers except for connections between units within each layer. DBNs can be considered as a hybrid multi-layered neural network, including directed and undirected connections. DBNs contains restricted Boltzmann machines (RBMs) which are trained in a greedy manner. Each RBM layer communicates with both the previous and subsequent layers [78, 85, 86]. This model is consists of a feed-forward network and several layers of restricted

Boltzmann machines or RBM as feature extractors [87]. A hidden layer and visible layer are only two layers of an RBM [88]. Figure 4 presents the architecture of the DBN method (Fig. 5 and Table 4).

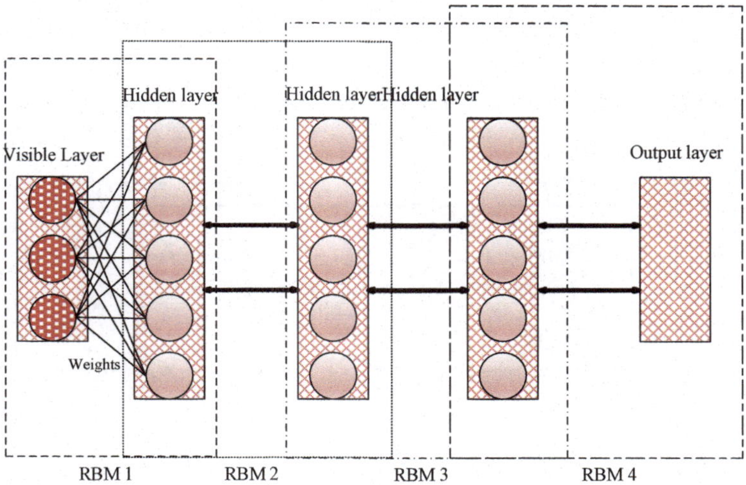

Fig. 5. DBN architecture

Table 4. The notable DBN applications

References	Application	Journal
Hassan et al. [89]	Human emotion recognition	Information Fusion
Cheng et al. [90]	Time series prediction	IEEE Internet of Things
Yu et al. [91]	wind speed prediction	IEEE Transactions on Electrical Engineering
Zheng et al. [92]	Exchange rate forecasting	Neural Computing and Applications
Ahmad et al. [93]	Automatic Liver Segmentation	IEEE Access
Ronoud et al. [94]	Breast cancer diagnosis	Soft Computing

DBN is one of the most reliable deep learning methods with high accuracy and computational efficiency. Thus, the application domains have been divers, including exciting application in a wide range of engineering and scientific problems. Human emotion detection, time series prediction, renewable energy prediction, economic forecasting, and cancer diagnosis have been among the public application domains.

2.5 Long Short-Term Memory (LSTM)

LSTM is an RNN method which benefits feedback connections to be used as a general-purpose computer. This method can of for both sequences and patterns recognition and image processing applications. In general, LSTM contains three central units, including input, output, and forget gates. LSTM can control on deciding when to let the input enter the neuron and to remember what was computed in the previous time step. One of the main strength of the LSTM method is that it decides all these based on the current input itself. Figure 6 presents the architecture of the LSTM method (Table 5).

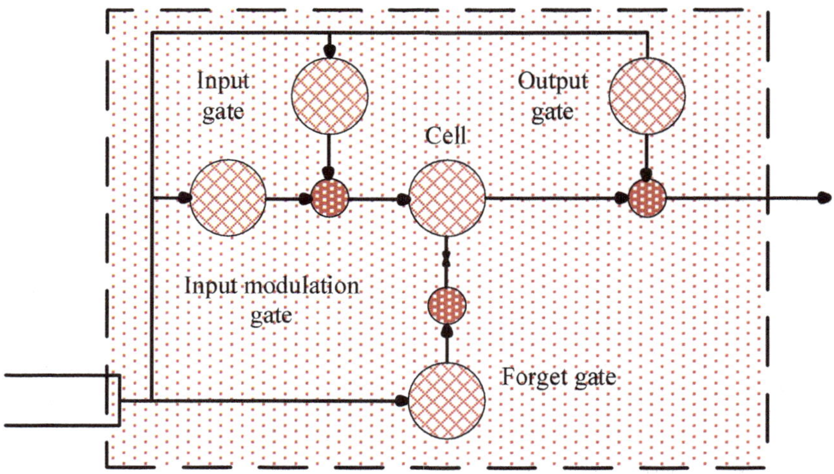

Fig. 6. LSTM architecture

Table 5. The notable applications of LSTM

References	Application	Journal
Ghimire et al. [95]	Solar radiation forecasting	Applied Energy
Liu [3]	Volatility forecasting	Expert Systems with Applications
Hong et al. [96]	Fault prognosis of battery systems	Applied Energy
Krishan [97]	Air quality prediction	Air Quality and Atmosphere
Zhang et al. [98]	Structural seismic prediction	Computers and Structures
Hua et al. [99]	Time Series Prediction	IEEE Communications
Zhang et al. [100]	Wind turbine power prediction	Applied Energy
Vardaan et al. [101]	Earthquake trend prediction	Electrical and Computer Engineering

LSTM has shown great potential in environmental applications, e.g., geological modeling, hydrological prediction, air quality, and hazard modeling. Due to the generalization ability of the LSTM architecture, it can be suitable for many application domains. Energy demand and consumption, wind energy industry, and solar power modeling are the other application domains of LSTM. Further investigation is essential to explore the new deep learning methods and explore the application domains, as it has been done for machine Learning methods [102–109].

3 Conclusions

Deep learning methods are fast-evolving. Some of them have advanced to be specialized in a particular application domain. However, there is a gap in research in introducing the deep learning methods and summarize the methods and application in a single paper. Consequently, this paper aims at providing a comprehensive list of the most popular deep learning methods and provide notable applications. CNN, RNN, DAE, DBNs, LSTM methods have been identified as the most popular deep learning method. The description of each deep learning method and the function of each building block of them is explained.

Acknowledgements. This publication has been supported by the Project: "Support of research and development activities of the J. Selye University in the field of Digital Slovakia and creative industry" of the Research & Innovation Operational Programme (ITMS code: NFP313010T504) co-funded by the European Regional Development Fund.

References

1. Diamant, A., et al.: Deep learning in head & neck cancer outcome prediction. Scientific Reports **9**(1) (2019)
2. Dong, Y., et al.: Bandgap prediction by deep learning in configurationally hybridized graphene and boron nitride. npj Comput. Mater. **5**(1) (2019)
3. Liu, Y.: Novel volatility forecasting using deep learning–long short term memory recurrent neural networks. Expert Syst. Appl. **132**, 99–109 (2019)
4. Ludwiczak, J., et al.: *PiPred* – a deep-learning method for prediction of π-helices in protein sequences. Scientific Reports **9**(1) (2019)
5. Matin, R., Hansen, C., Mølgaard, P.: Predicting distresses using deep learning of text segments in annual reports. Expert Syst. Appl. **132**, 199–208 (2019)
6. Nguyen, D., et al.: A feasibility study for predicting optimal radiation therapy dose distributions of prostate cancer patients from patient anatomy using deep learning. Scientific Reports **9**(1) (2019)
7. Shickel, B., et al.: DeepSOFA: a continuous acuity score for critically ill patients using clinically interpretable deep learning. Scientific Reports **9**(1) (2019)
8. Wang, K., Qi, X., Liu, H.: A comparison of day-ahead photovoltaic power forecasting models based on deep learning neural network. Appl. Energy **251** (2019)
9. Aram, F., et al.: Design and validation of a computational program for analysing mental maps: Aram mental map analyzer. Sustainability (Switzerland) **11**(14) (2019)

10. Asadi, E., et al.: Groundwater quality assessment for drinking and agricultural purposes in Tabriz Aquifer, Iran (2019)
11. Asghar, M.Z., Subhan, F., Imran, M., Kundi, F.M., Shamshirband, S., Mosavi, A., Csiba, P., Várkonyi-Kóczy, A.R.: Performance evaluation of supervised machine learning techniques for efficient detection of emotions from online content. Preprints 2019, 2019080019 https://doi.org/10.20944/preprints201908.0019.v1
12. Bemani, A., Baghban, A., Shamshirband, S., Mosavi, A., Csiba, P., Várkonyi-Kóczy, A.R.: Applying ANN, ANFIS, and LSSVM Models for Estimation of Acid Solvent Solubility in Supercritical CO2. Preprints 2019, 2019060055 https://doi.org/10.20944/preprints201906.0055.v2
13. Choubin, B., et al.: Snow avalanche hazard prediction using machine learning methods. J. Hydrol. **577** (2019)
14. Choubin, B., et al.: An ensemble prediction of flood susceptibility using multivariate discriminant analysis, classification and regression trees, and support vector machines. Sci. Total Environ. **651**, 2087–2096 (2019)
15. Dehghani, M., et al.: Prediction of hydropower generation using Grey wolf optimization adaptive neuro-fuzzy inference system. Energies **12**(2), 289 (2019)
16. Dineva, A., et al.: Review of soft computing models in design and control of rotating electrical machines. Energies **12**(6) (2019)
17. Dineva, A., et al.: Multi-label classification for fault diagnosis of rotating electrical machines (2019)
18. Farzaneh-Gord, M., et al.: Numerical simulation of pressure pulsation effects of a snubber in a CNG station for increasing measurement accuracy. Eng. Appl. Comput. Fluid Mech. **13**(1), 642–663 (2019)
19. Ghalandari, M., et al.: Investigation of submerged structures' flexibility on sloshing frequency using a boundary element method and finite element analysis. Eng. Appl. Comput.Fluid Mech. **13**(1), 519–528 (2019)
20. Ghalandari, M., et al.: Flutter speed estimation using presented differential quadrature method formulation. Eng. Appl. Comput. Fluid Mech. **13**(1), 804–810 (2019)
21. Karballaeezadeh, N., et al.: Prediction of remaining service life of pavement using an optimized support vector machine (case study of Semnan-Firuzkuh road). Eng. Appl. Comput. Fluid Mech. **13**(1), 188–198 (2019)
22. Menad, N.A., et al.: Modeling temperature dependency of oil—water relative permeability in thermal enhanced oil recovery processes using group method of data handling and gene expression programming. Eng. Appl. Comput. Fluid Mech. **13**(1), 724–743 (2019)
23. Mohammadzadeh, S., et al.: Prediction of compression index of fine-grained soils using a gene expression programming model. Infrastructures **4**(2), 26 (2019)
24. Mosavi, A., Edalatifar, M.: A Hybrid Neuro-Fuzzy Algorithm for Prediction of Reference Evapotranspiration, in Lecture Notes in Networks and Systems, pp. 235–243. Springer (2019)
25. Mosavi, A., Lopez, A., Varkonyi-Koczy, A.R.: Industrial applications of big data: state of the art survey, D. Luca, L. Sirghi, and C. Costin, Editors, pp. 225–232. Springer (2018)
26. Mosavi, A., Ozturk, P., Chau, K.W.: Flood prediction using machine learning models: literature review. Water (Switzerland) **10**(11) (2018)
27. Mosavi, A., Rabczuk, T.: Learning and intelligent optimization for material design innovation, D.E. Kvasov, et al., Editors, pp. 358–363. Springer (2017)
28. Mosavi, A., Rabczuk, T., Varkonyi-Koczy, A.R.: Reviewing the novel machine learning tools for materials design, D. Luca, L. Sirghi, and C. Costin, Editors, pp. 50–58. Springer (2018)

29. Mosavi, A., et al.: State of the art of machine learning models in energy systems, a systematic review. Energies **12**(7) (2019)
30. Mosavi, A., et al.: Prediction of multi-inputs bubble column reactor using a novel hybrid model of computational fluid dynamics and machine learning. Eng. Appl. Comput. Fluid Mech. **13**(1), 482–492 (2019)
31. Mosavi, A., Varkonyi-Koczy, A.R.: Integration of machine learning and optimization for robot learning, R. Jablonski and R. Szewczyk, Editors, pp. 349–355. Springer (2017)
32. Nosratabadi, S., et al.: Sustainable business models: a review. Sustainability (Switzerland) **11**(6) (2019)
33. Qasem, S.N., et al.: Estimating daily dew point temperature using machine learning algorithms. Water (Switzerland) **11**(3) (2019)
34. Rezakazemi, M., Mosavi, A., Shirazian, S.: ANFIS pattern for molecular membranes separation optimization. J. Mol. Liq. **274**, 470–476 (2019)
35. Riahi-Madvar, H., et al.: Comp. Anal. Soft Comput. Techn. RBF, MLP, ANFIS with MLR and MNLR Predict. Grade-control Scour Hole Geometry. Eng. Appl. Comput. Fluid Mech. **13**(1), 529–550 (2019)
36. Shabani, S., Samadianfard, S., Taghi Sattari, M., Shamshirband, S., Mosavi, A., Kmet, T., Várkonyi-Kóczy, A.R.: Modeling daily pan evaporation in humid climates using gaussian process regression. Preprints 2019, 2019070351 https://doi.org/10.20944/preprints201907.0351.v1
37. Shamshirband, S., Hadipoor, M., Baghban, A., Mosavi, A., Bukor J., Varkonyi-Koczy, A. R.: Developing an ANFIS-PSO model to predict mercury emissions in combustion flue gases. Preprints 2019, 2019070165 https://doi.org/10.20944/preprints201907.0165.v1
38. Shamshirband, S., et al.: Ensemble models with uncertainty analysis for multi-day ahead forecasting of chlorophyll a concentration in coastal waters. Eng. Appl. Comput. Fluid Mech. **13**(1), 91–101 (2019)
39. Shamshirband, S., Mosavi, A., Rabczuk, T.: Particle swarm optimization model to predict scour depth around bridge pier. arXiv preprint arXiv:1906.08863 (2019)
40. Taherei Ghazvinei, P., et al.: Sugarcane growth prediction based on meteorological parameters using extreme learning machine and artificial neural network. Eng. Appl. Comput. Fluid Mech. **12**(1), 738–749 (2018)
41. Torabi, M., et al.: A Hybrid clustering and classification technique for forecasting short-term energy consumption. Environ. Progr. Sustain. Energy **38**(1), 66–76 (2019)
42. Torabi, M., et al.: A Hybrid Machine Learning Approach for Daily Prediction of Solar Radiation, in Lecture Notes in Networks and Systems, pp. 266–274. Springer (2019)
43. Biswas, M., et al.: State-of-the-art review on deep learning in medical imaging. Front. Biosci. Landmark **24**(3), 392–426 (2019)
44. Bote-Curiel, L., et al.: Deep learning and big data in healthcare: a double review for critical beginners. Appl. Sci. (Switzerland) **9**(11) (2019)
45. Feng, Y., Teh, H.S., Cai, Y.: Deep learning for chest radiology: a review. Curr. Radiol. Reports **7**(8) (2019)
46. Griffiths, D., Boehm, J.: A Review on deep learning techniques for 3D sensed data classification. Remote Sens. **11**(12) (2019)
47. Gupta, A., et al.: Deep learning in image cytometry: a review. Cytom. Part A **95**(4), 366–380 (2019)
48. Ha, V.K., et al.: Deep learning based single image super-resolution: a survey. Int. J. Autom. Comput. **16**(4), 413–426 (2019)
49. Jiang, W., Zhang, C.S., Yin, X.C.: Deep learning based scene text detection: a survey. Tien Tzu Hsueh Pao/Acta Electronica Sinica **47**(5), 1152–1161 (2019)

50. Johnsirani Venkatesan, N., Nam, C., Shin, D.R.: Deep learning frameworks on apache spark: a review. IETE Techn. Rev. (Inst. Electron. Telecommun. Eng., India) **36**(2), 164–177 (2019)
51. Li, X., He, Y., Jing, X.: A survey of deep learning-based human activity recognition in radar. Remote Sens. **11**(9) (2019)
52. Litjens, G., et al.: State-of-the-art deep learning in cardiovascular image analysis. JACC: Cardiovasc. Imaging **12**(8P1), 1549–1565 (2019)
53. Liu, S., et al.: Deep learning in medical ultrasound analysis: a review. Engineering **5**(2), 261–275 (2019)
54. Mazurowski, M.A., et al.: Deep learning in radiology: an overview of the concepts and a survey of the state of the art with focus on MRI. J. Magn. Reson. Imaging **49**(4), 939–954 (2019)
55. Narendra, G., Sivakumar, D.: Deep learning based hyperspectral image analysis-a survey. J. Comput. Theor. Nanosci. **16**(4), 1528–1535 (2019)
56. Wang, H., et al.: A review of deep learning for renewable energy forecasting. Energy Convers. Manag. **198** (2019)
57. Wang, Y., et al.: Enhancing transportation systems via deep learning: a survey. Transp. Res. Part C: Emerg. Technol. **99**, 144–163 (2019)
58. Zhang, W., et al.: Deep learning-based multimedia analytics: a review. ACM Trans. Multimed. Comput. Commun. Appl. **15**(1s) (2019)
59. Zhou, J., et al.: Deep learning for aspect-level sentiment classification: survey, vision, and challenges. IEEE Access **7**, 78454–78483 (2019)
60. Zeiler, M.D., Fergus, R.: Visualizing and understanding convolutional networks. In: European Conference on Computer Vision. Springer (2014)
61. Szegedy, C., et al.: Going deeper with convolutions. In: Proceedings of the IEEE Conference on Computer Vision and Pattern Recognition (2015)
62. Simonyan, K., Zisserman, A.: Very deep convolutional networks for large-scale image recognition. arXiv preprint arXiv:1409.1556 (2014)
63. Krizhevsky, A., Sutskever, I., Hinton, G.E.: Imagenet classification with deep convolutional neural networks. In: Advances in Neural Information Processing Systems (2012)
64. He, K., et al.: Deep residual learning for image recognition. In Proceedings of the IEEE Conference on Computer Vision and Pattern Recognition (2016)
65. Kong, Z., et al.: Condition monitoring of wind turbines based on spatio-temporal fusion of SCADA data by convolutional neural networks and gated recurrent units. Renew. Energy **146**, 760–768 (2020)
66. Lossau, T., et al.: Motion estimation and correction in cardiac CT angiography images using convolutional neural networks. Computerized Med. Imaging Graphics **76** (2019)
67. Bhatnagar, S., et al.: Prediction of aerodynamic flow fields using convolutional neural networks. Comput. Mech. **64**(2), 525–545 (2019)
68. Nevavuori, P., Narra, N., Lipping, T.: Crop yield prediction with deep convolutional neural networks. Comput. Electron. Agric. 163 (2019)
69. Ajami, A., et al.: Identifying a slums' degree of deprivation from VHR images using convolutional neural networks. Remote Sens. **11**(11) (2019)
70. Min, S., Lee, B., Yoon, S.: Deep learning in bioinformatics. Brief. Bioinform. **18**(5), 851–869 (2017)
71. Zhu, S., et al.: Gaussian mixture model coupled recurrent neural networks for wind speed interval forecast. Energy Convers. Manag. **198** (2019)
72. Pan, B., Xu, X., Shi, Z.: Tropical cyclone intensity prediction based on recurrent neural networks. Electron. Lett. **55**(7), 413–415 (2019)

73. Bisharad, D., Laskar, R.H.: Music genre recognition using convolutional recurrent neural network architecture. Expert Syst. (2019)
74. Zhong, C., et al.: Inland ship trajectory restoration by recurrent neural network. J. Navig. (2019)
75. Jarrah, M., Salim, N.: A recurrent neural network and a discrete wavelet transform to predict the Saudi stock price trends. Int. J. Adv. Comput. Sci. Appl. **10**(4), 155–162 (2019)
76. Al Rahhal, M.M., et al.: Deep learning approach for active classification of electrocardiogram signals. Inf. Sci. **345**, 340–354 (2016)
77. Yin, Z., Zhang, J.: Cross-session classification of mental workload levels using EEG and an adaptive deep learning model. Biomed. Signal Process. Control **33**, 30–47 (2017)
78. Sun, W., Zheng, B., Qian, W.: Automatic feature learning using multichannel ROI based on deep structured algorithms for computerized lung cancer diagnosis. Comput. Biol. Med. **89**, 530–539 (2017)
79. Chen, Y., et al.: Indoor location method of interference source based on deep learning of spectrum fingerprint features in Smart Cyber-Physical systems. Eurasip J. Wirel. Commun. Netw. **2019**(1) (2019)
80. Liu, P., Zheng, P., Chen, Z.: Deep learning with stacked denoising auto-encoder for short-term electric load forecasting. Energies **12**(12) (2019)
81. Nicolai, A., Hollinger, G.A.: Denoising autoencoders for laser-based scan registration. IEEE Robot. Autom. Lett. **3**(4), 4391–4398 (2018)
82. Yue, L., et al.: Multiple auxiliary information based deep model for collaborative filtering. J. Comput. Sci. Technol. **33**(4), 668–681 (2018)
83. Roy, S.S., Ahmed, M., Akhand, M.A.H.: Noisy image classification using hybrid deep learning methods. J. Inf.Commun. Technol. **17**(2), 233–269 (2018)
84. Tan, Z., et al.: Denoised senone i-vectors for robust speaker verification. IEEE/ACM Trans. Audio Speech Lang. Process. **26**(4), 820–830 (2018)
85. Zhang, Q., et al.: Deep learning based classification of breast tumors with shear-wave elastography. Ultrasonics **72**, 150–157 (2016)
86. Wulsin, D., et al.: Modeling electroencephalography waveforms with semi-supervised deep belief nets: fast classification and anomaly measurement. J. Neural Eng. **8**(3), 036015 (2011)
87. Patterson, J., Gibson, A.: Deep Learning: A Practitioner's Approach. " O'Reilly Media, Inc." (2017)
88. Vieira, S., Pinaya, W.H., Mechelli, A.: Using deep learning to investigate the neuroimaging correlates of psychiatric and neurological disorders: methods and applications. Neurosci. Biobehav. Rev. **74**, 58–75 (2017)
89. Hassan, M.M., et al.: Human emotion recognition using deep belief network architecture. Inf. Fusion **51**, 10–18 (2019)
90. Cheng, Y., et al.: Deep belief network for meteorological time series prediction in the internet of things. IEEE Int. Things J. **6**(3), 4369–4376 (2019)
91. Yu, Y., et al.: Forecasting a short-term wind speed using a deep belief network combined with a local predictor. IEEJ Trans. Electr. Electron. Eng. **14**(2), 238–244 (2019)
92. Zheng, J., Fu, X., Zhang, G.: Research on exchange rate forecasting based on deep belief network. Neural Comput. Appl. **31**, 573–582 (2019)
93. Ahmad, M., et al.: Deep belief network modeling for automatic liver segmentation. IEEE Access **7**, 20585–20595 (2019)
94. Ronoud, S., Asadi, S.: An evolutionary deep belief network extreme learning-based for breast cancer diagnosis. Soft Comput. (2019)
95. Ghimire, S., et al.: Deep solar radiation forecasting with convolutional neural network and long short-term memory network algorithms. Appl. Energy (2019)

96. Hong, J., Wang, Z., Yao, Y.: Fault prognosis of battery system based on accurate voltage abnormity prognosis using long short-term memory neural networks. Appl. Energy (2019)
97. Krishan, M., et al.: Air quality modelling using long short-term memory (LSTM) over NCT-Delhi, India. Air Qual. Atmos. Health **12**(8), 899–908 (2019)
98. Zhang, R., et al.: Deep long short-term memory networks for nonlinear structural seismic response prediction. Comput. Struct. **220**, 55–68 (2019)
99. Hua, Y., et al.: Deep learning with long short-term memory for time series prediction. IEEE Commun. Mag. **57**(6), 114–119 (2019)
100. Zhang, J., et al.: Short-term forecasting and uncertainty analysis of wind turbine power based on long short-term memory network and Gaussian mixture model. Appl. Energy, 229–244 (2019)
101. Vardaan, K., et al.: Earthquake trend prediction using long short-term memory RNN. Int. J. Electr. Comput. Eng. **9**(2), 1304–1312 (2019)
102. Mesri Gundoshmian, T., Ardabili, S., Mosavi, A., Varkonyi-Koczy, A.: Prediction of combine harvester performance using hybrid machine learning modeling and re-sponse surface methodology, Preprints 2019
103. Ardabili, S., Mosavi, A., Varkonyi-Koczy, A.: Systematic review of deep learning and machine learning models in biofuels research, Preprints 2019
104. Ardabili, S., Mosavi, A., Varkonyi-Koczy, A.: Advances in machine learning modeling reviewing hybrid and ensemble methods, Preprints 2019
105. Ardabili, S., Mosavi, A., Varkonyi-Koczy, A.: Building Energy information: demand and consumption prediction with Machine Learning models for sustainable and smart cities, Preprints 2019
106. Ardabili, S., Mosavi, A., Dehghani, M., Varkonyi-Koczy, A., Deep learning and machine learning in hydrological processes climate change and earth systems a systematic review, Preprints 2019
107. Mohammadzadeh, D., Karballaeezadeh, N., Mohemmi, M., Mosavi, A., Varkonyi-Koczy A.: Urban train soil-structure interaction modeling and analysis, Preprints 2019
108. Mosavi, A., Ardabili, S., Varkonyi-Koczy, A.: List of deep learning models, Preprints 2019
109. Nosratabadi, S., Mosavi, A., Keivani, R., Ardabili, S., Aram, F.: State of the art survey of deep learning and machine learning models for smart cities and urban sustainability, Preprints 2019

Advances in Machine Learning Modeling Reviewing Hybrid and Ensemble Methods

Sina Ardabili[1], Amir Mosavi[2,3](✉), and Annamária R. Várkonyi-Kóczy[2,4]

[1] Institute of Advanced Studies Koszeg, Koszeg, Hungary
[2] Kalman Kando Faculty of Electrical Engineering, Obuda University, Budapest, Hungary
amir.mosavi@kvk.uni-obuda.hu
[3] School of the Built Environment, Oxford Brookes University, Oxford OX3 0BP, UK
[4] Department of Mathematics and Informatics, J. Selye University, Komarno, Slovakia

Abstract. The conventional machine learning (ML) algorithms are continuously advancing and evolving at a fast-paced by introducing the novel learning algorithms. ML models are continually improving using hybridization and ensemble techniques to empower computation, functionality, robustness, and accuracy aspects of modeling. Currently, numerous hybrid and ensemble ML models have been introduced. However, they have not been surveyed in a comprehensive manner. This paper presents the state of the art of novel ML models and their performance and application domains through a novel taxonomy.

Keywords: Machine learning · Deep learning · Ensemble models

Nomenclatures

ANN	Artificial neural network
ELM	Extreme learning machine
ML	Machine learning
SVM	Support vector machine
WNN	Wavelet neural networks
DL	Deep learning
ARIMA	Autoregressive integrated moving average
EE-ANT	Ensemble empirical with adaptive noise technology
DA-KF	Data assimilation Kalman filter-based
OSELM	Online sequential extreme learning machine
BAGNBT	Bagging-based naïve bayes trees
EEMD	Ensemble empirical mode decomposition
GOA	Grasshopper optimization algorithm
HybPAS	Hybrid of linear regression-deep neural network
TSM	Trauma Severity model
GBDT	Gradient boosting decision tree

A. R. Várkonyi-Kóczy (Ed.): INTER-ACADEMIA 2019, LNNS 101, pp. 215–227, 2020.
https://doi.org/10.1007/978-3-030-36841-8_21

EBFTM	Evidential belief function and tree-based models
DTFNN	Decision tree overfitting and neural network
ICEEMDMAN	Improved complete ensemble empirical mode decomposition method with adaptive noise
RF	Random forest

1 Introduction

Machine learning (ML) methods are reported to outperform most of the physical and statistical methods in predictive modeling in terms of accuracy, robustness, uncertainty analysis, data efficiency, simplicity, and computation cost. Thus, ML methods have gained massive popularity during the past few years in a diverse range of applications, energy, hydrology, hazard prediction, finance, economics, computational mechanics, etc. [1–9]. ML methods are numerous, and different classifications of methods have been recently given by researchers [10–13]. One of the popular classification methods is to divide the methods in three groups, i.e., single methods, hybrid methods, and ensembles [14–16].

The popular single ML methods which have been widely used include artificial neural networks (ANNs)-based methods [17–19], decision trees (DTs)-based methods [18, 20–23], support vector machines (SVM)-based methods [24–27], Bayesians-based [28–30], neuro-fuzzy-based [11, 12, 31, 32], classification and regression-based methods [33], and wavelet neural networks (WNNs)-based [12, 34]. Neuro-fuzzy methods and WNNs, although they are built upon two intelligent algorithms, have already been established as a single method. The ML methods are constantly progressing to hands-on higher performance algorithms [5, 9, 13, 35–46]. The hybrid and ensemble methods are often identified to outperform single ML methods [4, 47–56].

Ensemble and hybrid ML methods are the two major approaches toward more accurate, and reliable ML methods [57–59]. Hybrid ML models are made through integration of ML methods, with other ML methods, and/or with other soft computing, optimization techniques to improve the method in various aspects. While the ensemble methods are made using various grouping techniques such as bagging or boosting to use more than one ML classifier. It is suggested that the future success of ML highly depends on the advancement of novel ensemble and hybrids methods [32, 60–62]. Literature includes novel ML methods and various comparative analysis to identify the methods with the higher performance [63–65]. However, there is a gap in research in identifying the novel hybrid and ensemble ML methods and the applications they have been used in. Consequently, the contribution of this paper is to introduce these methods and highlight their applications.

2 Reviewing Ensemble and Hybrid ML Methods

The use of ML methods, including singles, ensembles, and hybrids, have been dramatically increasing. Figure 1 shows the ever fast-growing trend of ML methods used from 2009 up until now. The widespread application areas include engineering, mathematics, physics, astronomy, earth and planetary sciences, medicine, materials science, biochemistry, genetics and molecular biology, environmental science, social sciences, energy, chemistry, decision sciences, agricultural and biological sciences.

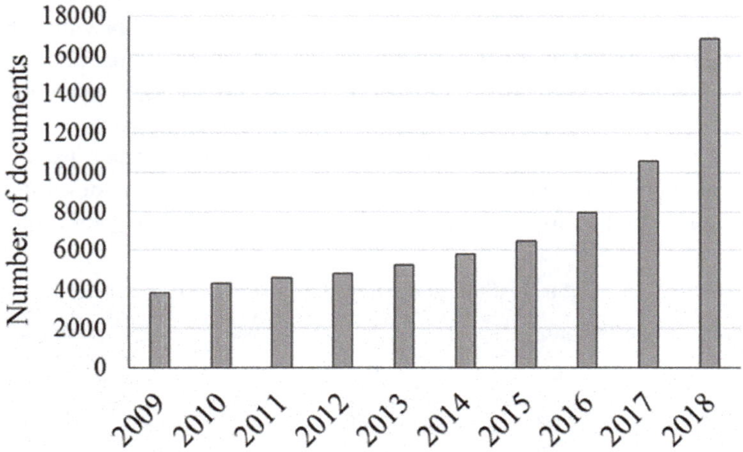

Fig. 1. The growing trend of ML methods the past decade (source: web of science)

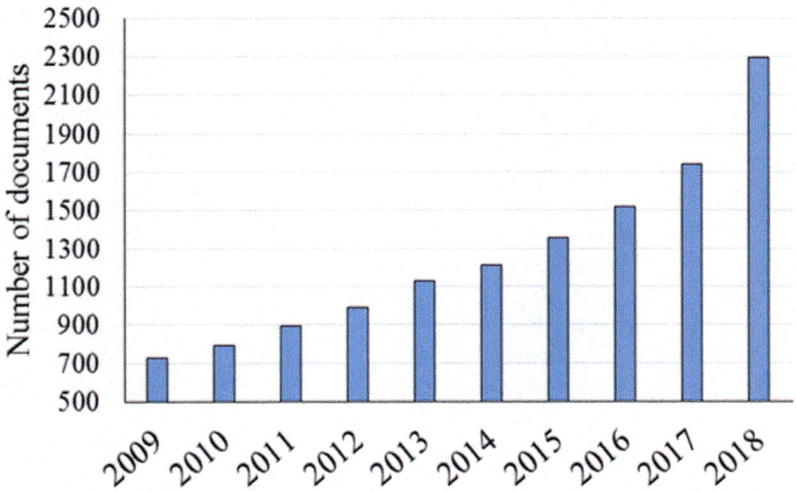

Fig. 2. The growing trend of hybrid and ensemble ML methods (source: web of science)

In the following, Fig. 2 shows the popularity of ensemble and hybrid ML models in advancing the novel method with higher performance.

2.1 Hybrid Methods

Hybrid methods combine two or more ML and/or soft computing methods for higher performance and optimum results. In fact, hybrid methods benefit from the advantage of two or more methods reach better performance. Sometimes, hybrid methods contain one unit for prediction and one unit for the optimization of the prediction unit for reaching an accurate output. Therefore, it can be claimed that hybrid methods contain different single methods and form a method with higher flexibility with a high capability compared with single methods. Hybrid methods have become more popular due to their high potential and capability. Hybrid methods are the same as a company with different employees with different expertise to achieve a single goal.

Table 1 presents the top six studies developed by hybrid methods. Table 1 contains four columns including reference, contribution of each study, the developed method and application domain for presenting the key-point of each study for a quick look.

Table 1. Studies developed by hybrid methods

References	Contributions	Methods	Application domains
[66]	To develop an adaptive hybrid methodology for the estimation of urban traffic flow	ARIMA-WNN	-Urban traffic flow -Advanced hybrid machine learning
[67]	To develop an innovative hybrid method for the estimation and optimization of wind energy	EE-ANT-WNN	-Wind power -Hybrid machine learning
[68]	To develop a novel hybrid multi-stage method to be applied in credit scoring	Hybrid multi-stage method	-Classification -Multi-stage hybrid model
[69]	To develop a novel hybrid bagging based method for the assessment of the Landslide susceptibility	Hybrid BAGNBT	-Landslide susceptibility -Hybrid machine learning
[70]	To develop a hybrid method for the estimation of electricity load	EEMD-ELM-GOA	-Electrical load -Hybrid machine learning
[71]	To develop a hybrid linear regression-based deep learning method for the estimation of poly (A) signals in DNA	HybPAS	-Hybrid machine learning -Signal processing

Hou et al. [66] developed a study in order to accurately estimate the urban traffic flow. The proposed method was an advanced hybrid wavelet neural network-integrated by autoregressive integrated moving average using a fuzzy method. The developed hybrid methods have been compared with the single form of each contributed methods in terms of mean absolute percentage error, and root mean square error. Results indicated about 60–70% improvement in the estimation accuracy of the hybrid method over the single methods. In another study, Du et al. [67] developed a novel hybrid method for the estimation and optimization of wind power. The method was including an integrated ensemble empirical with adaptive noise technology for eliminating noise and extracting the main features of original data followed by an optimized wavelet neural network to take a high estimation accuracy. Results have been compared using mean absolute percentage error. Based on the results, the hybrid method could increase the accuracy of the estimation as well as increasing the sustainability of the prediction and optimization process.

Zhang et al. [68] proposed a novel hybrid method based on feature and classifier selections in order to take an optimal classifier and feature subset in credit scoring task. Improving the accuracy of the estimation phase was performed by the use of classifier ensemble as well as using an enhanced multi-population niche genetic algorithm. Evaluations have been performed using accuracy and area under the curve factors. Based on the results, the proposed hybrid method could successfully cope with the estimation and optimization tasks over the single methods.

Pham and Prakash [69] developed a novel bagging-based naïve Bayes trees for the assessment of landslide susceptibility. The proposed hybrid method was compared with single methods including Rotation forest-based Naïve Bayes Trees, Naïve Bayes Trees, and SVM in terms of area under the curve and statistical indexes. Based on results, the proposed hybrid BAGNBT method could successfully increase the accuracy and could be introduced as the best alternative model for the assessment of landslide susceptibility over the single methods.

Wu et al. [70] developed a novel hybrid method for improving the accuracy of the electricity load forecasting. The proposed method was including an advanced integration of ELM, ensemble empirical mode decomposition, and grasshopper optimization algorithm. The hybrid method has been compared with the necessary methods by employing the test data sets in terms of root mean square error, mean absolute error and mean absolute percentage error. Based on the results, the proposed hybrid method has a higher performance and accuracy compared with the necessary methods. Albalawi et al. [71] developed a hybrid HybPAS including the integration of linear regression-deep neural network models for the estimation of ply (a) signals in DNA in the presence of sequence-based features and signal processing-based statistical as input values. Based on the results, the hybrid method could successfully increase the accuracy and performance by 30.29%.

As is clear from the above mentioned, brief literature, the hybrid methods are expanding and becoming popular due to their high potential and capability for increasing the estimation and optimizing performances. Table 2 represents a brief at the same time complete comparison for single and hybrid methods in terms of accuracy, reliability, and sustainability.

Table 2. The comparison results of Hybrid machine learning-based methods

Method	Application	Accuracy	Reliability	Sustainability	References
Hybrid WNN-ARIMA	Estimation	+++	+++	+++	[66]
WNN	Estimation	++	++	++	[66]
ARIMA	Estimation	++	+	+	[66]
Hybrid EE-ANT-WNN	Estimation	+++	+++	+++	[67]
Hybrid the optimized multi-stage method	Estimation	+++	+++	+++	[68]
BAGNBT	Estimation	+++	+++	+++	[69]
SVM	Estimation	++	+	+	[69]
NBT	Estimation	+	+	+	[69]
RFNBT	Estimation	++	++	++	[69]
EEMD-ELM-GOA	Estimation	+++	+++	+++	[70]
HybPAS	Estimation	+++	+++	+++	[71]

2.2 Ensemble Methods

Ensemble methods may use a series of ML classification trees instead of a single one. Through this technique, the accuracy of the model is substantially improved. Ensemble methods are considered as supervised learning algorithms. Ensemble methods benefit different training algorithms for increasing the training accuracy for reaching a higher testing accuracy. Ensemble method enables different training algorithms for making flexible training. Table 3 presents the top six studies developed by different Ensemble methods with different tasks.

Gorczyca et al. [72] developed a Trauma Severity model as an ensemble machine learning for risk estimation. This method has been compared with the Harborview Assessment for Risk of Mortality, Bayesian Logistic Injury Severity Score, and the Trauma Mortality Prediction Model in terms of accuracy and F-score values. Based on the results, the proposed ensemble method could successfully increase the accuracy compared with that of the base method. Results also indicated that trauma is an essential predictor for this task.

Wang et al. [73] developed an ensemble gradient boosting decision tree model for the estimation of customer churn and its relation with search Ads in the presence of two types of features including dynamic and static features. Bing Ads platform dataset was employed in order to evaluate the developed method. The results were highly promising and could successfully cope with the related task with high sustainability. Naghibi et al. [74] developed a rotation forest with decision trees as an ensemble methodology based on evidential belief function and tree-based models (EBFTM) for developing groundwater potential maps. The developed ensemble method has been compared with boosted regression tree, random forest, and classification and regression tree in terms of the performance factors receiver operating characteristics and area

Table 3. Studies developed by ensemble methods

References	Contributions	Method	Application domains
[72]	To develop an ensemble machine learning methodology for the estimation of risk	Ensemble TSM	-Risk prediction -Ensemble machine learning
[73]	To develop an ensemble model to estimate the churn in the relation of customers and search Ads	Ensemble GBDT	-Customer churn -Ensemble machine learning
[74]	To employ rotation forest with DT as an ensemble methodology based on EBF and tree-based models for developing GPM	Ensemble EBFTM	-Hydrogeology -Ensemble machine learning
[75]	To develop a novel ensemble machine learning method integrated by ELM for the estimation of significant wave height	Ensemble ICEEMDAN-ELM	-Wave height forecasting -Ensemble machine learning
[76]	To develop a novel Ensemble data assimilation Kalman filter-based for the estimation of parameters of the system's state	Ensemble DA-KF	-System's state -Ensemble machine learning
[77]	To develop an ensemble estimation model for forecasting the thyroid	Ensemble Bagging-Boosting	-Thyroid forecasting -Ensemble machine learning

under the curve. The highest performance was owned by the ensemble EBFTM method followed by random forest and boosted regression tree.

Ali and Prasad [75] developed a novel ensemble empirical mode decomposition method with adaptive noise integrated with extreme learning machine for accurately estimate the significant wave height. Evaluations have been performed geographically in the presence of the proposed method and ICEEMDAN-OSELM and ICEEMDAN-RF as the ensemble empirical mode decomposition method with adaptive noise integrated with online sequential extreme learning machine and random forest, respectively. Based on the results, the proposed ICEEMDAN-ELM method indicated the best performance over ICEEMDAN-OSELM and ICEEMDAN-RF with high accuracy and sustainability.

Yamanaka et al. [76] developed a novel ensemble method based on data assimilation-Kalman filter for the estimation of microstructure prediction using three-dimensional multi-phase-field as parameters of the system's state. Based on the results, the proposed method could successfully increase the accuracy with the lowest error, which indicates the capability of the model for directly applied to estimate the system parameters. Yadav and Pal [77] developed a novel ensemble method based on Bagging-Boosting for the estimation of women thyroid which is compared with decision tree overfitting and neural network (DTFNN) in the presence of root mean square error and mean absolute error. Based on results, the ensemble bagging-boosting method had about 65% higher accuracy over DTFNN method. As is apparent, the ensemble methods could successfully own a higher accuracy and sustainability, followed by higher attentions and trends. Therefore ensemble methods can be used by different policymakers. Accordingly, Table 4 presents a brief at the same time complete comparison for ensemble methods in terms of accuracy, reliability, and sustainability.

Table 4. The comparison results of Ensemble machine learning based methods

Method	Application	Accuracy	Reliability	Sustainability	References
Ensemble TSM	Estimation	+++	+++	+++	[72]
Ensemble GBDT	Estimation	+++	+++	+++	[73]
Ensemble EBFTM	Estimation	+++	+++	+++	[74]
RF	Estimation	+++	++	++	[74]
BRT	Estimation	++	++	++	[74]
ICEEMDAN-ELM	Estimation	+++	+++	+++	[75]
ICEEMDAN-OSELM	Estimation	++	+	+	[75]
ICEEMDAN-RF	Estimation	++	++	++	[75]
Ensemble KF-DA	Estimation	++	++	++	[76]
Ensemble bagging-boosting	Estimation	+++	+++	+++	[77]
DTFNN	Estimation	++	+	+	[77]

3 Conclusions

The ensemble and hybrid models are the new generations of machine learning. They provide higher accuracy and outperform most of the conventional machine learning models. This paper presented the state of the art of hybrids and ensembles and listed their most famous algorithms and application domains. Ensembles are currently limited to decision trees; however, it is expected to expand to other machine learning methods. Bagging and boosting methods are reported as the most popular technique to build ensembles. The hybrid models are not built through the integration of optimization and/or soft computing methods to optimize the method. Health, energy, climate change, urban informatics, and hydrology are the primary application domains of ensemble and hybrid models. Consequently, future research trends are devoted to the novel hybrid and ensemble methods [78–86].

Acknowledgments. This publication has been supported by the Project: "Support of research and development activities of the J. Selye University in the field of Digital Slovakia and creative industry" of the Research & Innovation Operational Programme (ITMS code: NFP313010T504) co-funded by the European Regional Development Fund.

References

1. Choubin, B., et al.: Snow avalanche hazard prediction using machine learning methods. J. Hydrol. **577** (2019)
2. Dehghani, M., et al.: Prediction of hydropower generation using Grey wolf optimization adaptive neuro-fuzzy inference system. Energies **12**(2) (2019)
3. Mosavi, A., Bathla, Y., Varkonyi-Koczy, A.: Predicting the future using web knowledge: state of the art survey. In: Luca, D., Sirghi, L., Costin, C. (eds.), pp. 341–349. Springer (2018)
4. Mosavi, A., Rabczuk, T.: Learning and intelligent optimization for material design innovation. In: Kvasov, D.E., et al. (eds.), pp. 358–363. Springer (2017)
5. Mosavi, A., Rabczuk, T., Varkonyi-Koczy, A.R.: Reviewing the novel machine learning tools for materials design. In: Luca, D., Sirghi, L., Costin, C. (eds.), pp. 50–58. Springer (2018)
6. Qasem, S.N., et al.: Estimating daily dew point temperature using machine learning algorithms. Water (Switzerland) **11**(3) (2019)
7. Taherei Ghazvinei, P., et al.: Sugarcane growth prediction based on meteorological parameters using extreme learning machine and artificial neural network. Eng. Appl. Comput. Fluid Mech. **12**(1), 738–749 (2018)
8. Torabi, M., et al.: A Hybrid clustering and classification technique for forecasting short-term energy consumption. Environ. Prog. Sustain. Energy **38**(1), 66–76 (2019)
9. Torabi, M., et al.: A hybrid machine learning approach for daily prediction of solar radiation. In: Lecture Notes in Networks and Systems, pp. 266–274. Springer (2019)
10. Dineva, A., et al.: Review of soft computing models in design and control of rotating electrical machines. Energies **12**(6) (2019)
11. Mosavi, A., Ozturk, P., Chau, K.W.: Flood prediction using machine learning models: literature review. Water (Switzerland) **10**(11) (2018)
12. Mosavi, A., et al.: State of the art of machine learning models in energy systems, a systematic review. Energies **12**(7) (2019)
13. Mosavi, A., Varkonyi-Koczy, A.R.: Integration of machine learning and optimization for robot learning. In: Jablonski, R., Szewczyk, R. (eds.), pp. 349–355. Springer (2017)
14. Cheng, L., Yu, T.: A new generation of AI: A review and perspective on machine learning technologies applied to smart energy and electric power systems. Int. J. Energy Res. **43**(6), 1928–1973 (2019)
15. Cheng, L., et al.: Machine Learning for Energy and Electric Power Systems: State of the Art and Prospects. Dianli Xitong Zidonghua/Autom. Electric Power Syst. **43**(1), 15–31 (2019)
16. Chou, J.S., Tran, D.S.: Forecasting energy consumption time series using machine learning techniques based on usage patterns of residential householders. Energy 709–726 (2018)
17. Chatterjee, B., et al.: RF-PUF: enhancing IoT security through authentication of wireless nodes using in-situ machine learning. IEEE Internet of Things J. **6**(1), 388–398 (2019)
18. Panesar, S.S., et al.: Machine learning versus logistic regression methods for 2-year mortality prognostication in a small, heterogeneous glioma database. World Neurosurg X **2** (2019)
19. Thomas, P.B.M., et al.: Feasibility of simple machine learning approaches to support detection of non-glaucomatous visual fields in future automated glaucoma clinics. Eye (Basingstoke) **33**(7), 1133–1139 (2019)
20. Alhajri, M.I., Ali, N.T., Shubair, R.M.: Classification of indoor environments for IoT applications: a machine learning approach. IEEE Antennas Wirel. Propag. Lett. **17**(12), 2164–2168 (2018)

21. Jamil, A., Bayram, B.: The delineation of tea gardens from high resolution digital orthoimages using mean-shift and supervised machine learning methods. Geocarto Int. (2019)
22. Maxwell, A.E., Warner, T.A., Fang, F.: Implementation of machine-learning classification in remote sensing: an applied review. Int. J. Remote Sens. **39**(9), 2784–2817 (2018)
23. Sehgal, V., et al.: Machine learning creates a simple endoscopic classification system that improves dysplasia detection in barrett's oesophagus amongst non-expert endoscopists. Gastroenterol. Res. Pract. **2018** (2018)
24. Azeem, M.I., et al.: Machine learning techniques for code smell detection: a systematic literature review and meta-analysis. Inf. Softw. Technol. **108**, 115–138 (2019)
25. Jabeen, A., Ranganathan, S.: Applications of machine learning in GPCR bioactive ligand discovery. Curr. Opin. Struct. Biol. **55**, 66–76 (2019)
26. Xu, C., Jackson, S.A.: Machine learning and complex biological data. Genome Biol. **20**(1) (2019)
27. Zhang, Z., Sejdić, E.: Radiological images and machine learning: trends, perspectives, and prospects. Comput. Biol. Med. **108**, 354–370 (2019)
28. Bock, F.E., et al., A review of the application of machine learning and data mining approaches in continuum materials mechanics. Front. Mater. **6** (2019)
29. Ekins, S., et al.: Exploiting machine learning for end-to-end drug discovery and development. Nat. Mater. **18**(5), 435–441 (2019)
30. Woldaregay, A.Z., et al.: Data-driven blood glucose pattern classification and anomalies detection: Machine-learning applications in type 1 diabetes. J. Med. Internet Res. **21**(5) (2019)
31. Najafzadeh, M., Ghaemi, A.: Prediction of the five-day biochemical oxygen demand and chemical oxygen demand in natural streams using machine learning methods. Environ. Monit. Assess. **191**(6) (2019)
32. Singh, H., Rana, P.S., Singh, U.: Prediction of drug synergy in cancer using ensemble-based machine learning techniques. Mod. Phys. Lett. B **32**(11) (2018)
33. Choubin, B., et al.: River suspended sediment modelling using the CART model: A comparative study of machine learning techniques. Sci. Total Environ. **615**, 272–281 (2018)
34. Zarkogianni, K., Athanasiou, M., Thanopoulou, A.C.: Comparison of machine learning approaches toward assessing the risk of developing cardiovascular disease as a long-term diabetes complication. IEEE J. Biomed. Health Inf. **22**(5), 1637–1647 (2018)
35. Aram, F., et al.: Design and validation of a computational program for analysing mental maps: aram mental map analyzer. Sustainability (Switzerland) **11**(14) (2019)
36. Asadi, E., et al.: Groundwater quality assessment for drinking and agricultural purposes in Tabriz Aquifer, Iran (2019)
37. Asghar, M.Z., Subhan, F., Imran, M., Kundi, F.M., Shamshirband, S., Mosavi, A., Csiba, P., Várkonyi-Kóczy, A.R.: Performance evaluation of supervised machine learning techniques for efficient detection of emotions from online content (2019), 2019080019. https://doi.org/10.20944/preprints201908.0019.v1
38. Bemani, A., Baghban, A., Shamshirband, S., Mosavi, A., Csiba, P., Várkonyi-Kóczy, A.R.: Applying ANN, ANFIS, and LSSVM models for estimation of acid solvent solubility in supercritical CO2 (2019), 2019060055. https://doi.org/10.20944/preprints201906.0055.v2
39. Mosavi, A., et al.: Prediction of multi-inputs bubble column reactor using a novel hybrid model of computational fluid dynamics and machine learning. Eng. Appl. Comput. Fluid Mech. **13**(1), 482–492 (2019)
40. Nosratabadi, S., et al.: Sustainable business models: a review. Sustainability (Switzerland) **11**(6) (2019)

41. Rezakazemi, M., Mosavi, A., Shirazian, S.: ANFIS pattern for molecular membranes separation optimization. J. Mol. Liq. **274**, 470–476 (2019)
42. Riahi-Madvar, H., et al.: Comparative analysis of soft computing techniques RBF, MLP, and ANFIS with MLR and MNLR for predicting grade-control scour hole geometry. Eng. Appl. Comput. Fluid Mech. **13**(1), 529–550 (2019)
43. Shabani, S., Samadianfard, S., Taghi Sattari, M., Shamshirband, S., Mosavi, A., Kmet, T., Várkonyi-Kóczy, A.R.: Modeling daily pan evaporation in humid climates using gaussian process regression (2019), 2019070351. (https://doi.org/10.20944/preprints201907.0351.v1)
44. Shamshirband, S., Hadipoor, M., Baghban, A., Mosavi, A., Bukor J., Varkonyi-Koczy, A. R.: Developing an ANFIS-PSO model to predict mercury emissions in combustion flue gases (2019), 2019070165. https://doi.org/10.20944/preprints201907.0165.v1
45. Shamshirband, S., et al.: Ensemble models with uncertainty analysis for multi-day ahead forecasting of chlorophyll a concentration in coastal waters. Eng. Appl. Comput. Fluid Mech. **13**(1), 91–101 (2019)
46. Shamshirband, S., Mosavi, A., Rabczuk, T.: Particle swarm optimization model to predict scour depth around bridge pier (2019). arXiv:1906.08863
47. Choubin, B., et al.: An ensemble prediction of flood susceptibility using multivariate discriminant analysis, classification and regression trees, and support vector machines. Sci. Total Environ. **651**, 2087–2096 (2019)
48. Dineva, A., et al.: Multi-label classification for fault diagnosis of rotating electrical machines (2019)
49. Farzaneh-Gord, M., et al.: Numerical simulation of pressure pulsation effects of a snubber in a CNG station for increasing measurement accuracy. Eng. Appl. Comput. Fluid Mech. **13**(1), 642–663 (2019)
50. Ghalandari, M., et al.: Investigation of submerged structures' flexibility on sloshing frequency using a boundary element method and finite element analysis. Eng. Appl. Comput. Fluid Mech. **13**(1), 519–528 (2019)
51. Ghalandari, M., et al.: Flutter speed estimation using presented differential quadrature method formulation. Eng. Appl. Comput. Fluid Mech. **13**(1), 804–810 (2019)
52. Karballaeezadeh, N., et al.: Prediction of remaining service life of pavement using an optimized support vector machine (case study of Semnan-Firuzkuh road). Eng. Appl. Comput. Fluid Mech. **13**(1), 188–198 (2019)
53. Menad, N.A., et al.: Modeling temperature dependency of oil-water relative permeability in thermal enhanced oil recovery processes using group method of data handling and gene expression programming. Eng. Appl. Comput. Fluid Mech. **13**(1), 724–743 (2019)
54. Mohammadzadeh, S., et al.: Prediction of compression index of fine-grained soils using a gene expression programming model. Infrastructures **4**(2), 26 (2019)
55. Mosavi, A., Edalatifar, M.: A hybrid neuro-fuzzy algorithm for prediction of reference evapotranspiration. In: Lecture Notes in Networks and Systems, pp. 235–243. Springer (2019)
56. Mosavi, A., Lopez, A., Varkonyi-Koczy, A.R.: Industrial applications of big data: state of the art survey. Luca, D., Sirghi, L., Costin, C. (eds.), pp. 225–232. Springer (2018)
57. Bui, D.T., et al.: Shallow landslide prediction using a novel hybrid functional machine learning algorithm. Remote Sens. **11**(8) (2019)
58. Pham, B.T., et al.: Landslide susceptibility modeling using Reduced Error Pruning Trees and different ensemble techniques: hybrid machine learning approaches. Catena **175**, 203–218 (2019)
59. Zhang, X., Mahadevan, S.: Ensemble machine learning models for aviation incident risk prediction. Decis. Support Syst. **116**, 48–63 (2019)

60. Jaiswal, A., Malhotra, R.: Software reliability prediction using machine learning techniques. Int. J. Syst. Assur. Eng. Manag. **9**(1), 230–244 (2018)
61. Khagi, B., Kwon, G.R., Lama, R.: Comparative analysis of Alzheimer's disease classification by CDR level using CNN, feature selection, and machine-learning techniques. Int. J. Imaging Syst. Technol. (2019)
62. Kumari, M., et al.: Comparative analysis of machine learning based QSAR models and molecular docking studies to screen potential anti-tubercular inhibitors against InhA of mycobacterium tuberculosis. Int. J. Comput. Biol. Drug Des. **11**(3), 209–235 (2018)
63. Bataineh, A.A.: A comparative analysis of nonlinear machine learning algorithms for breast cancer detection. Int. J. Mach. Learn. Comput. **9**(3), 248–254 (2019)
64. Manzoor, S.I., Singla, J.: A comparative analysis of machine learning techniques for spam detection. Int. J. Adv. Trends Comput. Sci. Eng. **8**(3), 810–814 (2019)
65. Odugu, K., Rajasekar, B.: Comparative analysis on supervised machine learning models for future wireless communication networks. Int. J. Innov. Technol. Explor. Eng. **8**(6), 721–723 (2019)
66. Hou Q, et al.: An adaptive hybrid model for short-term urban traffic flow prediction. Phys. A Stat. Mech. Appl. **527** (2019)
67. Du, P., et al.: A novel hybrid model for short-term wind power forecasting. Appl. Soft Comput. J. **80**, 93–106 (2019)
68. Zhang, W., He, H., Zhang, S.: A novel multi-stage hybrid model with enhanced multi-population niche genetic algorithm: An application in credit scoring. Expert Syst. Appl. **121**, 221–232 (2019)
69. Pham, B.T., Prakash, I.: A novel hybrid model of Bagging-based Naïve Bayes Trees for landslide susceptibility assessment. Bull. Eng. Geol. Env. **78**(3), 1911–1925 (2019)
70. Wu, J., et al.: A new hybrid model to predict the electrical load in five states of Australia. Energy, 598–609 (2019)
71. Albalawi, F., et al.: Hybrid model for efficient prediction of poly(A) signals in human genomic DNA. Methods (2019)
72. Gorczyca, M.T., Toscano, N.C., Cheng, J.D.: The trauma severity model: an ensemble machine learning approach to risk prediction. Comput. Biol. Med. **108**, 9–19 (2019)
73. Wang, Q.F., Xu, M., Hussain, A.: Large-scale ensemble model for customer churn prediction in search ads. Cogn. Comput. **11**(2), 262–270 (2019)
74. Naghibi, S.A., et al.: Application of rotation forest with decision trees as base classifier and a novel ensemble model in spatial modeling of groundwater potential. Environ. Monit. Assess. **191**(4) (2019)
75. Ali, M., Prasad, R.: Significant wave height forecasting via an extreme learning machine model integrated with improved complete ensemble empirical mode decomposition. Renew. Sustain. Energy Rev. 281–295 (2019)
76. Yamanaka, A., Maeda, Y., Sasaki, K.: Ensemble Kalman filter-based data assimilation for three-dimensional multi-phase-field model: Estimation of anisotropic grain boundary properties. Mater. Des. **165** (2019)
77. Yadav, D.C., Pal, S.: To generate an ensemble model for women thyroid prediction using data mining techniques. Asian Pac. J. Cancer Prev. **20**(4), 1275–1281 (2019)
78. Ardabili, S., Mosavi, A., Mahmoudi, A., Mesri Gundoshmian, T., Nosratabadi, S., Varkonyi-Koczy, A.: Modelling temperature variation of mushroom growing hall using artificial neural networks (2019)
79. Mesri Gundoshmian, T., Ardabili, S., Mosavi, A., Varkonyi-Koczy, A.: Prediction of combine harvester performance using hybrid machine learning modeling and response surface methodology (2019)

80. Ardabili, S., Mosavi, A., Varkonyi-Koczy, A.: Systematic review of deep learning and machine learning models in biofuels research (2019)
81. Ardabili, S., Mosavi, A., Varkonyi-Koczy, A.: Advances in machine learning model-ing reviewing hybrid and ensemble methods (2019)
82. Ardabili, S., Mosavi, A., Varkonyi-Koczy, A.: Building Energy information: demand and consumption prediction with Machine Learning models for sustainable and smart cities (2019)
83. Ardabili, S., Mosavi, A., Dehghani, M., Varkonyi-Koczy, A.: Deep learning and machine learning in hydrological processes climate change and earth systems a systematic review (2019)
84. Mohammadzadeh D., Karballaeezadeh, N., Mohemmi, M., Mosavi, A., Varkonyi-Koczy A.: Urban train soil-structure interaction modeling and analysis (2019)
85. Mosavi, A., Ardabili, S., Varkonyi-Koczy, A.: List of deep learning models (2019)
86. Nosratabadi, S., Mosavi, A., Keivani, R., Ardabili, S., Aram, F., State of the art survey of deep learning and machine learning models for smart cities and urban sustainability (2019)

State of the Art Survey of Deep Learning and Machine Learning Models for Smart Cities and Urban Sustainability

Saeed Nosratabadi[1], Amir Mosavi[2,3](✉), Ramin Keivani[2], Sina Ardabili[4], and Farshid Aram[5]

[1] Institute of Business Studies, Szent Istvan University, Godollo 2100, Hungary
[2] School of the Built Environment, Oxford Brookes University, Oxford, UK
amir.mosavi@kvk.uni-obuda.hu
[3] Kalman Kando Faculty of Electrical Engineering, Obuda University, Budapest, Hungary
[4] Institute of Advanced Studies Koszeg, Koszeg, Hungary
[5] Escuela Técnica Superior de Arquitectura, Universidad Politécnica de Madrid-UPM, 28040 Madrid, Spain

Abstract. Deep learning (DL) and machine learning (ML) methods have recently contributed to the advancement of models in the various aspects of prediction, planning, and uncertainty analysis of smart cities and urban development. This paper presents the state of the art of DL and ML methods used in this realm. Through a novel taxonomy, the advances in model development and new application domains in urban sustainability and smart cities are presented. Findings reveal that five DL and ML methods have been most applied to address the different aspects of smart cities. These are artificial neural networks; support vector machines; decision trees; ensembles, Bayesians, hybrids, and neuro-fuzzy; and deep learning. It is also disclosed that energy, health, and urban transport are the main domains of smart cities that DL and ML methods contributed in to address their problems.

Keywords: Deep learning · Machine learning · Smart cities · Urban sustainability · Cities of future · Internet of things (IoT) · Data science · Big data

1 Introduction

Global urbanization is growing at a fast pace [1]. In the near future, major population of the world will be moving to the cities [2]. This trend will be extremely challenging for the land use management, sustainable urban development, food supply, safety, security, and human well-being in general [3, 4].

The emerging technologies and novel concepts for smart cities have been very promising to encourage a brighter future in dealing with the cities of the future. The artificial intelligence applications such as internet of things (IoT) [5], machine learning (ML) [6–9], deep learning (DL) [10] and big data [7, 11–15], have been essential in supporting the smart cities evolution and technological advancement. Among them,

A. R. Várkonyi-Kóczy (Ed.): INTER-ACADEMIA 2019, LNNS 101, pp. 228–238, 2020.
https://doi.org/10.1007/978-3-030-36841-8_22

ML methods have been contributing to various application domains with promising results in, e.g., mobility management and monitoring, city planning, resource allocation, energy demand and consumption prediction, food supply and production prediction, air pollution monitoring and prediction, etc. [16–21].

Literature includes an adequate number of state of the art review papers and comparative analysis on the general applications of ML and DL methods [22–35]. The trends of the advancement of ML and DL methods are reported to be hybrid and ensemble methods [36–46]. Considering the smart cities research, although, there have been various surveys on the applications of artificial intelligence, ML and DL methods, an insight into the popular methods, classification of the methods, and future trend in the advancement of novel methods are not given yet [47–54]. Thus, the current research aims to fill this gap through providing the state-of-the-art of ML and DL methods used for smart cities toward a more sustainable approach for urban sustainability. To do so, a novel classification is used to identify the most popular ML and DL models and review them in individual groups according to the methods used. This paper further contributes to identifying future trends in the advancement of learning algorithms for smart cities. Unlike in other fields, e.g., atmospheric sciences and hydrology were hybrids, and ensemble ML models have increased in popularity, in the smart city domain DL applications are dominant.

2 ML and DL Models for Smart Cities

To identify the most relevant literature in the realm of using ML and DL methods for smart cities and sustainable urban development we explored the web of science (WoS) and Scopus with the following search keywords: "smart cities" or "sustainable urban development" and all the existing ML and DL methods [12]. Figure 1 represents the overall research results showing the exponential growth in using the ML and DL methods for smart cities and sustainable urban development. The search in the major

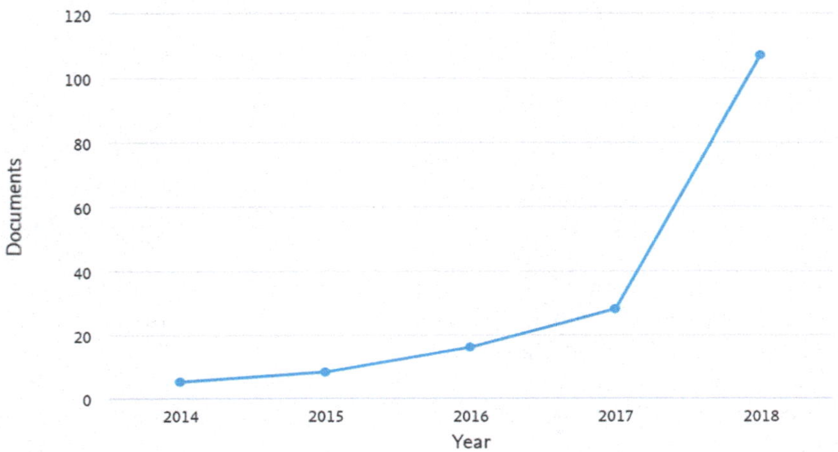

Fig. 1. The exponential growth in using ML and DL methods for smart cities and sustainable urban development (source: web of science)

research databases revealed that the popularity of machine learning is still limited in only a handful of ML and DL methods.

In the following the popular ML and DL methods are identified and reviewed in the classified tables based on the methods used.

2.1 Artificial Neural Networks in Smart Cities

As it is summarized in Table 1, several research papers have applied artificial neural networks (ANNs) in the context of smart cities. ANNs have many applications in smart cities, including hazard detection, water supply, energy, and urban transport. Ullah et al. [55], for instance, develop infrastructures for smart lightning detection system aided by ANNs. Yuan et al. [56] provide an approach to improve the stability of wind turbines in smart cities. Rojek and Studzinski [57] design a method to detect water leaks in smart cities utilizing neural networks. Pan et al. [58] and Vlahogianni et al. [59] have applied ANNs to provide solutions for urban transport in smart cities.

Table 1. Notable applications and contributions of ANNs in the smart cities

Literature references	Contributions	Application domains
Ullah et al. [55]	Utilizing artificial neural network to develop Infrastructures for the smart lightning detection system	Hazard detection
Yuan et al. [56]	Increasing the stability of wind turbines in smart cities	Energy
Rojek and Studzinski [57]	Providing a solution for water leaks detection utilizing neural networks in smart cities	Water supply and energy
Pan et al. [58]	Network traffic prediction for the smart cities using DE-BP neural network	Urban transport
Vlahogianni et al. [59]	Developing a prediction system for real-time parking in smart cities	Urban transport

2.2 Support Vector Machines

Support vector machines (SVMs) is another machine learning method that can be used to deal with smart cities' problems. SVMs have been applied in different aspects of a smart city such as water supply, Energy, evaluation and management of smart, and health domains. Table 2 elaborates the contribution and domain application of SVMs in the smart cities, where Livingston et al. [60] provide solutions to improve water utilization in smart cities utilizing SVM. Chen and Zhang [61] providing a model to evaluate green smart cities in China. Chui et al. [62] propose an innovative approach to evaluate energy sustainability in smart cities. Ultimately, Aborokbah et al. [63] and Muhammad et al. [64] utilize SVMs to provide solutions for the health industry of smart cities.

Table 2. Notable applications and contributions of SVMs in the smart cities

Literature references	Contributions	Application domains
Livingston et al. [60]	Improving water utilization in smart cities using SVM	Water supply
Chen and Zhang [61]	Providing a model to assess green smart cities in China	Evaluation and management of smart cities
Chui et al. [62]	Proposing a novel approach to evaluate energy sustainability in smart cities	Energy
Aborokbah et al. [63]	Designing an adaptive context-aware decision computing paradigm for intensive health care delivery in smart cities	Health
Muhammad et al. [64]	Designing a facial-expression monitoring system to improve healthcare in smart cities	Health

2.3 Tree-Based Models (Decision Trees)

Another ML method which has been applied to solve problems of different aspects of smart cities is Decision trees (DTs) method. As it is detailed in Table 3, researchers have applied DTs to address the issues related to businesses, air pollution, urban transport, and food to develop a smart city. Ilapakurti et al. [65] formulate adaptive edge analytics for creating a memorable customer experience and venue brand engagement for smart cities. Orlowski et al. [66] design business models to measure air quality for smart cities for using IoT (Internet of Things) and SaaS (software as a service). Mei et al. [67] design an incentive framework for an intelligent traffic system based on initiative game-theory. Vuppalapati et al. [68] design a smart dairy model using IoT sensor network measuring cattle's health issues, milk production prediction, and productivity improvement.

Table 3. Notable applications and contributions of Decision trees in the smart cities

Literature references	Contributions	Application domains
Ilapakurti et al. [65]	Formulating adaptive edge analytics for creating a memorable customer experience and venue brand engagement for smart cities	Businesses
Orlowski et al. [66]	Designing business models to measure air quality for smart cities for using IoT (Internet of Things) and SaaS (Software as a Service)	Air Pollution
Mei et al. [67]	Designing an incentive framework for an intelligent traffic system based on initiative game-theory	Urban transport
Vuppalapati et al. [68]	Designing a smart dairy model using IoT sensor network measuring cattle's health issues, milk production prediction and productivity improvement	Food

2.4 Ensembles, Bayesian, Hybrids, and Neuro-Fuzzy

In addition to ANNs, SVMs, and DTs, which have a remarkable contribution to smart cities; ensembles, Bayesian, hybrids, and neuro-fuzzy have been applied to address the issues in the domains such as energy, urban governance, evaluation and management of smart cities, and health for smart cities. Table 4 summarizes the articles that have used ensembles, Bayesian, hybrids, and neuro-fuzzy to deal with different problems in smart cities. Where Nguyen et al. [69] design a sustainable model for urban landscape evolution city. Taveres-Cachat et al. [70] propose a framework to build a zero-emission neighborhood using responsive building envelope. Ju et al. [71] design a framework to apply citizen-centered big data for governance intelligence in smart cities. Tan et al. [72] develop an adaptive neuro-fuzzy inference system approach for urban sustainability assessment. Finally, Sajjad et al. [73] provide a quality computer-aided blood analysis system to discover and count the white blood cells in blood samples. Their approach contributes to making the healthcare industry smart in the smart city (see Table 4).

Table 4. Notable applications and contributions of ensembles, Bayesian, hybrids, and neuro-fuzzy in the smart cities

Literature references	Contributions	Application domains
Nguyen et al. [69]	Designing a sustainable model for urban landscape evolution city	Evaluation and management of smart cities
Taveres-Cachat et al. [70]	Designing a framework to build a zero-emission neighborhood using responsive building envelope	Energy
Ju et al. [71]	Proposing a framework to apply citizen-centered big data for governance intelligence in smart cities	Evaluation and management of smart cities
Tan et al. [72]	Developing an adaptive neuro-fuzzy inference system approach for urban sustainability assessment	Evaluation and management of smart cities
Sajjad et al. [73]	Providing a quality computer-aided blood analysis system to the discover and count the white blood cells in blood samples	Health

2.5 Deep Learning

Deep learning methods have had numerous and various applications in developing smart cities. The body of research, aided by such methods, has contributed to different aspects of a smart city such as energy sector, health, transportation, and even management of smart cities. As it is presented in Table 5, Luo et al. [74] and Vázquez-Canteli et al. [75] utilize deep learning to provide solutions in the energy sector for the smart cities. Where Luo et al. [74] design a system for a short-term energy prediction for a smart city. Vázquez-Canteli et al. [75] develop an integrated simulation environment to manage energy intelligently. Baba et al. [76] provide a sensor network for violence detection in smart cities. Reddy and Mehta [77] propose a system for smart traffic management for smart cities. Muhammed et al. [48] and Obinikpo and Kantarci [78] applied deep learning to deal with the concerns in the health sector. Finally, Madu et al. [79] propose a framework to evaluate urban sustainability utilizing deep learning.

Table 5. Notable applications and contributions of deep learning in the smart cities

Literature references	Contributions	Application domains
Luo et al. [74]	Crafting a system for a short-term energy prediction for smart city	Energy
Baba et al. [76]	Designing a sensor network for violence detection in smart cities	Security
Reddy and Mehta [77]	Proposing a system for Smart traffic management in smart cities utilizing reinforcement learning algorithm	Urban transport
Vázquez-Canteli et al. [75]	Developing an integrated simulation environment to manage energy intelligently in smart cities	Energy
Muhammed et al. [48]	Providing a ubiquitous healthcare framework, utilizing edge computing, deep learning, big data, high-performance computing (HPC), and the Internet of Things (IoT)	Health
Obinikpo and Kantarci [78]	Providing all the applications of deep learning methods used in sensed data for prediction in smart health services	Health
Madu et al. [79]	Providing a framework to evaluate urban sustainability utilizing deep learning	Evaluation and management of smart cities

3 Discussion and Conclusions

The current paper provided a comprehensive state-of-the-art of ML and DL methods used for smart cities toward urban sustainability. A novel classification is used to identify the most popular ML and DL model and review them in individual groups according to the methods used. The ANNs, SVMs, DTs, Ensembles, Bayesians, and neuro-fuzzy methods have been seen as the most used machine learning methods. The finding revealed that ML and DL methods had remarkable contributions to the development of smart cities. It is also shown that energy, health, urban transport, evaluation and management of smart cities, water supply, businesses, air pollution, food, urban governance, security, and hazard detection are different domains of smart cities have borrowed ML and DL methods to deal with the similar problems.

Energy was the sector which has most leveraged the ML and DL methods in development of smart cities as four out of the 5 methods concerned in this study (artificial neural networks, support vector machines (SVMs), Ensembles, Bayesians, hybrids, and neuro-fuzzy, deep learning) are applied to provide different solutions for this sector. After energy, health, urban transport, and evaluation and management of smart cities are the other smart cities domains that have had most attention by the researchers in the standard fields of ML and DL methods and smart cities where at least three different ML and DL methods are applied to address their research questions. This paper also reveals an unexpected result, i.e., the immense popularity of DL methods. The DL methods have been seen dramatically popular in smart city applications mainly published in 2018 and 2019. This paper further identified future trends in the advancement of learning algorithms for smart cities. The trend in smart cities have shown to follow the trend in the overall trend which is a shift toward the advancement of the more sophisticated hybrid, ensemble and deep learning models, as also shown in [80–88].

References

1. Haase, D., et al. Global Urbanization. In: The Urban Planet: Knowledge Towards Sustainable Cities, vol. 19 (2018)
2. Galea, S., Ettman, C.K., Vlahov, D.: The Present and Future of Cities. Urban Health, p. 1 (2019)
3. Wang, S.J., Moriarty, P.: Urban health and well-being challenges. In: Big Data for Urban Sustainability, pp. 23–43. Springer (2018)
4. Nosratabadi, S., et al.: Sustainable business models: a review. Sustainability **11**(6), 1663 (2019)
5. Alavi, A.H., et al.: Internet of Things-enabled smart cities: state-of-the-art and future trends. Measurement **129**, 589–606 (2018)
6. Dineva, A., et al.: Review of soft computing models in design and control of rotating electrical machines. Energies **12**(6) (2019)
7. Ghalandari, M., et al.: Investigation of submerged structures' flexibility on sloshing frequency using a boundary element method and finite element analysis. Eng. Appl. Comput. Fluid Mech. **13**(1), 519–528 (2019)
8. Qasem, S.N., et al.: Estimating daily dew point temperature using machine learning algorithms. Water (Switzerland) **11**(3) (2019)
9. Taherei Ghazvinei, P., et al.: Sugarcane growth prediction based on meteorological parameters using extreme learning machine and artificial neural network. Eng. Appl. Comput. Fluid Mech. **12**(1), 738–749 (2018)
10. Vargas, R., Mosavi, A., Ruiz, R.: Deep learning: a review. In: Advances in Intelligent Systems and Computing (2017)
11. Mosavi, A., Lopez, A., Varkonyi-Koczy, A.R.: Industrial applications of big data: state of the art survey. In: Luca, D., Sirghi, L., Costin, C. (eds.), pp. 225–232. Springer (2018)
12. Mosavi, A., Ozturk, P., Chau, K.W.: Flood prediction using machine learning models: literature review. Water (Switzerland) **10**(11) (2018)
13. Mosavi, A,. Rabczuk, T.: Learning and intelligent optimization for material design innovation. In: Kvasov, D.E., et al. (eds.), pp. 358–363. Springer (2017)

14. Mosavi, A., Varkonyi-Koczy, A.R.: Integration of machine learning and optimization for robot learning. In: Jablonski, R., Szewczyk, R. (eds.), pp. 349–355. Springer (2017)
15. Torabi, M., et al.: A Hybrid clustering and classification technique for forecasting short-term energy consumption. Environ. Prog. Sustain. Energy **38**(1), 66–76 (2019)
16. Audu, A.R.A., et al.: An intelligent predictive analytics system for transportation analytics on open data towards the development of a smart city. In: Hussain, F.K., Barolli, L., Ikeda, M. (eds.) pp. 224–236. Springer (2020)
17. Rebelo, F., Noriega, P., Oliveira, T.: Evaluation of the concept of a smart city gamification from a user centered design perspective. In: Soares, M.M., Rebelo, F. (eds.), pp. 207–219. Springer (2020)
18. Ruzina, E.I.: From information city to smart city: Russian experience of state entrepreneurship. In: Solovev, D.B. (ed.), pp. 419–430. Springer Science and Business Media Deutschland GmbH (2020)
19. Sharifi, A.: A critical review of selected smart city assessment tools and indicator sets. J. Clean. Prod. **233**, 1269–1283 (2019)
20. Valdeolmillos, D., Mezquita, Y., Ludeiro, A.R.: Sensing as a service: An architecture proposal for big data environments in smart cities. In: Novais, P., et al. (eds.), pp. 97–104. Springer (2020)
21. Wataya, E., Shaw, R.: Measuring the value and the role of soft assets in smart city development. Cities **94**, 106–115 (2019)
22. Aram, F., et al.: Design and validation of a computational program for analysing mental maps: Aram mental map analyzer. Sustainability (Switzerland) **11**(14) (2019)
23. Asadi, E., et al.: Groundwater quality assessment for drinking and agricultural purposes in Tabriz Aquifer, Iran (2019)
24. Asghar, M.Z., Subhan, F., Imran, M., Kundi, F.M., Shamshirband, S., Mosavi, A., Csiba, P., Várkonyi-Kóczy, A.R.: Performance evaluation of supervised machine learning techniques for efficient detection of emotions from online content (2019), 2019080019. https://doi.org/10.20944/preprints201908.0019.v1
25. Bemani, A., Baghban, A., Shamshirband, S., Mosavi, A., Csiba, P., Várkonyi-Kóczy, A.R.: Applying ANN, ANFIS, and LSSVM models for estimation of acid solvent solubility in supercritical CO2. (2019), 2019060055. https://doi.org/10.20944/preprints201906.0055.v2
26. Choubin, B., et al.: Snow avalanche hazard prediction using machine learning methods. J. Hydrol. **577** (2019)
27. Choubin, B., et al.: An ensemble prediction of flood susceptibility using multivariate discriminant analysis, classification and regression trees, and support vector machines. Sci. Total Environ. **651**, 2087–2096 (2019)
28. Dehghani, M., et al.: Prediction of hydropower generation using Grey wolf optimization adaptive neuro-fuzzy inference system. Energies **12**(2) (2019)
29. Dineva, A., et al.: Multi-label classification for fault diagnosis of rotating electrical machines (2019)
30. Farzaneh-Gord, M., et al.: Numerical simulation of pressure pulsation effects of a snubber in a CNG station for increasing measurement accuracy. Eng. Appl. Comput. Fluid Mech. **13**(1), 642–663 (2019)
31. Ghalandari, M., et al.: Flutter speed estimation using presented differential quadrature method formulation. Eng. Appl. Comput. Fluid Mech. **13**(1), 804–810 (2019)
32. Karballaeezadeh, N., et al.: Prediction of remaining service life of pavement using an optimized support vector machine (case study of Semnan-Firuzkuh road). Eng. Appl. Comput. Fluid Mech. **13**(1), 188–198 (2019)

33. Menad, N.A., et al.: Modeling temperature dependency of oil - water relative permeability in thermal enhanced oil recovery processes using group method of data handling and gene expression programming. Eng. Appl. Comput. Fluid Mech. **13**(1), 724–743 (2019)
34. Mohammadzadeh, S., et al.: Prediction of compression index of fine-grained soils using a gene expression programming model. Infrastructures **4**(2), 26 (2019)
35. Mosavi, A., Edalatifar, M.: A hybrid neuro-fuzzy algorithm for prediction of reference evapotranspiration. in Lecture Notes in Networks and Systems, pp. 235–243. Springer (2019)
36. Mosavi, A., Rabczuk, T., Varkonyi-Koczy, A.R.: Reviewing the novel machine learning tools for materials design. In: Luca, D., Sirghi, L., Costin, C. (eds.), pp. 50–58. Springer (2018)
37. Mosavi, A., et al.: State of the art of machine learning models in energy systems, a systematic review. Energies **12**(7) (2019)
38. Mosavi, A., et al.: Prediction of multi-inputs bubble column reactor using a novel hybrid model of computational fluid dynamics and machine learning. Eng. Appl. Comput. Fluid Mech. **13**(1), 482–492 (2019)
39. Nosratabadi, S., et al.: Sustainable business models: a review. Sustainability (Switzerland) **11**(6) (2019)
40. Rezakazemi, M., Mosavi, A., Shirazian, S.: ANFIS pattern for molecular membranes separation optimization. J. Mol. Liq. **274**, 470–476 (2019)
41. Riahi-Madvar, H., et al.: Comparative analysis of soft computing techniques RBF, MLP, and ANFIS with MLR and MNLR for predicting grade-control scour hole geometry. Eng. Appl. Comput. Fluid Mech. **13**(1), 529–550 (2019)
42. Shabani, S., Samadianfard, S., Taghi Sattari, M., Shamshirband, S., Mosavi, A., Kmet, T., Várkonyi-Kóczy, A.R.: Modeling daily pan evaporation in humid cli-mates using gaussian process regression (2019), 2019070351. https://doi.org/10.20944/preprints201907.0351.v1
43. Shamshirband, S., Hadipoor, M., Baghban, A., Mosavi, A., Bukor J., Varkonyi-Koczy, A. R.: Developing an ANFIS-PSO model to predict mercury emissions in combustion flue gases (2019), 2019070165. https://doi.org/10.20944/preprints201907.0165.v1
44. Shamshirband, S., et al.: Ensemble models with uncertainty analysis for multi-day ahead forecasting of chlorophyll a concentration in coastal waters. Eng. App. Comput. Fluid Mech. **13**(1), 91–101 (2019)
45. Shamshirband, S., Mosavi, A., Rabczuk, T.: Particle swarm optimization model to predict scour depth around bridge pier (2019). arXiv:1906.08863
46. Torabi, M., et al.: A hybrid machine learning approach for daily prediction of solar radiation. In: Lecture Notes in Networks and Systems, pp. 266–274. Springer (2019)
47. de Souza, J.T., et al.: Data mining and machine learning to promote smart cities: a systematic review from 2000 to 2018. Sustainability (Switzerland) **11**(4) (2019)
48. Muhammed, T., et al.: UbeHealth: A personalized ubiquitous cloud and edge-enabled networked healthcare system for smart cities. IEEE Access **6**, 32258–32285 (2018)
49. Nagy, A.M., Simon, V.: Survey on traffic prediction in smart cities. Pervasive Mobile Comput. **50**, 148–163 (2018)
50. O'Dwyer, E., et al.: Smart energy systems for sustainable smart cities: current developments, trends and future directions. Appl. Energy 581–597 (2019)
51. Soomro, K., et al.: Smart city big data analytics: An advanced review. In: Data Mining and Knowledge Discovery. Wiley Interdisciplinary Reviews (2019)
52. Usman, M., et al.: A survey on big multimedia data processing and management in smart cities. ACM Comput. Surv. **52**(3) (2019)
53. Zhao, L., et al.: Routing for crowd management in smart cities: A deep reinforcement learning perspective. IEEE Commun. Mag. **57**(4), 88–93 (2019)

54. Aram, F., et al.: Design and validation of a computational program for analysing mental maps: aram mental map analyzer. Sustainability **11**(14), 3790 (2019)
55. Ullah, I., et al.: Smart lightning detection system for smart-city infrastructure using artificial neural network. Wirel. Pers. Commun. **106**(4), 1743–1766 (2019)
56. Yuan, Z., Wang, W., Fan, X.: Back propagation neural network clustering architecture for stability enhancement and harmonic suppression in wind turbines for smart cities. Comput. Electr. Eng. **74**, 105–116 (2019)
57. Rojek, I., Studzinski, J.: Detection and localization of water leaks in water nets supported by an ICT system with artificial intelligence methods as away forward for smart cities. Sustainability (Switzerland) **11**(2) (2019)
58. Pan, X., et al.: Prediction of network traffic of smart cities based on DE-BP neural network. IEEE Access **7**, 55807–55816 (2019)
59. Vlahogianni, E.I., et al.: A real-time parking prediction system for smart cities. J. Intell. Trans. Syst. Technol. Plan. Oper. **20**(2), 192–204 (2016)
60. Livingston, S.J., et al.: A hybrid approach for water utilization in smart cities using machine learning techniques. Int. J. Innov. Technol. Explor. Eng. **8**(6), 488–493 (2019)
61. Chen, L., Zhang, H.: Evaluation of green smart cities in china based on entropy weight-cloud model. Xitong Fangzhen Xuebao/J Syst Simul. **31**(1), 136–144 (2019)
62. Chui, K.T., Lytras, M.D., Visvizi, A.: Energy sustainability in smart cities: artificial intelligence, smart monitoring, and optimization of energy consumption. Energies **11**(11) (2018)
63. Aborokbah, M.M., et al.: Adaptive context aware decision computing paradigm for intensive health care delivery in smart cities—A case analysis. Sustain. Cities Soc. **41**, 919–924 (2018)
64. Muhammad, G., et al.: A facial-expression monitoring system for improved healthcare in smart cities. IEEE Access **5**, 10871–10881 (2017)
65. Ilapakurti, A., et al.: Adaptive edge analytics for creating memorable customer experience and venue brand engagement, a scented case for Smart Cities. Institute of Electrical and Electronics Engineers Inc. (2018)
66. Orlowski, C., et al.: Decision processes based on IoT data for sustainable smart cities. In: Lecture Notes in Computer Science (including subseries Lecture Notes in Artificial Intelligence and Lecture Notes in Bioinformatics), pp. 136–146. Springer (2018)
67. Mei, H., Poslad, S., Du, S.: A game-theory based incentive framework for an intelligent traffic system as part of a smart city initiative. Sensors (Switzerland) **17**(12) (2017)
68. Vuppalapati, J.S., et al.: Smart dairies-enablement of smart city at gross root level. Institute of Electrical and Electronics Engineers Inc. (2017)
69. Nguyen, T.A., et al.: Toward a sustainable city of tomorrow: a hybrid Markov-Cellular Automata modeling for urban landscape evolution in the Hanoi city (Vietnam) during 1990–2030. Environ. Dev. Sustain. **21**(1), 429–446 (2019)
70. Taveres-Cachat, E., et al.: Responsive building envelope concepts in zero emission neighborhoods and smart cities-a roadmap to implementation. Build. Environ. **149**, 446–457 (2019)
71. Ju, J., Liu, L., Feng, Y.: Citizen-centered big data analysis-driven governance intelligence framework for smart cities. Telecommun. Policy **42**(10), 881–896 (2018)
72. Tan, Y., et al.: Adaptive neuro-fuzzy inference system approach for urban sustainability assessment: a China case study. Sustain. Dev. **26**(6), 749–764 (2018)
73. Sajjad, M., et al.: Leukocytes classification and segmentation in microscopic blood smear: a resource-aware healthcare service in smart cities. IEEE Access **5**, 3475–3489 (2017)
74. Luo, H., et al.: A short-term energy prediction system based on edge computing for smart city. Future Gener. Comput. Syst. **101**, 444–457 (2019)

75. Vázquez-Canteli, J.R., et al.: Fusing TensorFlow with building energy simulation for intelligent energy management in smart cities. Sustain. Cities Soc. **45**, 243–257 (2019)
76. Baba, M., et al.: A sensor network approach for violence detection in smart cities using deep learning. Sensors (Switzerland) **19**(7) (2019)
77. Reddy, D.V.S., Mehta, R.V.K.: Smart traffic management system for smart cities using reinforcement learning algorithm. Int. J. Recent Technol. Eng. **7**(6), 12–15 (2019)
78. Obinikpo, A.A., Kantarci, B.: Big sensed data meets deep learning for smarter health care in smart cities. J. Sens. Actuator Netw. **6**(4) (2017)
79. Madu, C.N., Kuei, C.H., Lee, P.: Urban sustainability management: A deep learning perspective. Sustain. Cities Soc. **30**, 1–17 (2017)
80. Ardabili, S., Mosavi, A., Mahmoudi, Gundoshmian, T.M., Nosratabadi, S., Varkonyi-Koczy, A.: Modelling temperature variation of mushroom growing hall using artificial neural networks (2019)
81. Gundoshmian, T.M., Ardabili, S., Mosavi, A., Varkonyi-Koczy, A., Prediction of combine harvester performance using hybrid machine learning modeling and response surface methodology (2019)
82. Ardabili, S., Mosavi, A., Varkonyi-Koczy, A.: Systematic review of deep learning and machine learning models in biofuels research (2019)
83. Ardabili, S., Mosavi, A., Varkonyi-Koczy, A., Advances in machine learning model-ing reviewing hybrid and ensemble methods (2019)
84. Ardabili, S., Mosavi, A., Varkonyi-Koczy, A.: Building Energy information: demand and consumption prediction with Machine Learning models for sustainable and smart cities (2019)
85. Ardabili, S., Mosavi, A., Dehghani, M., Varkonyi-Koczy, A.: Deep learning and machine learning in hydrological processes climate change and earth systems a systematic review (2019)
86. Mohammadzadeh D., Karballaeezadeh, N., Mohemmi, M., Mosavi, A., Varkonyi-Koczy A.: Urban train soil-structure interaction modeling and analysis (2019)
87. Mosavi, A., Ardabili, S., Varkonyi-Koczy, A., List of deep learning models (2019)
88. Nosratabadi, S., Mosavi, A., Keivani, R., Ardabili, S., Aram, F.: State of the art survey of deep learning and machine learning models for smart cities and urban sustainability (2019)

Improvement of Abstract Reasoning in Teaching Computer Science at Higher Education Level

Teréz A. Várkonyi(✉), Tibor Gregorics, and András Nagy

Faculty of Informatics, Eötvös Loránd University, Budapest, Hungary
{treszka, gt, nagyandras95}@inf.elte.hu

Abstract. Nowadays, abstract reasoning is a key competence in computer science. It strongly affects mathematical and programming ability and it needs to be improved. The improvement, at higher education level, is usually connected to some knowledge transfer where present tendencies and needs are introduced. In programming, worldwide trends are object orientation and component-based programming. In this paper, the authors introduce new features of a template class library which has already been applied successfully to teach object-oriented programming. The library requires abstract understanding of problems and helps teachers to show relevant programming techniques to the students.

Keywords: Abstract reasoning · Object orientation · Component-based · Library · Programming · Teaching computer science

1 Introduction

Abstract reasoning has a very important role in computer science. It strongly influences mathematical and programming skills. It helps to discover and recognize new things and conduces to structured way of thinking [1]. Because of its importance, it is worthy to start its improvement in childhood, and continue also in adulthood. At the beginning of adulthood, higher level education may contribute much to the improvement though indirectly: by tasks that also help to get concrete knowledge in a certain area.

To help students in this manner, in computer science, there are several possibilities, but in this paper two of them are mainly emphasized: teaching object-oriented programming [2] and component-based programming [3].

In component-based programming, pre-written libraries and templates are given to be used. The libraries help to spare much time and effort, that is why they are widely used [4–6], but they can also limit the freedom of the programmer. For using them, it is necessary to understand the problem to be solved and recognize where the components of e.g. a library could be applied effectively in a solution. This understanding needs abstraction already at the level of planning.

The second possibility to improve abstract reasoning at higher education level is teaching object-oriented programming. This type of programming is used worldwide and is an integral part of the curriculum. The central idea of object orientation is

A. R. Várkonyi-Kóczy (Ed.): INTER-ACADEMIA 2019, LNNS 101, pp. 239–248, 2020.
https://doi.org/10.1007/978-3-030-36841-8_23

classification of objects based on their common properties and operations. This classification needs abstract understanding.

The two above possibilities (component-based and object-oriented programming) alone help to improve abstract reasoning. At the same time, in many cases, libraries are too complex to come up in the classroom, especially at the beginning of the curriculum. Fortunately, by creating a simpler class library, which is easily understandable, both possibilities can be introduced to the students.

At Eötvös Loránd University, object orientation and libraries have already been joined in a template class library [7, 8] with which the authors aim to introduce both topics for the students and try to point at the necessity of abstract reasoning, too. With the template class library, any algorithm which contains loop may be restructured into a sequence of basic algorithmic patterns (e.g. summation, linear search, counting, etc.) [9] and any task might be described in an abstract level. The only question that has to be answered for every problem is: which algorithmic patterns it needs for the solution. The usage of the library is already taught, but from programming standpoint, it still needs corrections and extensions to make it capable to solve wider variety of task.

In this paper, authors introduce new parts and modifications for the library and show how they influence problem solving. The paper is organized as follows. Section 2 overviews the template class library. Section 3 explains the logic of it. Section 4 presents the new parts of the library. Section 5 concludes the paper.

2 The Template Class Library

Though the basic algorithmic patterns (maximum search, summation, counting, linear search, and selection) differ in detail, their general algorithm is very similar. Thus, the general algorithm can be put in an abstract base class called *Procedure* and the classes of the concrete algorithmic patterns are inherited from it. As all the patterns use loop and they process some kind of collection, another class group is used. This class group is responsible for traversing collections through enumerators. Enumerators are special iterators that cannot modify the items of the collection [10]. They have four main operations: *first()*, *next()*, *current()*, and *end()*. The four main operations are present in an interface class (*Enumerator*) as virtual methods. Then the concrete enumerator types which are inherited from the interface class override and concretize them. Two main enumerator types are introduced in [7], one for arrays and one for sequential input files.

The two class groups (patterns and enumerators) are connected with association [2], as class *Procedure* contains a pointer of *Enumerator* and uses its operations. Through runtime polymorphism [2], *Procedure* can handle the enumerator by calling the proper operations. The UML class diagram of the library can be seen in Fig. 1.

The operation of the library relies on the *Template method*, the *Strategy*, and the *Iterator* design patters [11]: the general algorithm of the algorithmic patterns is implemented in one main method called *run()*. This main method contains one loop through which the collection to be traversed is enumerated. Before the loop, method *run()* calls virtual method *init()* to initialize the parameters of the concrete pattern. It also calls method *first()* to set the first element of the collection to be the current item of the enumerator. As in some cases the algorithmic patterns have to use an already started

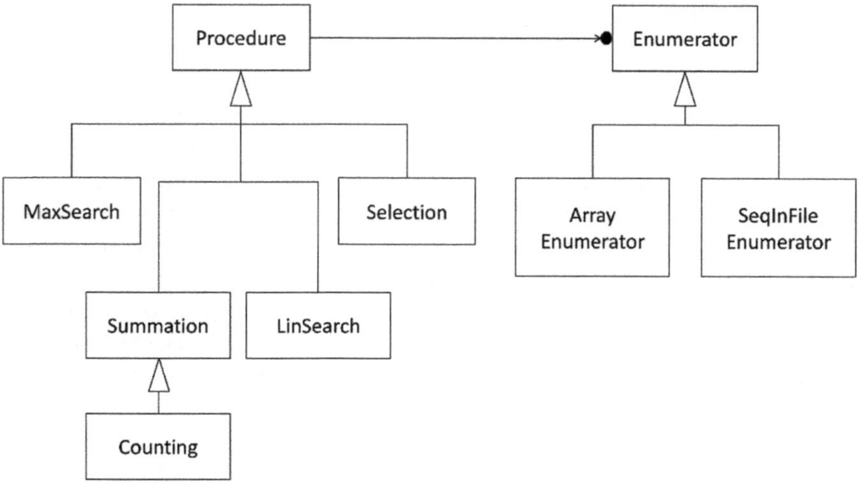

Fig. 1. UML class diagram of the template class library.

enumerator, method *first()* has to be virtual to be able to override its algorithm to continue the enumeration from the actual element instead of restarting it. After initializing the parameters, method *run()* gets into the loop. The loop runs as long as the enumerator has not ended and some other condition (described in virtual method *whileCond()*) is satisfied. In the loop, virtual method *body()* is called to run the typical algorithm of the concrete algorithmic pattern. In the loop, the next element of the collection is also selected by the enumerator.

Method *run()* uses several virtual methods which are pattern-dependent: they can be overridden in the inherited classes so that the patterns could be implemented correctly. The C++ algorithm of procedure *run()* is the following:

```
init();
first();
while (!enumerator.end() &&
        whileCond(enumerator.current()))
{
  body(enumerator.current());
  enumerator.next();
}
```

An important remark for the library: it should be used for any type of data. Thus, all the classes have to be templates. A template parameter has to be given to define what type of data the enumerator enumerates, otherwise the classes of the algorithmic patterns cannot handle the enumerators.

3 Abstract Reasoning with the Library

In this section, the logic of the template class library is explained through an example and then, its possible effects on the abstract reasoning are presented with some other advantages. Consider the following problem. Peter stores his grades of different courses in a text file, one course per line. A grade consists of a one-word comment and the actual mark. Peter would like to print out the average of his marks for each course.

For the solution, two things have to be realized: what kind of algorithmic pattern and what kind of enumerator are needed. At the highest level, courses are enumerated from a text file by a sequential input file enumerator. The enumerator enumerates two important data for each course: its name and the average of the marks (they are stored together in a record called *Statistics*). From the enumerator, the elements get to the console with the help of the Summation algorithmic pattern, as initially the console is empty and some text is added to it in each step of the main loop. The only problem is that in the file, not averages are stored but marks with comments (the name of the course can be extracted very easily).

To get one average from several grades, the reading operator of record *Statistics* has to be overloaded. In the reading operator, the grades of that line have to be summed up and counted, and at the end, the two numbers have to be divided. The adding and counting are done together by another Summation and the result is stored in a record called *Result* (it has two attributes). This Summation uses a new enumerator type, called *StringStreamEnumerator*, which is going to be explained in the next section. The enumerator enumerates items of record *Grade* (with a text and a numeric attribute), but in the calculation, only its numeric part is used.

As a summary, for the solution, two Summations are needed. The first one processes the items of a sequential input file enumerator of course names and mark averages. The second one calculates with grades and results in two numbers: the sum and the count of the numeric part of the grades.

As the above explanation shows, for the library-based solution, the whole program has to be structured and the components (algorithmic patterns) in the structure have to be planned before starting the actual coding.

It is important to emphasize this, because at the beginning of the studies at a university, in many cases, the students do not look at a programming tasks like this. As the years of teaching experience shows, most of the students just start to code and they discover the next steps of the solution during working on it. (Discovering the next steps includes fixing errors when they occur.) It makes their solution unique, but unfortunately, it may result in wrong or inefficient programs. On the other hand, by forcing the pre-planning of the components, they learn how to see through a problem quickly and as they have an idea of the solution, the time of the coding reduces. They can also recognize many possible weaknesses of their programs before starting to write it, as all of the algorithmic patters have typical vulnerabilities: e.g. the algorithm of the Maximum search cannot be used for empty enumerators. Another important advantage is that by learning the correct and most efficient algorithm of the patterns (through the library), the students start to create correct and more efficient programs.

Overall, the template class library can be considered as a tool to improve the programming skills of the students. It requires the abstract reasoning to see through the programming task. As a result, the students may solve tasks more quickly and their solution may become more trustworthy. They might also learn three important steps of creating programs: discovering the main components, handling their weaknesses, and applying correct algorithms.

4 Results

In this section, the new elements and modifications of the template class library are presented.

First of all, the structure of class *Summation* has been modified. Previously, the class was too general, method *add()* was not constant, thus any code could be written inside. The initial value of the result of the summation had to be defined by the user instead of that the summation itself set the result to the neutral element of the adding operator through method *init()*. It was not also concretized by separate method how to transform the elements of the enumerator to add them to the result. The deficiencies made the library insecure, because it could be used out of its purpose like this. In the new version, three new virtual methods are given that can be overridden by the user. Method *neutral()* is responsible for setting the initial value of the result. *func()* determines how to transform the current element of the enumerator to be added to the result. Finally, *add()* shows how two values can be added to each other to calculate the result. It is necessary, because in general, the adding operator is not defined for non-numeric data. The C++ definition of the class is the following:

```
template < typename Item , typename Value = Item >
class Summation : public Procedure<Item , Value>
{
 private :
  Value _result ;
 protected :
  void init() final override { _result = neutral(); }
  void body(const Item& e) final override {
     if (cond(e)) _result = add(result, func(e));
  }
  virtual Value func(const Item& e) const = 0;
  virtual Value neutral() const = 0;
  virtual Value add(const Value& a, const Value& b) const = 0;
  virtual bool cond(const Item& e) const { return true; }
 public :
  Summation() {}
  Summation(const Value &v): _result(v) {}
  Value result() const { return _result; }
};
```

As second modification of the library, two new enumerator types have been added to the already existing ones. The first one is an interval enumerator. It enumerates integers between two bounds. It is very useful, because in several cases, programs calculate with integers and it is unnecessary to store them in a container: only the upper and lower bounds have to be defined. A simple example where an interval enumerator might be useful is the task of calculating the factorial of a natural number. In this case, the given number is the upper bound of the interval while the lower bound equals 2. For the calculation, *Summation* algorithmic pattern is needed, where the adding operator is the multiplication, the neutral number of the operator is number 1, and the elements of the enumerator are not transformed (method *func()* does nothing).

The other new enumerator type deals with stringstreams. Stringstreams are C++ specific objects containing sequence of characters. They are very useful when, for example, varying-length data stored line-by-line in a text file is to be processed. It helps to work with files used as smaller databases. With the help of this enumerator, after getting one line of the file through method *getline()*, and putting it into a stringtream, the enumerator can extract the data in it one by one. Short example for the usage of the *StringStreamEnumerator* is mentioned in the previous section. Now, the C++ solution for that problem is shown.

In the solution, first, record *Grade* has to be created with a reading operator.

```
struct Grade {
  std::string comment;
  int mark;
  friend std::istream& operator>>(std::istream& in, Grade &gr);
};
std::istream& operator>>(std::istream& in, Grade &gr ) {
  in >> gr.comment >> g.mark;
  return in;
}
```

After that, as two things are calculated in one line (sum and count), another structure (*Result*) is needed with two corresponding attributes and the necessary constructors. The other important step is creating the custom class which executes the concrete Summation (*SumUp*). For class *SumUp*, two template parameters have to be declared: *Grade* which is the type of the enumerator and *Result* as the type of the result of the summation. In the class definition, only the three main virtual methods have to be overridden.

```
struct Result {
   int sum;
   int count;
   Result(int s, int c): sum(s), count (c) {}
   Result(): sum(0), count (0) {}
};
class SumUp : public Summation<Grade, Result> {
   Result func(const Grade& e) const override {
      return Result (e.mark, 1);
   }
   Result neutral() const override { return Result(); }
   Result add(const Result& a, const Result& b) const override {
      return Result(a.sum + b.sum, a.count + b.count);
   }
};
```

The last step for processing one line of the file is a last record (*Statistics*) which describes the useful information of one line in the file: the name of the course and the average of the marks. This record needs a reading operator, too, which after getting the name of the course, puts the rest of the line in a stringstream, creates an enumerator, and through class *SumUp* processes the enumerated grades for the average calculation. For the *StringStreamEnumerator*, one template parameter is needed (*Grade*) which denotes the type of its elements.

```
struct Statistics
{
   std::string name;
   double avg;
   friend std::istream& operator>>(std::istream& in,
            Statistics &st);
};
std::istream& operator>>(std::istream& in, Statistics &st ) {
   in >> st .name;
   std::string line;
   getline(in, line, '\n ');
   std::stringstream ss(line);
   StringStreamEnumerator<Grade> ss_enor (ss);
   SumUp sup;
   sup.addEnumerator(&ss_enor);
   sup.run();
   st.avg = double(sup.result().sum)/sup.result().count;
   return in;
}
```

Other details of the solution are going to be explained together with the next modification of the library.

The third main modification for the library affects class *Summation* again. This class mainly deals with summing up data, where only the adding operator and its neutral element have to be defined. Unfortunately, there are special cases, when these two properties cannot be implemented so easily. One such case is writing a sequence to an output stream (e.g. to the console or to a text file), the other one is building up a vector from the enumerated data. For these two cases, template specializations of class *Summation* have been created. Template specialization means that for special template parameters, separately defined algorithm is used. For the vector specialization, the adding operator is the *push back()* method of the vectors, the neutral element of which is the empty vector (but is does not need to be declared as in C++ every vector is empty initially). The only virtual function which has to be overridden for the concrete tasks is method *func()*.

The other template specialization is about writing sequences to an output stream. This specialization needs a parametrized constructor to know where to write the desired data. Otherwise, its adding operator is the default writing operator of C++, neutral element is not needed, and method *func()* always transforms the elements of the enumerator to string. By this specialization, the solution of the above example with Peter and his grades can be given as follows. A custom class inherited from *Summation* has to be given (*Print*) which is responsible for writing the statistics to the console. It needs one constructor which calls the parametrized constructor of the template specialization with the memory address of the console. In the class description, only method *func()* is overridden. For using the codes of the template specialization, *std:: ostream* has to be given as second template parameter:

```
class Print : public Summation<Statistics, std::ostream>
{
   std::string func(const Statistics& e) const override {
      return e.name + " : " + to_string(e.avg) + "\n";
   }
 public :
   Print() : Summation<Statistics, std::ostream>(&cout) {}
};
```

The last step of the solution is the main program, where the path of the input file is given and the main classes are instantiated. The enumerator object needs the proper template parameter again:

```
int main()
{
   SeqInFileEnumerator<Statistics> enor("input.txt");
   Print pr;
   pr.addEnumerator(&enor);
   pr.run();
   return 0;
}
```

The fourth modification of the library affects the inheritance of the classes. In the previous version, there were no restrictions for the inheritance, thus it was possible to create new children of class *Procedure* with any new methods. It made possible to use the library out of its purpose. To avoid this weakness, the constructor of *Procedure* has become private and the classes of the algorithmic patterns have become friend classes. Like this, the patterns can call the constructor of Procedure, even if it is private, but other classes that are not marked as friends cannot. If they cannot call the constructor of their parent, then they cannot be created.

The next modifications of the library are about inheritance again. First of all, to avoid users to override those methods that are responsible for the main logic of the library, keyword *final* has been introduced. The keyword does not effect the visibility of the methods in the parent classes, but it blocks the children classes so that they cannot override them. Second, to make the inheritance more secure, keyword *override* has been applied, too. If *override* is not used and the method description in a child does not match the method description of the parent, the compiler handles the method as a new function or procedure. If the keyword is used, and description in the parent and the child are not the same, the compiler handles it as an error and makes the inheritance more secure. This keyword has an important educational purpose. It would not be necessary, as the library is written correctly, but helps the students to get used to its usage.

The last modification is about safer compiling. Before the C++11 standard, there was not any notion for pointers pointing nowhere. Constant 0 had to be used, which the compiler could identify as number 0. With the new standard, keyword *nullptr* has been introduced which helps the compiler to compile the code more unambiguously.

5 Conclusions

In this paper, extensions and modifications are presented for a template class library. The library has already been used successfully at higher level education. It helps to improve the abstract reasoning of students while introduces relevant knowledge, like object orientation. The extensions make sure that the library can be used to solve wider variety of tasks. The modifications make the usage more secure and help the students to program more elegantly.

Acknowledgement. The project has been supported by the European Union, co-financed by the European Social Fund (EFOP-3.6.3-VEKOP-16-2017-00001).

References

1. Hansen, N., Koudenburg, N., Hiersemann, R., Tellegen, P.J., Kocsev, M., Postmes, T.: Laptop usage affects abstract reasoning of children in the developing world. Comput. Educ. **59**, 989–1000 (2012)
2. Rumbaugh, J., Blaha, M., Premerlani, W., Eddy, F., Lorensen, W.: Object-Oriented Modeling and Design. Prentice-Hall, Englewood Cliffs, NJ (1991)
3. Szyperski, C., Gruntz, D., Mure, S.: Component Software: Beyond Object-oriented Programming. ACM Press - Pearson Educational, London (2002)
4. Kirk, B.S., Peterson, J.W., Stogner, R.H.: libMesh: a C++ library for parallel adaptive mesh refinement/coarsening simulations. Eng. Comput. **22**(3–4), 237–254 (2006)
5. Blumenthal, D.B., Bougleux, S., Gamper, J., Brun, L.: GEDLIB: A C++ Library for graph edit distance computation. In: Conte, D., Ramel J.-Y., Foggia, P. (eds.) Graph-Based Representations in Pattern Recognition 2019. Springer International Publishing (2019)
6. Carpentier, J., Saurel, G., Buondonno, G., Mirabel, J., Lamiraux, F., Stasse, O., Mansard, N.: The Pinocchio C++ library: a fast and flexible implementation of rigid body dynamics algorithms and their analytical derivatives. In: 2019 IEEE/SICE International Symposium on System Integration (SII), pp. 614–619. IEEE Press, New York (2019)
7. Gregorics, T.: Analogous programming with a template class library. Teach. Math. Comput. Sci. **10**(1), 135–152 (2012)
8. The template class library. https://people.inf.elte.hu/gt/oep/library.zip
9. Ginat, D., Menashe, E., Taya, A.: Novice difficulties with interleaved pattern composition. In: Proceedings 6th International Conference on Informatics in Schools: Situation, Evolution, and Perspectives, pp. 57–67. Springer, Heidelberg (2013)
10. Kreher, D.L., Stinson, D.R.: Combinatorial Algorithms: Generation, Enumeration, and Search. CRC Press, London (1998)
11. Gamma, R., Helm, R., Johnson, R., Vlissides, J., Booch, G.: Design Patterns: Elements of Reusable Object-Oriented Software. Addison-Wesley Professional Boston (1994)

Measurement, Monitoring, and Identification

About the Size of Photon (Again)

Ryszard Jablonski(✉)

Faculty of Mechatronics, Warsaw University of Technology, Warsaw, Poland
yabu@mchtr.pw.edu.pl

Abstract. Single photons are essential for development of modern technologies and there are still many fundamental features investigated now, related to the measurement of photons. Photons are often referred to as particles, but they are light quanta and depending on the situation, sometimes the particle aspect is useful and sometimes the wave aspect is. As a photon gets closer to measured object, the chance of it interacting becomes greater.

The first step of our research was to generate a stable photon beam. This photon beam was used as a measuring tool for measuring photon transmission through micro-gaps and micro-bores versus their dimension. Photon' cross section varies depending on what it is interacting with, also depends upon the size of the object.

Then, the experiments on the heterodyning the light beams were carried out. For the intensity of photon beam so low that the distance between photons exceeds wavelength, such photons should not overlap, but in fact beatings were recorded. This test allows to assume that actual length of photon is much larger than wavelength.

The above considerations refer to 2D space, but real objects are 3D and usually of very compound character. All experiments on photons are of great value. Usually they are of very specialized character, but nevertheless, they always bring us closer to the general theory of interaction between photon and object.

Keywords: Photon size. photon beam · Single photon · Heterodyning · Cross section · Elementary particle

1 Introduction

Light is almost involved in all fields of science and daily life of everyone, and yet light's true nature has eluded us for century. Single photons are essential for fundamental study of quantum mechanics and the development of photonic quantum technologies, but again the advances in understanding of quantum mechanics, occurring over the past years, show that there are still fundamental features related to the measurement of photons, that have been under investigated until now.

Most interpretations of quantum mechanics are focused on explaining the phenomenon of the wave function collapse, ignoring not only the usefulness of such interpretations in applying to actual systems, but also a multitude of other non-trivial phenomena observable in different scientific researches.

A. R. Várkonyi-Kóczy (Ed.): INTER-ACADEMIA 2019, LNNS 101, pp. 251–259, 2020.
https://doi.org/10.1007/978-3-030-36841-8_24

Single-photon sources and detectors are widely described in review article [1]. Research in this field is progressing rapidly, but in engineering applications, rather photon beam is used and as a detector, in most cases, Photo Multiplier Tube (PMT). Knowing the photon structure is very important, but in metrology (at the present stage of knowledge) the wave form of the photon is of the greatest importance. Photon is a measuring tool and either the energy represented by the number of photons or phase relationships are of the greatest importance. In the nano- and microstructural studies, the most important is the interaction of the photon beam with the object, photon transmission and phase relations.

2 Photon and Photon Beam

"The photon is a type of elementary particle, the quantum of the electromagnetic field and carries energy proportional to the radiation frequency but has zero rest mass" [2]. Going deeper into the issues, we can find that photon always moves at the speed of light within a vacuum and that photon is having bosonic statistics, weak electromagnetic, gravity interactions, stable lifetime, 0 electric charge, spin 1, parity -1, C parity -1 and also that is a force carrier (in particle physics).

Does this definition help in determining the size of the photon, so that the obtained result can be used in photonic measurements of micro and nanostructures? My answer is: not much, but it indicates that photon can behave differently, depending on the environment.

In scientific publications there is a multitude of papers about "how big is the photon", and the quantitative answers vary very much, depending on approach. Let me start with defining what we mean by photon (1) and a photon beam (2).

As an author/co-author of several works [3–10] on the properties of photons and their use as a measurement tool, I have adopted a practical approach:

(1) It is generally accepted that light, including laser beam, consists of photons. At relatively high intensity light can be considered as a classical electromagnetic wave. The registration of the light of low intensity with a broadband photodetector gives short electrical impulses, which usually are compared with the time interval of registration of an individual photon. Hence, the use of the word "photon" is quite consistent with the generally used scientific and technical terminology.
(2) Although the concept of a single photon was proposed by Plank, a true single-photon source was created by utilizing a cascade transition within mercury atoms [11]. From then, this term is often used in many articles concerning "single photon" subject. Its meaning is simple and natural: light beams of intensity so small, that the average distance between photons considerably exceeds the size of the photon in the Bohr model, which means that the length of the wave (and photons) do not overlap. This interpretation raises our doubts.

The performed experiments have shown that in the laboratory conditions "a single photon" mode have been achieved and we claim that photon has a needle-like shape.

We began our research with the generation of a stable photon beam. It was obtained from laser sources by strong attenuation of their intensity, either by using pinholes and diaphragms or polarizers. The photons are counted by PMT working in Single Photon Counting mode, as it is shown in Figs. 1 and 2 respectively.

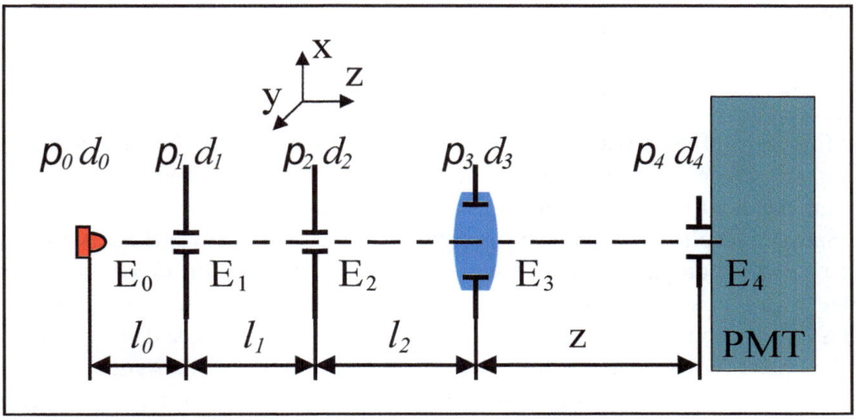

Fig. 1. Photon beam generator (attenuation about 1017 times); p0 - light source (laser), d0 - light beam diameter, p1, p2 – pinholes (diameters d1 = d2 = 5um), p3 – diaphragm of adjustable diameter, l0, l1, l2, z - distances (10–100 mm, as shown on figure), d4 - entrance pupil of PMT diameter, E0 ÷ E3 - photon field amplitude in respective part of generator, E4 - photon field amplitude on the photo cathode of PMT. Coordinate z is coaxial with the photon beam.

With photon beam generator we could obtain the pulses containing from 50 to 516 photons (s = 2%, for pulses 0.16 s and sampling rate 0.0032 s). Figure 2 presents photon distribution for pulses 516 photons, duration 0.16 s and 0.0032 s sampling rate [4, 6].

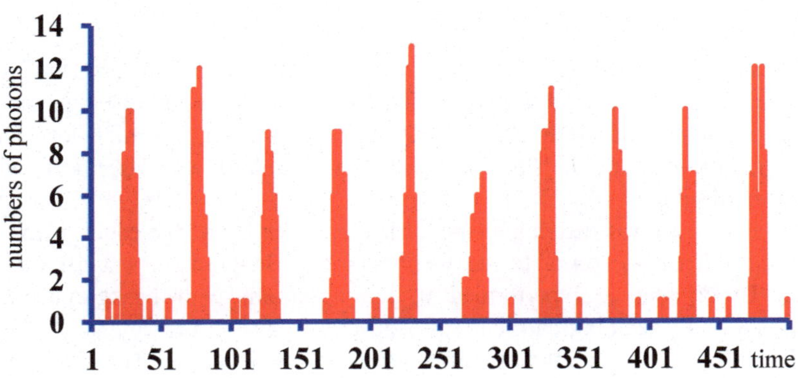

Fig. 2. Photon distribution for pulses 516 photons, 0.16 s and sampling time 0,0032 s

The photon beam was also obtained with the Mach–Zender interferometer shown in Fig. 6. The intensity of reference beam Ir and heterodyne beam Ih were regulated independently over the wide range by attenuators (polarizers P). With this interferometer single photons were registered (see p. 3.2).

3 Experiments on Photon Size

3.1 Microstructures Measurement

The first stage of the research concerned the use of single photon beam as a measuring probe for nondestructive testing of microcracs (microgaps), then extended to the measurements of microbores and other microstructures.

Special samples were prepared for microgap measurement [8]. Each sample consists of two silicon plates: one of them is in the shape of Johnson's type angular block gauge with one angle 40 arcsec (nominally) and the second one is rectangular. The plates adhere each other on their lateral sides, as shown in Fig. 3, and the inclination of the first plate forms wedge with the other one. The volume of the wedge is treated as measurement object. The samples of different thickness has been used: t = 0, 1; 0, 35 and 0, 5 mm.

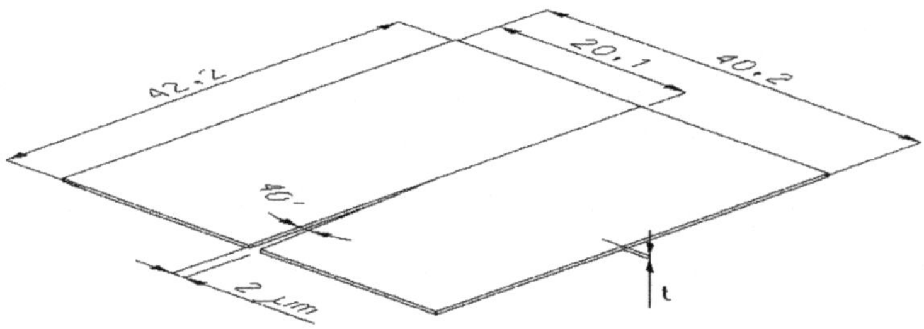

Fig. 3. The set of adhered silicon plates forming a wedge slot.

Transmission efficiency is understood as a relation between outgoing and incoming light energy. As the energy was normalized in time, in practice the light power (in number of photons) has been measured. Outgoing power, for particular slot width, was determined by scanning the slot transversely and angularly with laser beam and integrating the results.

The slot width was determined from triangle arisen by wedge edges, using optical measuring microscope (prior to measurement). The He–Ne laser was used as a light source. Experiment was performed in the darkroom with background 5 – 10 photons/second (including dark current of PMT).

The series of experiment was carried out for on each sample. The obtained results of the dependence of light transmittance versus slot width and sample thickness are shown in Fig. 4. The characteristic of efficiency versus slot width is non-linear in the width range between (1.5 + 0.1) μm and (1.8 + 0.1) μm.

The results presented above proved the measurement possibility of 0.2 μm gaps in silicon samples. Non-zero light transmission efficiency for gaps smaller than 0.2 μm indicates that the method can be applied for detection of objects far below 0.2 μm. However at present, it is impossible to determine exactly the lower limit of slot size, we can only claim it to be about 0.1 μm. Measurements exhibit the existence of non-linearity in transmission function (power transmission vs. slot width) in the slot range between 0.3 and 1.8 μm, however it is possible to approximate it with two linear functions.

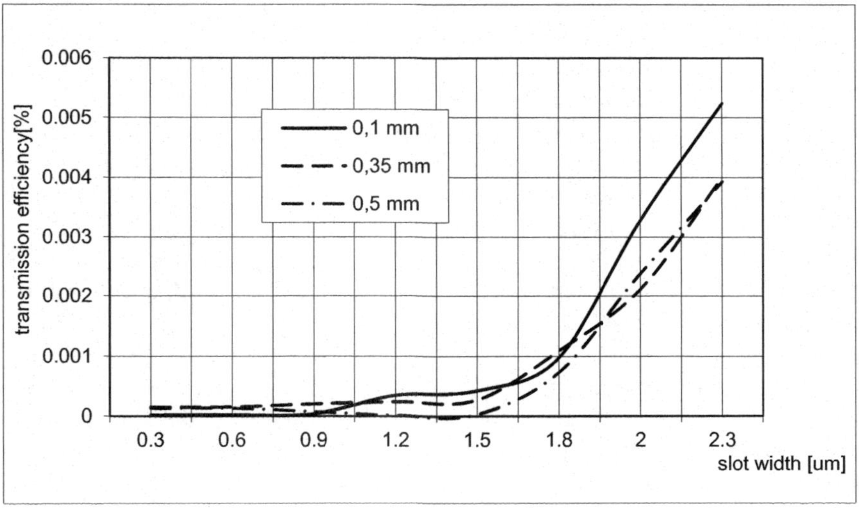

Fig. 4. The efficiency of light transmission vs. slot width, for sample thickness range 0.1–0.5 mm.

The experiments with microbores were performed with the master of adjustable microbore formed of mutually accurately adjusted stack of 20 brass plates, each 0.1 mm thick. Each bore had diameter 100 μm [3, 5, 7].

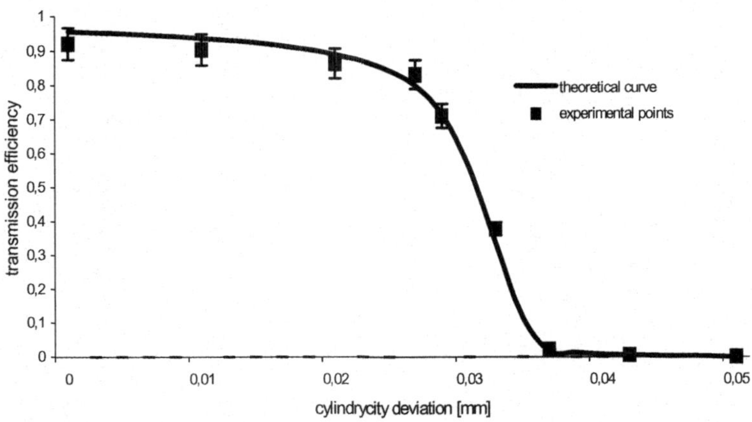

Fig. 5. Light transmission efficiency versus cylindricity (bend) deviation. Microbore diameter – 100 μm, length – 2 mm.

Figure 5 shows the exemplary transmission efficiency for bended microbore. The measurement set up is analogous to the scheme used for microgap measurement. The theoretical curve was determined on the basis of statistics, taking into account only multiple reflections.

3.2 Heterodyning the Light Beams

At high level intensities, in the classical representation of light waves, the process of heterodyning is intelligible. With the propagation in the linear medium the beams do not interact between themselves, both preserve their structures even if they overlap in the space and in the time. However, the fields of these beams linearly interfere, being summarized in the places of overlap. If the directions of propagation of beams and space location coincide, then as a result interference causes the amplitude modulation of combined traveling wave. The frequency modulation is equal to a difference in the frequencies of the waves of initial beams. When this wave reaches photoreceiver, at its output appears the signal of beatings. It is clear that the appearance of beatings during heterodyning is the purely interference process, which is possible only during the superposition of the waves of different frequency.

This special feature of the heterodyning process can be used for evaluating the length of the wave packet of photon [9, 10]. If we decrease the intensity of the light beam, so that the interval between the single photon pulses at the output PM would exceed the period of light wave (i.e. that the distance between the photons exceeds wavelength), then photons with the length λ must not overlap in the space. Accordingly, no signal of beatings can appear. However, practically the signal of beatings is recorded. It proves the interference; the wave packets of the photons of differed frequency overlap in the space, but for overlapping, the wave packets of photons in the space, in such conditions, must have large length. It is natural to assume that this length is determined by the spectral width of the line of the emission $l_0 = c/\Delta\omega$. For example, if the width of the line of emission is of the order of 1 MHz, then photon must have length of the order of 30 m. In stabilized lasers, the width of line is smaller, respectively making the length of photon longer.

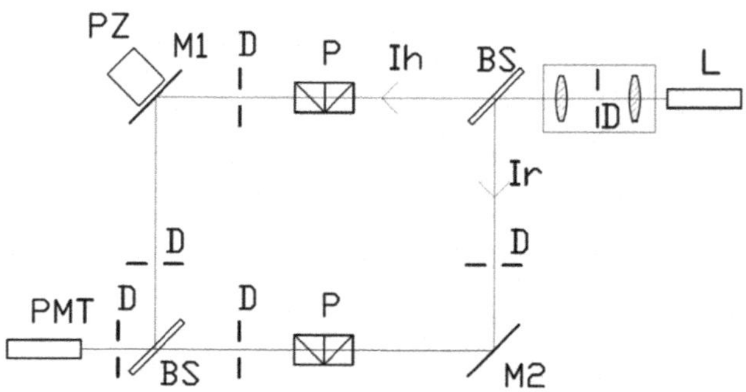

Fig. 6. Schematic representation of the experimental setup. L – He–Ne laser; PZ - piezoceramic actuator; PMT - photomultiplier tube; BS - beam splitter; M1, M2 - mirrors; P - beam attenuator; D - aperture; Ir - reference and Ih - heterodyne light beams.

An experimental setup is designed according to the schematic of Mach-Zender interferometer (Fig. 6). The He–Ne laser (λ = 632.8 nm) was used as the source of light (L) with an output power of 4 mW. The expanded beam was then splitted at the beam splitter (BS). One of the mirrors (M1) was mounted on the piezoceramic actuator and oscillating at the frequency 20 Hz. The intensity of reference beam I_r and heterodyne beam I_h were regulated independently over wide limits by attenuators (P). The photomultiplier tube (PMT) signal was recorded with the PC equipped with acquisition card.

In our experiment the statistics of single photon pulses at the output of PMT were investigated. The intensities of both beams were equaled. The average number of recorded photons was 30 000 counts/s. If beams I_r and I_h were directed to the detector in turns, then the distribution of single photon pulses at the output had random nature dictated by poissonian probability distribution. But if both beams were detected, then statistics changed fundamentally. A group of pulses of higher population density distinctly separated from that of smaller density was observed. The periodicity of groups coincides with the period of the oscillations of mirror M1.

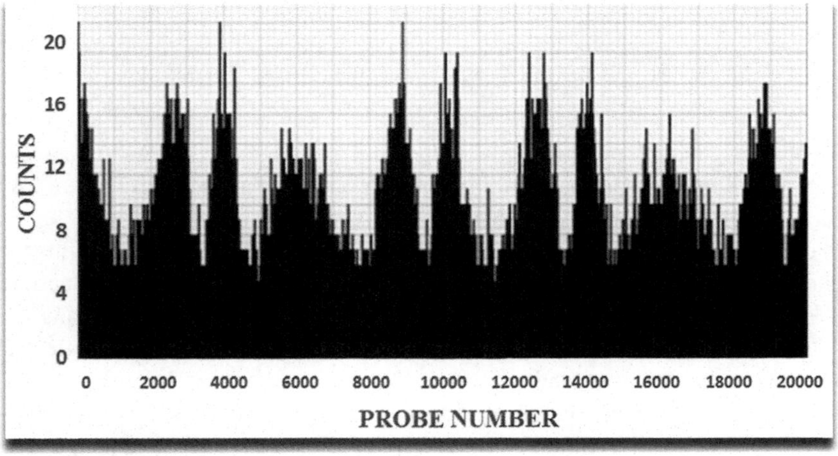

Fig. 7. Oscillogram of signal obtained with PMT. signal for one passage of sweep, Is = Ih. The total number of photons per second 30 000 counts/s.

This illustrates Fig. 7 presenting the oscillogram of single photon pulses for one passage of sweep. The results of experiment clearly show that the interference occurs even when the distance between photons greatly exceeds λ.

Experimental results confirmed the assumption that the length of photon considerably exceeds λ, reaching at least the order of meters. Within the framework of semi-classical model of photon (with assumed photon length of the order of λ) such results, until now, were not obtained with satisfactory explanation. The presented model of the photon (of large length) also makes possible to interpret results of multiple other "single photon" experiments.

4 Conclusion

It is said that photons and all elementary particles are weird and this is reflected in the mystery of entire quantum mechanics. Only what we are sure of is that time and space are mutually related through the constant c (speed of light), and in accordance with Heisenberg principle of uncertainty ($\Delta E \cdot \Delta t = h$; $\Delta p \cdot \Delta x = h$) we can present it in the form: Fig. 7.

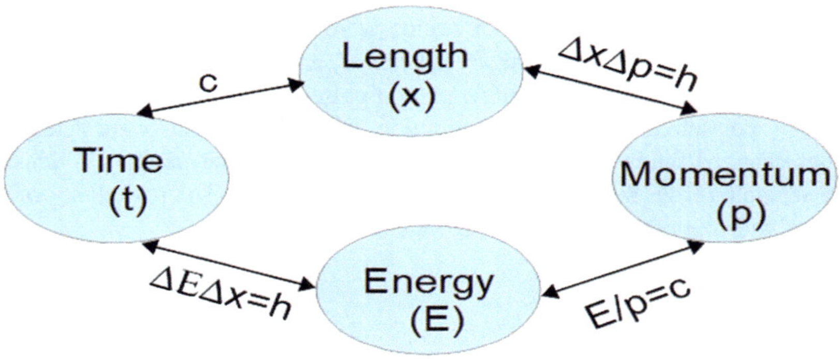

Fig. 8. The relationship between time, momentum, length and energy

From this mutual relationship we can find that the uncertainty of c results in uncertainty of length, and for a fractional *uncertainty* of c in vacuum of 3.5×10^{-9}, an uncertainty of 1 m in position amounts to an uncertainty in time 3.3 ns.

Following [12] we can obtain the same order of magnitude when calculating the size of photon wave from He–Ne laser (for 4.7•1014 Hz and λ = 628 nm we get about 9 m).

With regard to this, the result achieved in p. 3.2 should not be surprising.

Our experiments have shown that in the laboratory conditions "a single photon" mode have been achieved and we claim that photon has a needle-like shape. It is desired to have the interpretation of the experimental results based on both the semi-classical approach and quantum electrodynamics. However, many results, presented in a number of works in the 30-year period of "a single photon" experiments, unfortunately, still do not have a clear interpretation in a form of the model and also in the framework of the quantum theory. The proposed model of photon, based on modification of semi-classical Bohr model, did not lead to the understanding of observed effects. Interpretations at the level of quantum electrodynamics theories do not exist yet. So far, only the proposed by us "needling-like photon" model allowed to qualitatively interpret these effects.

In the experiment, we used a laser beam at a distance of about 2 m from laser, and after weakening by 15 orders of magnitude. There is no doubt that the photons can be regarded as propagating in free space. The feedback between these photons and laser practically does not exist.

Photon is an elementary bosons, the quantum of electromagnetic field, with zero invariant mass. So, they are light quanta, and depending on the situation their wave aspect or particle aspect can be used.

In measurements, when a photon gets closer to the measuring object the mutual interaction takes place. The area of this interaction is often represented as a "cross section", and is related to the effective transverse size of colliding particles [13].

So, does "cross section" represent the size of photon? Definitely not, because cross section varies depending on what it is interacting with, also depends on frequency, polarization and physical phenomena on which measurements are based. Mostly all the above considerations are theoretical and refer to 2D space, but real objects are 3D and usually of very compound character. Due to weirdness of photons, all the experiments are of great importance. It is not important that they are fragmentary and of very specialized character and refer to very specific conditions. However, thanks to them we are getting closer to developing a general theory of interaction between photon and object. This theory can be given in a form of general equation containing all influential quantities.

References

1. Eisaman, M.D., Fan, J., Mugdall, A., Polyakov, S.V.: Single-photon sources and detectors (Invited Review Article). Rev Sci Instrum **82**, 071101 (2011)
2. Wikipedia: https://en.wikipedia.org/wiki/Photon
3. Jablonski R (2000) Measurement of extremely long microbores by application of laser metrology. Measurement 28(3):139–145
4. Baszak J, Jabłoński R Photon beam forming device. In: Proceedings of Mechatronics 2000, vol 2, pp 367–372
5. Jablonski R, Orzechowski P (2004) Shape measurement of long microbores applying photon counting method. In: 2004 Proceedings of European international conference, 2004, pp 332–334
6. Baszak J, Jablonski R (2005) An application of LD and photon counting techniques for fluid flow measurement. J Phys 13:86–89
7. Jablonski R, Orzechowski P (2006) Shape measurement of long microbores applying photon counting method. Precis Eng 30(2):180–181
8. Jablonski R (2013) Single photon beam as a tool for measuring micro- and nanoobjects. In: Proceedings of ISMQC 2013, Publ. IMEKO, pp 355–359
9. Jablonski R, Al-Tabich A (2014) On the size of photon. Adv Mater Res Trans Tech Publ 1117:69–74
10. Jablonski R, Al-Tabich A, Rysakov V (2018) Classical interpretation of ultra-low intensity optical heterodyning as a pragmatic approach to photon size determination. In: Mechatronics 2017. Springer, Berlin, pp 111–118
11. Clauser, John F.: Experimental distinction between the quantum and classical field-theoretic predictions for the photoelectric effect. Phys Rev D **9**(4), 853–860 (1974)
12. Feynman RP Lectures I-32–3
13. Hubbel JH (1969) Photon cross sections, attenuation coefficient, and energy absorption coefficient, from 10 keV to 100 GeV. J. H. Hubbel, U.S. Department of Commerce, National Bureau of Standards, NSRDS-NBS 29

Development of Plasma Driven Permeation Measurement System for Plasma Facing Materials

Mingzhong Zhao[1(✉)], Shota Yamazaki[2], Moeko Nakata[2], Fei Sun[3], Takuro Wada[2], Ayaka Koike[2], Yoji Someya[4], Kenji Tobita[4], and Yasuhisa Oya[2]

[1] Graduate School of Science and Technology, Shizuoka University, Shizuoka 422-8529, Japan
zhao.mingzhong.17@shizuoka.ac.jp
[2] Graduate School of Integrated Science and Technology, Shizuoka University, Shizuoka 422-8529, Japan
[3] Faculty of Science, Shizuoka University, Shizuoka 422-8529, Japan
[4] National Institutes for Quantum and Radiological Science and Technology, Aomori 039-3212, Japan

Abstract. To study the hydrogen isotopes plasma driven permeation (PDP) behavior in plasma facing materials, a linear Radio Frequency (RF) plasma device has been constructed in the radiation controlled area at Shizuoka University. The deuterium (D) plasma is generated by injecting RF power with the frequency of 13.56 MHz through a copper antenna and confined by DC magnetic field. The sample is sealed by gold (Au) coated O-ring and one side (upstream side) of sample is exposed to the D plasma. The other side of sample, named as downstream side, is pumped out by a turbo molecular pump and a rotary pump. The permeated D through the sample is monitored by a quadrupole mass spectrometer (QMS) which is connected to the downstream chamber. Infrared heater is adopted to control the sample temperature. The PDP experiments under different plasma parameters show that the permeation process agrees with RD regime. The D recombination coefficient on upstream surface of W is obtained.

Keywords: Hydrogen isotopes · Plasma driven permeation · Tungsten

1 Introduction

Plasma facing materials (PFMs) for fusion reactor will be subjected to both of deuterium (D)-tritium (T) plasma and heat flux. In addition, the volume of fusion power plant with high fusion gain will be larger than 1000 m^3 [1]. Taking them into account, the evolution of radioactive T permeation through PFMs is important for the safety operation of fusion reactor.

Tungsten (W) will be used as the PFMs in divertor of International Thermonuclear Experimental Reactor (ITER). W is also the most suitable candidate PFMs for future fusion reactors attribute to its favorable properties, such as high melting point, low

A. R. Várkonyi-Kóczy (Ed.): INTER-ACADEMIA 2019, LNNS 101, pp. 260–268, 2020.
https://doi.org/10.1007/978-3-030-36841-8_25

physical sputtering yield and low T retention [2]. Hence, the evaluation of hydrogen isotopes permeation in W is important for the development of fusion reactor. Gas driven permeation (GDP) [3–5] and ion driven permeation (IDP) [6–8] experiments have been carried out to study the hydrogen isotope permeation behavior in W. The hydrogen isotopes GDP process in W consists absorption, dissociation and dissolution on the feed side, diffusion in material, recombination and release from back side [9]. The relevant hydrogen isotopes permeation parameters, solubility, diffusivity and permeability can be obtained by GDP method. In the case of IDP experiment, electron is stripped and energetic hydrogen ions directly inject into the material. The effect of impurity gas on the deuterium (D) permeation behavior in W have been studied by IDP method. Reduction in D permeation flux are observed by seeding helium (He), neon (Ne) and argon (Ar) in D ion beam [7, 10]. However, enhancement in D permeation flux are observed in the carbon (C) and nitrogen (N) seeding experiment [11, 12]. These IDP results provide good references for estimating hydrogen isotopes retention and permeation behavior in W. However, the IDP experiments are carried out with high incident energy of hydrogen isotope ions. In the fusion reactor, the energy of hydrogen isotope particles ranges from eV to several keV [13]. To data, the hydrogen isotope plasma driven permeation behavior in W is not sufficient understood, especially the hydrogen permeation behavior in W under fusion relevant high temperature condition.

The objective of the present study is to examine the hydrogen isotopes plasma driven permeation behavior in W under fusion relevant high temperature. Hence, a linear plasma device combining with an infrared heater is constructed in the radiation controlled area at Shizuoka University. The preliminary PDP experiments with W material are carried out using the present developed device.

2 Experiment

The schematic of plasma driven permeation measurement device is given in Fig. 1. The device consists of two vacuum chambers, namely upstream chamber and downstream chamber, which are separated by the sample. The upstream chamber and downstream chamber are evacuated to a base pressure lower than 5×10^{-5} Pa by the combination of turbo-molecular pump (TMP) and rotary pump (RP), respectively. The plasma discharge is formed in quartz tube by a copper antenna and confined by magnetic fields. A 13.56 MHz–3 kW radio frequency (RF) source with matching box is used to energize the antenna. The D_2 gas is supplied to the upstream chamber to maintain the discharge pressure. The plasma parameters are measured by Langmuir probe which locates in front of sample. The sample is sealed by gold (Au) coated O-ring and electrically insulated with vacuum chamber by ceramic insulator. Hence, bias can be applied to the sample by a DC power supply. The sample is heated from the rear side (downstream side) by an infrared heater. The sample can be heated up to 1073 K which is limited by the working temperature of O-ring. A K-type thermocouple fixed into the sample flange is used to measure the sample temperature. Under sample temperature at 1073 K, the tube for quartz rod temperature is lower than 473 K. In consideration of the discharge pressure is within 0.2 Pa, the effect of GDP from stainless steel tube can be ignored. The permeation signal through the sample is measured by a quadrupole mass spectrometer (QMS) connecting to the downstream vacuum. In the PDP experiment, the mass 3 (HD) and mass 4 (D_2) are recorded by QMS which is calibrated by D_2 standard leak bottles. The sensitivity of mass 3 is taken as an average of mass 2 and mass 4 sensitivities.

In the present study, polycrystalline W wafer purchased from Allied Material (A.L. M.T) Corp. Ltd. is used. The sample size is 10 mm in diameter and 0.5 mm in thickness. Both side of the sample are polished as a mirror finish. To release the internal stress, W sample is annealed under 1173 K for 30 min in vacuum. The annealed sample is then sealed by O-rings. Thereafter, the PDP experiments are conducted in the temperature range from 749 K to 813 K.

The W sample surface condition is analyzed by X-ray photoelectron spectroscopy (XPS). Argon ion with the energy of 3 keV is used for sputtering the sample surface.

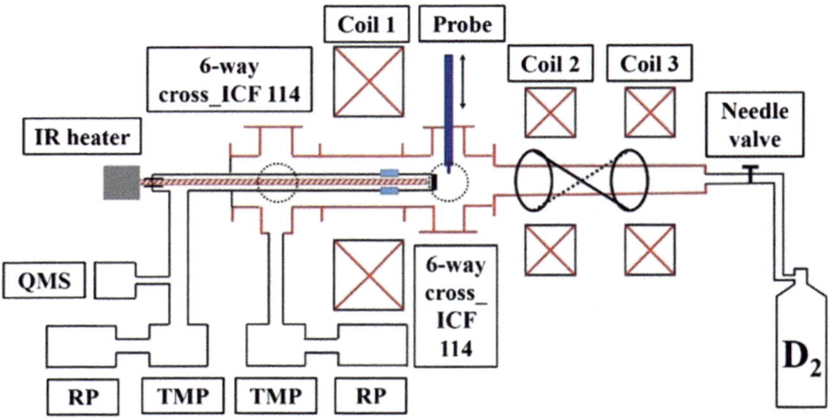

Fig. 1. Schematic of the RF- plasma driven permeation measurement device.

3 Results and Discussion

3.1 Plasma Parameters

Plasma experiments are performed with two different input powers of 500 W and 700 W. In the case of 500 W, D_2 pressure in the upstream chamber is maintained to be 0.16 Pa. For 700 W, D_2 pressure of 0.05 Pa is used. The plasma parameters at the center of D plasma are measured by a Langmuir probe. The current-voltage characteristics for 500 W and 700 W discharge are given in Fig. 2. The calculated plasma parameters, namely electron temperature (T_e), electron density (n_e) and ion flux, are summarized in Table 1., where the ion flux is calculated by applying the Bohm criterion [14]. The n_e of 700 W is obvious higher than that of 500 W under the same magnetic field. Such jump in plasma density is due to that the helicon mode is excited with increasing the input power to 700 W [15].

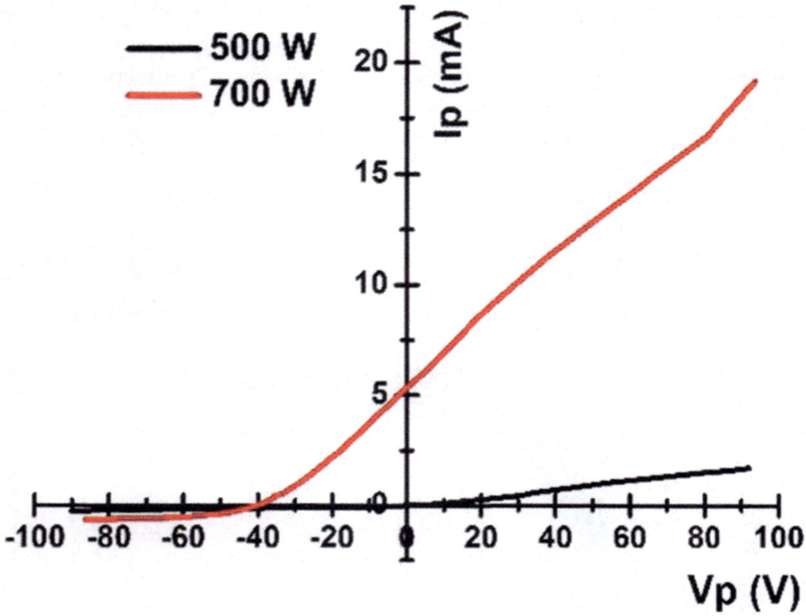

Fig. 2. The current-voltage characteristics measured by Langmuir probe.

Table 1. Plasma parameters under various discharge condition.

	T_e (eV)	n_e (m^{-3})	Ion flux (m^{-2} s^{-1})
500 W	12.4	4.4×10^{14}	1.1×10^{19}
700 W	7.5	1.2×10^{16}	2.3×10^{20}

3.2 The Plasma Driven Permeation Behavior in W

The influence of bias on the PDP flux under the discharge power of 500 W is shown in Fig. 3. The permeation flux increases gradually and reaches to the steady state with time after the plasma ignited at time 0 s. When the negative bias is applied to the W sample against to the plasma, the PDP steady state flux decreases. Such reduction of PDP flux by negative bias has also been observed in PDP experiment with nickel (Ni) material as reported in Ref. [16]. The possible reason is that the electrons dissociate molecules to atoms near the upstream surface and increase the incident flux of D atoms. In the case of negative bias, electrons will be repelled near the sample surface leading to the lower PDP flux.

As shown in Fig. 3, a hump is observed in the permeation curve without bias. This phenomenon is associated with the sample surface conditions. Impurities, carbon (C) and oxygen (O), are observed on the initial W sample surface by XPS as shown in Fig. 4, and the impurities on W sample can be removed after Ar ion sputtering 10 s. The implanted D on the upstream surface will recombine with each other and escape from the upstream surface or diffuse toward the downstream surface. The existence of impurities on the W sample surface retards the recombination of D on the upstream surface leading to the increment in D concentration at upstream surface. Such an increase of D concentration at the upstream surface causing the increment in permeation flux at the initial stage. With the increasing of plasma exposure time, the impurities are gradually removed, leading to the decrease of permeation flux as observed in Fig. 3 [17].

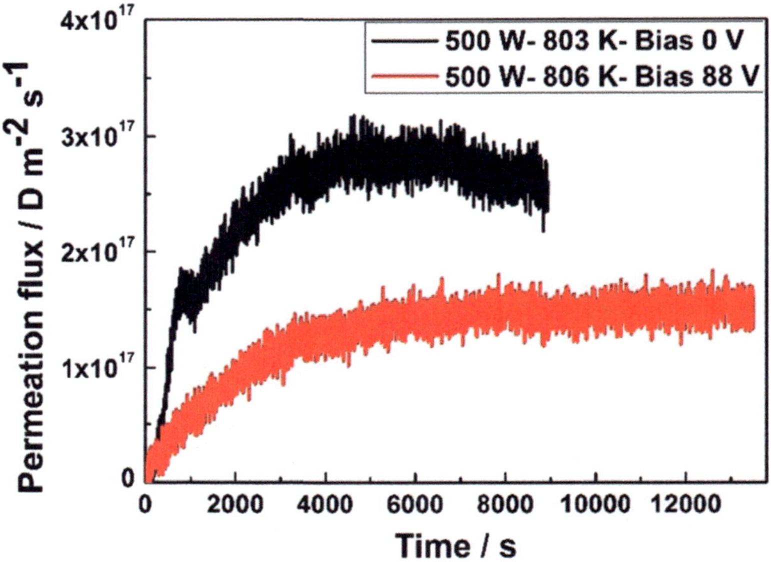

Fig. 3. The permeation curves with and without bias under discharge power of 500 W.

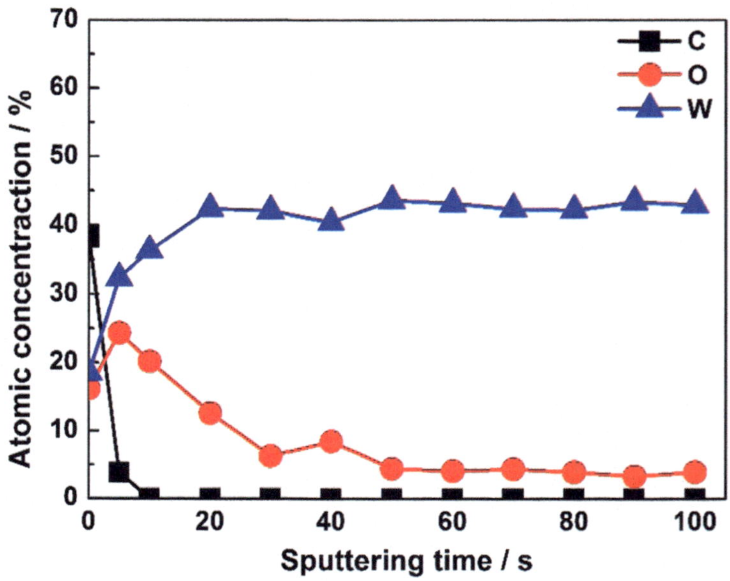

Fig. 4. The atomic concentrations as a function of Ar ion sputtering time.

As shown in Fig. 5, the influence of temperature on the D permeation behavior in W is studied with the discharge power 700 W. Under the temperature of 803 K, the steady state permeation flux for the discharge power 700 W and 500 W are 1.5×10^{18} D m^{-2} s^{-1} and 2.6×10^{17} D m^{-2} s^{-1}, respectively. As discussed in Sect. 3.1, the ion flux of discharge power 700 W is about 20 times as high as that of 500 W.

There are three regimes for the steady state plasma driven permeation process [18]: (1) the release of hydrogen isotopes from the upstream and downstream surfaces are limited by diffusion process, named as DD-regime, (2) release of hydrogen isotopes from the upstream surface and downstream surface are limited by the recombination of hydrogen isotopes and diffusion of hydrogen isotopes, named as RD-regime, (3) the release of hydrogen isotopes from both surfaces are limited by the recombination process, named as RR-regime. The relationship of steady state permeation flux (J_p) and the implantation flux (J_i) in different regime can be expressed as:

$$J_p = \frac{R}{L} J_i \text{ (DD-regime)} \tag{1}$$

$$J_P = \frac{D}{L}\sqrt{\frac{J_i}{K_U}} \text{ (RD-regime)} \tag{2}$$

$$J_P = \frac{K_U}{K_U + K_D} J_i \text{ (RR-regime)}, \tag{3}$$

where, R, L and D represent the ion projected range, sample thickness and the bulk diffusion coefficient. K_U and K_D are the recombination coefficient on upstream surface and downstream surface. In the present experiment, the steady state permeation flux is approximate proportional to the square root of the ion flux indicating that permeation takes place in the RD regime.

The steady state permeation flux is plotted as function of inverse temperature as shown in Fig. 5(b). For the RD regime, the D recombination coefficient (K_U) at the upstream surface can be obtained by the following equation:

$$K_U = \frac{D^2 J_i}{L^2 J_P^2} \tag{4}$$

The diffusivity measured by Frauenfelder [3] is used as the bulk diffusion coefficient. Because the diffusivity in Ref. [3] is measured at elevated temperature and less likely to be effected by defects. The calculated K_U as a function of inverse temperature is shown in Fig. 6 in which the literature data from Takagi [19] and Anderl [20] are also plotted. The available hydrogen isotopes recombination coefficient in W material is real scarce at present. The possible reasons for the difference in recombination coefficient are that (1) the recombination coefficient is obtained by different experiment method, (2) the recombination coefficient is sensitive to the sample surface condition. In Ref. [19], the recombination coefficient is obtained according to the D concentration which is measured by nuclear reaction analysis (NRA). In Ref. [20], the recombination coefficient is determined by the IDP experiment with 3 keV D_3^+ ions.

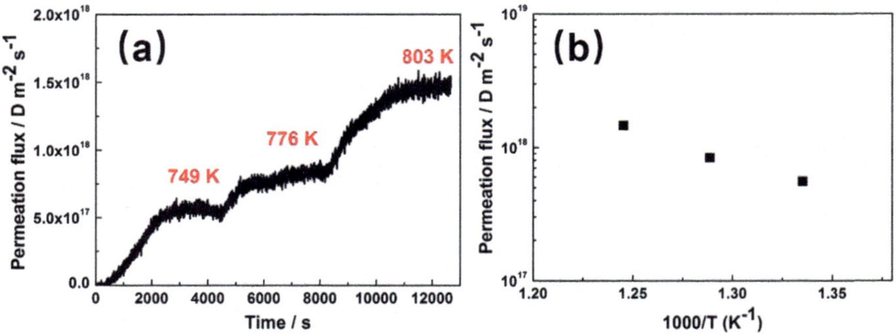

Fig. 5. The PDP results under different temperature, (a) permeation curve, (b) steady state permeation flux as a function of inverse temperature.

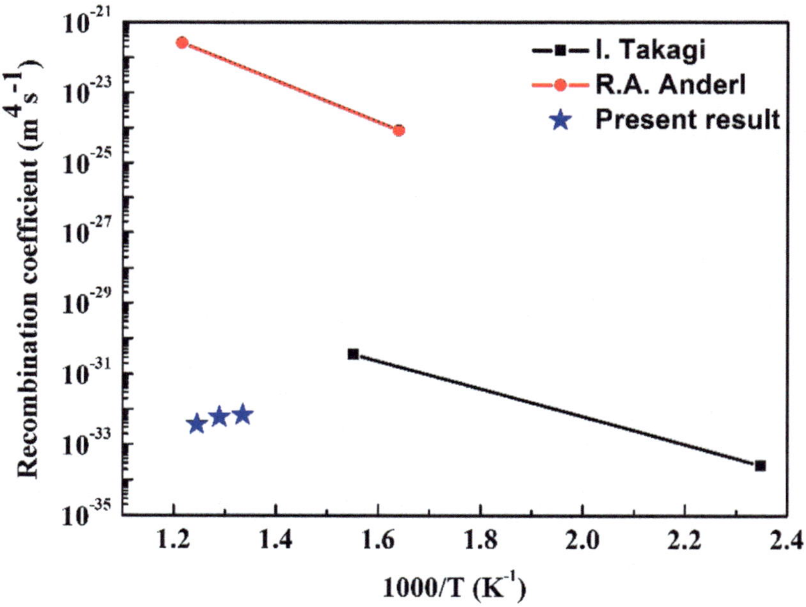

Fig. 6. The recombination coefficient of D for W sample.

4 Conclusion

RF-Plasma device for the PDP study is constructed in the radiation controlled area at Shizuoka University. The D plasma parameters are measured by Langmuir probe. The ion flux can be reached to be 1.1×10^{19} m^{-2} s^{-1} and 2.3×10^{20} m^{-2} s^{-1} under the discharge power of 500 W and 700 W. The PDP experiments are carried out under different discharge power, bias condition and sample temperature. The negative bias can reduce the steady state D permeation flux as the electrons at the upstream side are repelled by the negative bias. In addition, the PDP results under different discharge power show that the permeation process in W agrees with the RD regime. The recombination coefficient, which is an important parameter for the estimation of the hydrogen isotopes permeation behavior in fusion reactor, is obtained by the RD regime.

References

1. Costley, A., Hugill, J., Buxton, P.: On the power and size of tokamak fusion pilot plants and reactors. Nucl. Fusion **55**, 033001 (2015)
2. Philipps, V.: Tungsten as material for plasma-facing components in fusion devices. J. Nucl. Mater. **415**, S2 (2011)
3. Frauenfelder, R.: Solution and diffusion of hydrogen in tungsten. J. Vac. Sci. Technol. **6**, 388 (1969)
4. Zakharov, A., Sharapov, V., Evko, E.: Hydrogen permeability of polycrystalline and monocrystalline molybdenum and tungsten. Mater. Sci. **9**, 149 (1975)
5. Uemura, Y., et al.: Effect of helium irradiation on deuterium permeation behavior in tungsten. J. Nucl. Mater. **490**, 242 (2017)
6. Lee, H., Markina, E., Ohtsuka, Y., Ueda, Y.: Deuterium ion-driven permeation in tungsten with different microstructures. Phys. Scripta **2011**, 014045 (2011)
7. Lee, H., Tanaka, H., Ohtsuka, Y., Ueda, Y.: Ion-driven permeation of deuterium through tungsten under simultaneous helium and deuterium irradiation. J. Nucl. Mater. **415**, S696 (2011)
8. Gasparyan, Y., Rasinski, M., Mayer, M., Pisarev, A., Roth, J.: Deuterium ion-driven permeation and bulk retention in tungsten. J. Nucl. Mater. **417**, 540 (2011)
9. Nguyen, T.H., Mori, S., Suzuki, M.: Hydrogen permeance and the effect of H_2O and CO on the permeability of $Pd_{0.75}Ag_{0.25}$ membranes under gas-driven permeation and plasma-driven permeation. Chem. Eng. J. **155**, 55 (2009)
10. Ishida, M., Lee, H., Ueda, Y.: The influence of neon or argon impurities on deuterium permeation in tungsten. J. Nucl. Mater. **463**, 1062 (2015)
11. Lee, H., Ishida, M., Ohtsuka, Y., Ueda, Y.: The influence of nitrogen on deuterium permeation through tungsten. Phys. Scripta **2014**, 014021 (2014)
12. Peng, H., Lee, H., Ohtsuka, Y., Ueda, Y.: Ion-driven permeation of deuterium in tungsten by deuterium and carbon-mixed ion irradiation. Phys. Scripta **2011**, 014046 (2011)
13. Joachim, R., Klaus, S.: Hydrogen in tungsten as plasma-facing material. Phys. Scripta **2011**, 014031 (2011)
14. Amemiya, H.: Sheath formation criterion and ion flux for non-Maxwellian plasma. J. Phys. Soc. Jpn. **66**, 1335 (1997)
15. Barada, K.K., Chattopadhyay, P., Ghosh, J., Kumar, S., Saxena, Y.: A linear helicon plasma device with controllable magnetic field gradient. Rev. Sci. Instrum. **83**, 063501 (2012)
16. Hatano, Y., Nakamura, H., Furuya, H., Sugisaki, M.: Influence of surface impurities on plasma-driven permeation of deuterium through nickel. J. Vac. Sci. Technol. Vac. Surf. Films **16**, 2078 (1998)
17. Liu, H.-D., et al.: Deuterium plasma driven permeation behavior in a Chinese reduced activation martensitic/ferritic steel CLF-1. J. Nucl. Mater. **514**, 109 (2019)

18. Zhou, H., Hirooka, Y., Ashikawa, N., Muroga, T., Sagara, A.: Gas-and plasma-driven hydrogen permeation through a reduced activation ferritic steel alloy F82H. J. Nucl. Mater. **455**, 470 (2014)
19. Takagi, I., et al.: Deuterium recombination coefficients on tungsten exposed to RF plasma. J. Nucl. Mater. **417**, 564 (2011)
20. Anderl, R., et al.: Deuterium transport and trapping in polycrystalline tungsten. Fusion Technol. **21**, 745 (1992)

Monitoring the Environmental Quality of Marine Waters Through the Analysis of Biomineralization in Bivalve Shells

Iuliana Motrescu[1,2](✉), Anca Elena Calistru[2,3], Gerard Jitareanu[2,3], and Liviu Dan Miron[4]

[1] Science Department, University of Agricultural Sciences and Veterinary Medicine of Iasi, Iasi, Romania
imotrescu@yahoo.com
[2] Research Institute for Agriculture and Environment, University of Agricultural Sciences and Veterinary Medicine of Iasi, Iasi, Romania
[3] Pedotechnics Department, University of Agricultural Sciences and Veterinary Medicine of Iasi, Iasi, Romania
[4] Clinics Department, University of Agricultural Sciences and Veterinary Medicine of Iasi, Iasi, Romania

Abstract. Bivalve shells, such as *Mytilus*, offer great potential as environmental proxies. Analysis of the biomineralization process with determination of elemental composition gives information about the quality of environment and reflects the possible safety issues related to mollusk consumption because high pollutant quantities in shells indicate high pollutants presence in the consumed parts. In this work we study the biomineralization process in some bivalve shells and identify the presence of pollutants such as Pb, U, and other heavy metals using scanning electron microscopy (SEM) coupled with energy dispersive spectroscopy (EDAX). Through the obtained results, this methodology proved to be very reliable and fast for this purpose. We also show a correlation of the biomineralization results with the environmental conditions where the shells developed, such as estimation of water temperature by the Sr/Ca ratios, all the results proving the ability of bivalve shells of providing information about the environment quality.

Keywords: *Mytilus* · Environmental quality · Environmental monitoring · Biomineralization

1 Introduction

Climate change is impacting marine life, increasing temperature and acidity of water [1]. The chemical composition of the adjacent fluids directly influences the biomineralization process. Studying the microstructure and composition of marine shells would offer information about the environment through the biomineralization process. Because molluscan have the ability to integrate aquatic pollutants, the biomineralization of molluscan shells is also an indicator of aquatic pollutants as a result of accumulation of elements during mineralization [1–3]. The findings regarding the

A. R. Várkonyi-Kóczy (Ed.): INTER-ACADEMIA 2019, LNNS 101, pp. 269–277, 2020.
https://doi.org/10.1007/978-3-030-36841-8_26

biomineralization process might be even applied to the fabrication of composite materials.

Researchers have been suggesting since 1950s that bivalve shells could be collecting and storing unwanted chemical species. Some found that contaminants such as lead are mainly mineralized in the inner part of the shell, the nacre [4]. They also found that the concentration in the tissue is closely related to the one in the nacre, so the latter can not only give an indication about the quality of the environment but also rise alarms towards the risks that might be involved in their consumption.

The methods that have been commonly used so far are mostly chemical, time consuming, invasive, and some are using toxic chemicals. In this work we put to use environmental friendlier methods that can give fast information on the morphology and chemistry of shells: environmental scanning electron microscopy (ESEM) coupled with energy dispersive spectroscopy (EDAX).

We focus here on the shells belonging to *Mytilus* spp., a very common bivalve, with an important role of filter feeder, removing bacteria and toxins from the water. The minerals in the shells are also taken from the environment along with contaminants. Along with the phytoplankton from the water, bivalves filter and integrate aquatic pollutants during the biomineralizarion of the shells. Thus, *Mytilus* are bivalve shells that have been believed to be environmental proxies. In this work we focus on the full characterization of the shells collected from different environments and draw conclusions on the environmental impact and on the individual diversity in the same genus.

2 Materials and Methods

The samples used in analysis in this work were commercially procured and they origin in different locations indicated in Fig. 1. From each site only samples having the same ontogenetic stage were chosen (about 4 cm length, corresponding with shells of 24–26 months). The difference between the environments is expected to be mirrored in the biomineralization process for the individuals belonging to a specific area. The environmental trace elements and anthropogenic enhancement has potential to study the effects of environmental factors on the mineralogy of the shells.

To study the biomineralization mechanism, a detailed analysis from macro to nanoscale of the biomineralized skeletal materials was performed. Environmental scanning electron microscopy (ESEM) was performed using FEI-Quanta 450. The shells were analyzed without any preparation (no covering with conductive material) since ESEM allows such imaging for non-conductive samples. The samples were directly mounted on stubs using adhesive carbon tape. The imaging was done under high vacuum conditions (working pressure was about $4 \cdot 10^{-4}$ Pa) and using electrons accelerated at 12.5 kV. These conditions were considered the best for imaging different structures in the shells and are in good agreement with those reported in the literature. Microscopy measurements were coupled with elemental analysis using energy dispersive spectrometry (EDS) using an EDAX detector. The analysis of the composition gives important information about the organic-inorganic character of the biomineralized materials.

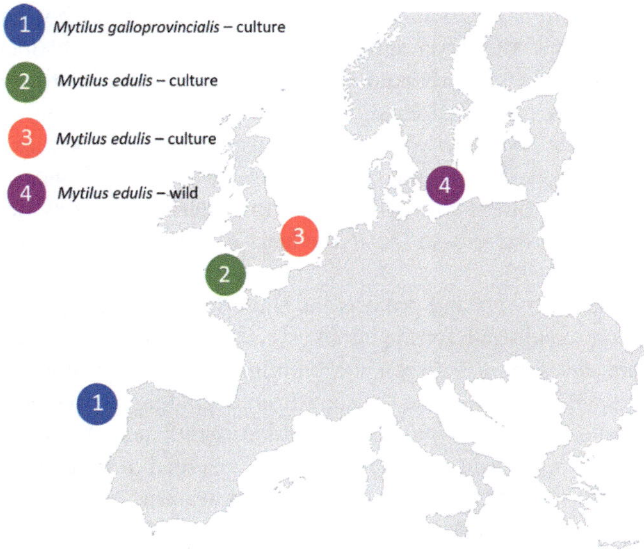

Fig. 1. Location of *Mytilus* sampling.

3 Results and Discussions

At macroscopic level, the differences between individuals in the same genus are insignificant. Detailed statistical analysis and modelling have been reported by Telesca et al. [6]. In our case only slightly different biomineralization in the nacre layer is observed, probably due to environmental conditions. Thus, it is necessary to solve the structures at microscopic level from both the morphological and composition point of views. ESEM analysis was performed for the outer part of the shells – called periostracum, inner part – nacre, and also the longitudinal section was imaged to analyze the structure of the prism layer under the periostracum.

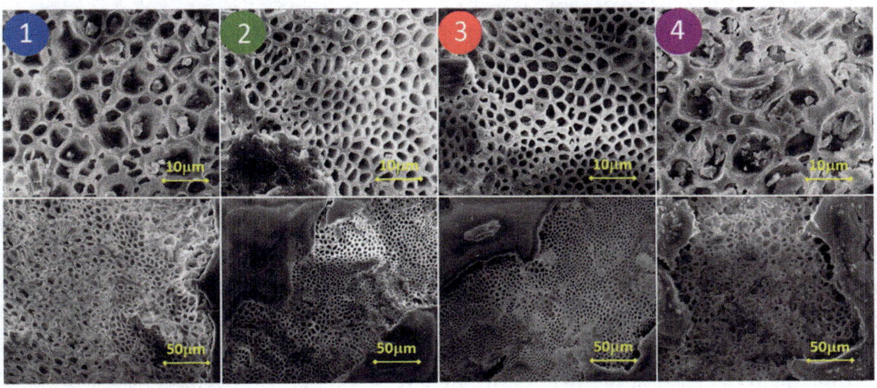

Fig. 2. Comparison of the ESEM images of the periostacum for the *Mytilus* samples.

The shell of *Mytilus edulis* is comprised of two calcium carbonate polymorphs: prismatic calcite in the outer layer and aragonite in the nacre. The element distribution varies between the calcium carbonate forms in different ontogenetic stages and also depending on the environmental conditions that influenced the biomineralization process. The most outer layer, periostracum, is mostly made of complex proteins. The minerals and biopolymeric parts are organized in many types of microstructures with different textures. The internal morphology of the shells is probably developed as an adaptation to the environment so the individuals to survive in certain environmental conditions [7–10].

The periostracum covers and protects the shell. It is an organic layer consisting of proteins that act as a substrate for the initial crystal nucleation of the shell. The organic components are developing between the biomineral structures and are hydrophobic macromolecules able to prevent the structure from degrading in aqueous environments. This organic material form as an intercrystalline matrix and it consists of polysaccharides and proteins. These components are also assumed to influence the nucleation of the carbonate crystals as well as the crystallization form.

Figure 2 presents a comparison of the periostracum appearance for the four samples. It is very clear that the matrix presents different structuring with a denser and more regular network in the case of samples 2 and 3 and larger and less regular matrix for sample 4. Since these parts of the shells are those mostly exposed to the environment, we believe this appearance is a direct indication of the environmental impact on the biomineralization.

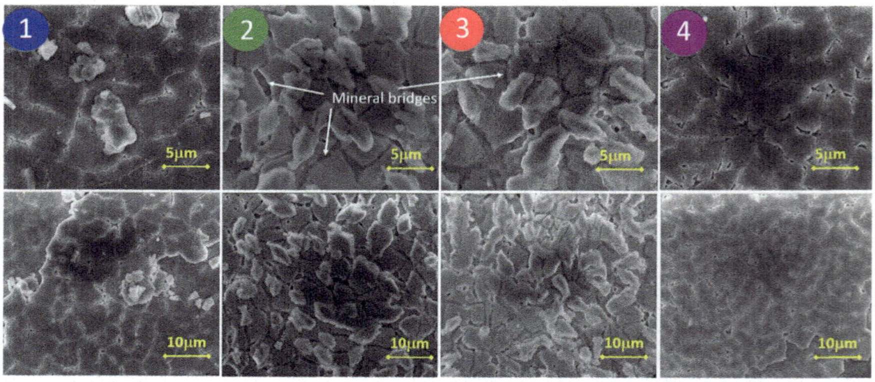

Fig. 3. Comparison of nacre layer appearance for the samples imaged with ESEM.

The images in Fig. 3 indicate that the nacre formation corresponds to Schaffer model showing growth through mineral bridges across the pre-existing organic interlamellar sheets by precipitation of the $CaCO_3$. Differences between samples consists in the different size of the mineral bridges and different micro-structuring in all layers. Analyzing the inside of the shells, namely the nacre layer, similar findings can be reported as in the case of periostracum. Images of the nacre presented in Fig. 3 indicate similarities between samples 2 and 3, with discernable and regular calcite crystal tablets. In samples 1 and 4 the mineral bridges seem thinner than for samples 2 and 3, while the mineralized tablets look less differentiated and more bound together.

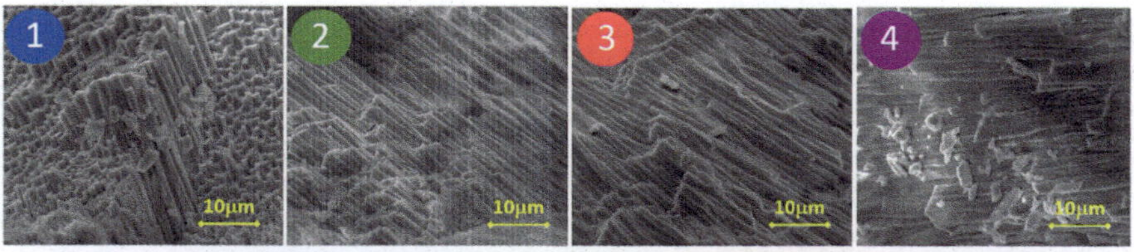

Sample	1	2	3	4
Width (μm)±3%	1.38	1.89	2.20	2.61

Fig. 4. ESEM images of the prism layer in all samples and average values of the layering.

Figure 4 shows images of the prism layer obtained by fracturing the shells. The main difference detected is in the size of the layer thickness, for which average values are shown under the images. The largest thickness was measured for sample 4, about 50% higher than for sample 1, indicating a strong difference in the biomineralization process between the samples, probably own to the environment where the shells developed since the shells are on the same ontogenetic stage. The pictures are representative for the samples, the distribution of the prismatic structures is the same along the length of the shell.

Chemical analysis of the minerals biologically produced by the mollusks can yield critical information about the biomineralization process and the interactions with the surrounding environment. Ca, C, O are the major elements in the minerals comprising the shells. Other elements are incorporated in the system as a result of their presence in the environment where the shell develops and where the biomineralization takes place. These trace elements ($\sim$0.1 wt%) incorporated in the biogenic carbonate system along with the major elements, do not form crystal phases but are believed to influence the biomineralization process [10–12].

The elemental composition of the samples was determined with energy dispersive spectrometry, the measurements being performed in the same time as electron microscopy imaging. The analysis was directed towards detecting all present elements in the mineral structure and trace contaminants filtered and fixed from the environment (e.g. Pb, U). Electron beam accelerated at various voltages (20 kV, 15 kV, and 10 kV) were used to analyze the nacre in layers corresponding to about 10 μm, 7 μm, and 5 μm, respectively. These depths were estimated considering the chemical composition of the shells.

Magnesium (Mg) is believed to be important in the growth of $CaCO_3$ because Mg^{2+} and other divalent cations can substitute Ca^{2+} in the crystal lattice, determining the crystal growth [10, 11]. In the calcite layer younger specimens were reported to have higher Mg, S, and Sr concentration, while older ones have higher concentrations. Among the inorganic components, the presence of Mg is believed to be an inhibitor of carbonate crystal growth and stabilizing the amorphous calcium carbonate structures. In our case, the maximum ratio of Mg/Ca was found in sample 3, with close value as sample 2. These values were measured for the nacre and are consistent with the similarities found in micro-imaging of the nacre previously discussed. Sample 2 also shown highest ratio in the case of Na/Ca. Na^+ incorporation has been known to be related to interstitial mineral bridges, although its presence is still debated.

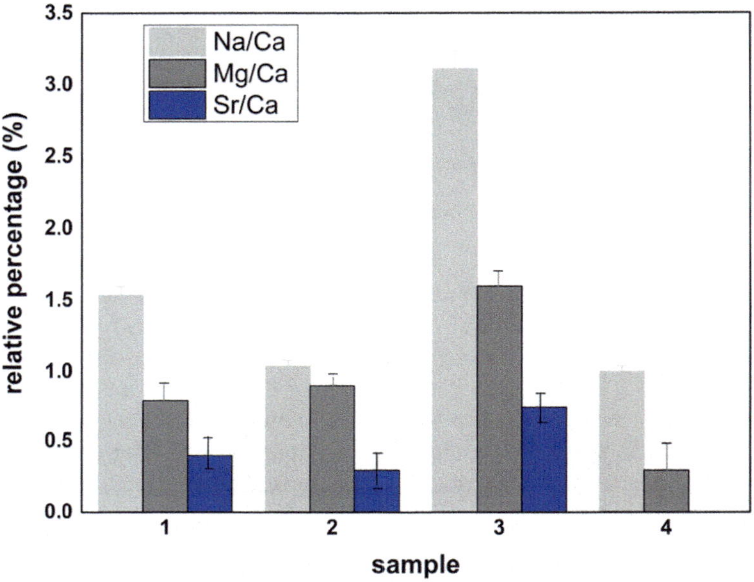

Fig. 5. Comparison of Na/Ca, Mg/Ca, and Sr/Ca ratios from EDAX measurements (statistic data from 20 kV electron beam conditions measurements) for all the samples.

Figure 5 also shows the Sr/Ca ratios for the samples. Except for sample 4 in which Sr was not detected in any point or conditions, for the other samples the values are similar. Sr is believed to affect the kinetics of calcite growth depending on its concentration and it is also an indicator of the temperature of carbonate formation [13, 14]. The temperature of the environment during shell development can be estimated as shown by Wanamaker et al. the relationship between Mg/Ca ratio and temperature being a linear one [13]. The ratios as presented in Fig. 5 are similar for samples 1 and 2 (about 8 °C ± 10%), which developed in similar conditions thus similar average temperatures. The environmental temperature for sample 3 is found to be slightly higher (14 °C ± 10%) probably due to lower current and wave activity as compared to areas where samples 1 and 2 were collected. Sample 4 corresponds with the lower temperature (calculated from the ratios to about (about 3 °C ± 10%), which is in agreement with the cooler sea where in area 4. The values are based on Wanamaker formula and are slightly smaller than the real averages. Even if the accuracy is not very high, these results support the reported findings in the literature which connect the Mg/Ca ratio with the environmental temperature.

For all samples, the Na/Ca and Mg/Ca ratios increased with the depth, indicating the increase of the amount of ACC (amorphous calcium carbonate) and the presence of crystallin forms in the vicinity of the nacre surface. The amount was the highest for sample 2. EDS analysis also found that the organic matrix, which proved to contain more than 48 proteins [15], is denser towards the inside of the nacre, by the amount of nitrogen and sulfur detected. The highest amount of sulfur was found in sample 2.

Table 1. Elemental weight percentage in the samples estimated from EDAX measurements (highest values measured are indicated in bold characters).

Element	Sample 1	Sample 2	Sample 3	Sample 4
Mn	0.66	0.83	**1.73**	0.44
Fe	1.26	1.39	**1.86**	0.39
Cu	0.40	0.51	**0.66**	0.06
Zn	**0.16**	0.13	0.06	0.07
Al	0.17	**2.10**	0	0
Pb	0	**0.58**	0.50	0.07
U	0.47	0.79	**1.53**	0

Except the above-mentioned main elements that form the shell structure, trace toxic elements are found, fixed during the biomineralization from the environment [16–20]. The elements found depend on the environmental conditions. The trace elements measured in our samples are presented in Table 1 with average values measured for 20 kV accelerated beams. The percentages generally increase with the measured depth and is different between the samples, indicating different environmental conditions. Highest amounts found are represented in bold characters and red color.

Measurements made in different points on the nacre layer indicate a relatively uniform distribution of the trace elements. Their presence was first believed to be due to passive adsorption but after discrete analyses of the structural components of the shells it was found that the elements are filtered by the mollusks and fixed through the biomineralization. Our measurements shown that the concentration of toxic elements on the outer side of the shell, in the periostracum, is much lower than at nacre level, proving accumulation during biomineralization. These results support the fact that the nacreous layer in *Mytilus* shells can be used as a monitoring tool for pollution and also provides the means for assessing possible health risks of mussels consumption [18–20].

4 Conclusions

Mytilus shells grown in different environmental conditions were analyzed using non-invasive physical methods. Differences in the structures at micro-level were detected as well as in the chemical composition of the samples in the same genus, leading to indications about the environmental influence on the biomineralization process in *Mytilus* spp. shells. The results give a strong basis for a further genotype characterization.

ESEM coupled with EDAX proves to be fast and accurate method to analyze the details of the biomineralization process and correlate it with the environmental conditions. The biomineralization of potential toxigenic elements (e.g. Pb, U) recommends a veterinary-sanitary examination of the products and should rise alarms towards public advisory regarding the possible harms of mollusk consumption in some areas. Considering the presented results, *Mytilus* can be successfully used as monitoring tool for the marine environmental conditions.

Acknowledgments. This work was supported by Joint Institute for Nuclear Research Grant (04-4-1121-2015/2020 RO-JINR item 34).

References

1. Rodolfo-Metalpa, R., Houlbreque, F., Tambutte, E., Boisson, F., Baggini, C., Patti, F.P., Jeffree, R., Fine, M., Foggo, A., Gattuso, J.-P., Hall-Spencer, J.M.: Coral and mollusk resistance to ocean acidification adversely affected by warming. Nat. Clim. Change **1**, 308–312 (2011)
2. Hubbard, F.H., Al-Dabbas, M.A.M., McManus, J.: Environmental influences on the shell mineralogy of *Mytilus edulis*. Geo. Mar. Lett. **1**, 267–269 (1981)
3. Fitzer, S.C., Zhu, W., Tanner, K.E., Phoenix, V.R., Kamenos, N.A., Cusack, M.: Ocean acidification alters the material properties of *Mytilus edulis* shells. J. R. Soc. Interface (103):20141227 (2014)
4. Boutgoin, B.P.: *Mytilus edulis* shell as bioindicator of lead pollution: considerations of bioavailability and variability. Mar. Ecol. Prog. Ser. **61**, 253–262 (1990)
5. Krzysztof, R.B., Szopa, K.: Morphological diversity of microstructures occurring in selected recent bivalve shells and their ecological implications. Contemp. Trends Geosci. **5**(2), 104–112 (2016)
6. Telesca, L., Michalek, K., Sanders, T., Peck, L.S., Thyrring, J., Harper, E.M.: Blue mussel shell shape plasticity and natural environments: a quantitative approach. Sci. Rep. **8**, 2865 (2018)
7. Innes, D.J., Bates, J.A.: Morphological variations of *Mytilus edulis* and *Mytilus trossulus* in eastern Newfoundland. Mar. Biol. **133**, 691–699 (1999)
8. Moschino, V., Bressan, M., Cavaleri, L., Da Ros, L.: Shell-shaped and morphometric variability in *Mytilus galloprovinciallis* from micro-tidal environments: responses to different hydrodynamic drivers. Mar. Ecol. **36**(4), 1–14 (2015)
9. Alunno-Bruscia, M., Bourget, E., Frechette, M.: Shell allomertry and length-mass-density relationships for *Mytilus edulis* in an experimental food-regulated situation. Mar. Ecol. Prog. Ser. **219**, 177–188 (2001)
10. Meldrum, F.C., Hyde, S.T.: Morphological influence of magnesium and organic additives on the precipitation of calcite. J. Cryst. Growth **231**, 544–558 (2001)
11. Sugawara, A., Kato, T.: Aragonite $CaCO_3$ thin-film formation by cooperation of Mg^{2+} and organic polymer matrices. Chem. Commun. 487–488 (2000)
12. Rosenthal, Y., Boyle, E.A., Slowey, N.: Temperature control on the incorporation of magnesium, strontium, fluorine, and cadmium into benthic foraminiferal shells from Little Bahama Bank: Prospects for thermocline paleoceanography. Geochim. Cosmochim. Acta **61** (17), 3633–3643 (1997)

13. Wanamaker, A.D., Kreutz, K.J., Wilson, T., Borns Jr., H.W., Introne, D.S., Feindel, S.: Experimentally determined Mg/Ca and Sr/Ca ratios in juvenile bivalve calcite for *Mytilus edulis:* implications for paleotemperature reconstructions. Geo-Mar. Lett. **28**(5–6), 359–368 (2008)
14. Jacob, D.E., Soldati, A.L., Wirth, R., Huth, J., Wehrmeister, U., Hofmeister, W.: nanostructure, composition and mechanisms of bivalve shell growth. Geochim. Cosmochim. Acta **72**, 5401–5415 (2008)
15. Gao, P., Liao, Z., Wang, X., Bao, L., Fan, M., Li, X., Wu, C., Xia, S.: Layer-by-layer proteomic analysis of *Mytilus galloprovincialis* shell. PLoS ONE **10**(7), e0133913 (2015)
16. Piwoni-Piorewicz, A., Kuklinski, P., Strekopytov, S., Humphreys-Williams, E., Najorka, J., Iglikowska, A.: Size effect on the mineralogy and chemistry of *Mytilus trossulus* shells from the southern Baltic Sea: implications for environmental monitoring. Environ. Monit. Assess **189**, 197 (2017)
17. Vander Putten, E., Dehairs, F., Keppens, E., Baeyens, W.: High resolution distribution of trace elements in the calcite shell layer of modern *Mytilus edulis*: environmental and biological controls. Geochim. Cosmochim. Acta **64**, 997–1011 (2000)
18. Yigit, M., Celikkol, B., Yimaz, S., Bulut, M., Ozalp, B., Dwyer, R., Maita, M., Kizilkaya, B., Yigit, U., Ergun, S., Gurses, K., Buyukates, Y.: Bioaccumulation of trace metals in Mediterranean mussels (*Mytilus galloprovincialis)* from fish farm with copper-alloy meh pens and potential risk assessment. Hum. Ecol. Risk Assess. **24**(2), 465–481 (2018)
19. Jovic, M., Onjia, A., Stankovic, S.: Toxic metal health risk by mussel consumption. Environ. Chem. Lett. **10**(1), 69–77 (2012)
20. Tanaskovki, B., Jovic, M., Mandic, M., Pezo, L., Degetto, S., Stankovic, S.: Elemental analysis of mussels and possible health risks arising from their consumption as a food: the case of Boka Kotorska Bay, Adriatic Sea. Ecotoxicol. Environ. Saf. **130**, 65–73 (2016)

Utilization of Atmospheric Plasmas for Agricultural Applications

Iuliana Motrescu[1,2(✉)], Manuela Filip[2], Florin Petrian Herciu[2], and Gerard Jitareanu[2,3]

[1] Science Department, University of Agricultural Sciences and Veterinary Medicine of Iasi, Iasi, Romania
imotrescu@yahoo.com
[2] Research Institute for Agriculture and Environment, University of Agricultural Sciences and Veterinary Medicine of Iasi, Iasi, Romania
[3] Pedotechnics Department, University of Agricultural Sciences and Veterinary Medicine of Iasi, Iasi, Romania

Abstract. Atmospheric pressure plasma fit great in agricultural applications due to their reduced complexity and to their chemical reactivity, being produced in air. In this work we present some of our results regarding agricultural applications of plasmas, obtained with seeds and soil exposed in atmospheric pressure plasmas conditions. The treatment of seeds shown a non-linear behavior with the exposure time and voltage for seed germination and development. The effects are strongly dependent on the type of seeds. Radish seeds were stimulated with lower voltages plasma and shorter exposures as compared to broccoli. In some conditions plasma exposure inhibited the growth, with lower germination rates than un-exposed samples and smaller size of the sprouts. For soil treatment we found the possibility to increase the nitrogen content of soil when tuning plasma treatment conditions, and we believe it is due to the reaction between reactive nitrogen species produced in plasma and organic components in soil.

Keywords: Plasma agriculture · Atmospheric pressure plasma applications · Seed germination · Spouting stimulation · Nitrogen content of soil · Soil treatment

1 Introduction

Plasma agriculture is a fast emerging field of plasma applications, with reported results such as stimulated germination and improved physical characteristics of the plants [1–3], inactivation of microorganisms [3–5], decreased harvest period and increased number of seeds and total seed weight [5], or all of the above [6, 7]. However some of these studies use plasma produced at low pressure [2, 9] which involved a vacuum system and expensive equipment, or make use of rare gases [7], expensive as well. Moreover, most of the studies present only some modifications of the physical properties of seeds and resulting plants [1, 9, 10] without any biochemical analysis or explanation of the mechanisms or obtained results.

A. R. Várkonyi-Kóczy (Ed.): INTER-ACADEMIA 2019, LNNS 101, pp. 278–287, 2020.
https://doi.org/10.1007/978-3-030-36841-8_27

Sprouts are very popular all around the world, in some countries especially from Asia. In others, like Romania, they are only recently on the market. Sprouts contain a wide range of bioactive compounds that are believed to be benefic for human health [11–13]. From the plasma agricultural point of view, the focus of performing plasma treatment of seeds is to reduce the costs of sprout production, and increase the safety and viability of the products on the shelf. In this study we focus on sprouts and study the effects of a non-thermal surface dielectric barrier discharge (sDBD) on the properties and viability of some plasma treated seeds. The sDBD used here has a simple and flexible configuration, with low operational costs, functioning in atmospheric pressure air. A similar configuration proved to be useful in inactivating *Geobacillus steraothermophilus* spores from the surface of medical equipment [14] and other pathogens like *Listeria, E. coli* and *Salmonella* from the surface of meat products [15].

In this work we study the effects of such sDBD on the treatment of two types of sprouting seeds, discussing the direct effects on the seeds and also indirect ones regarding the properties after sprouting in order to get an insight on the mechanisms and determine the optimum conditions of plasma treatment necessary to fulfill our goal of germination stimulation.

Another agricultural application approached by this work tests the possibility of soil fertilization using atmospheric plasma treatment. We propose this alternative to chemical fertilization methods used to increase the nitrogen content of soils, based on the fact that atmospheric pressure plasmas contain a lot of reactive nitrogen species (RNS) that could interact with the soil and fix the nitrogen in a plant-accessible form. Electrical discharges have been previously used in the industry of nitrogen fertilizers since the end of eighteen century, to aid the production of ammonia [16]. Using direct atmospheric plasma treatment, the use of chemical would be avoided, as well as other collateral possible issues of nitrogen fertilizers such as the increase of gas emission, and reduction of plant species diversity. Plasma treatment of soil was performed before [17] showing the influence of ozone on the chemical and biological properties of soil but without any focus on the nitrogen content.

2 Materials and Methods

The setup used for both seeds and soil treatments is schematically presented in Fig. 1. It consists of a flexible electrode configuration similar with the one reported by Eto et al. [14]. For the ease of treatments, instead of the polymer sheet we use a Petri dish with a metallic mesh attached to the inside as driving electrode, and a conductive aluminum tape on the outside acting as ground electrode. The discharge is ignited using 10 kHz high voltage provided by a variable power supply 0–18 kV (HV18K603AC). The signals are monitored on a RIGOL DS2072A oscilloscope. For the seed treatment, the seeds were placed on the grid electrode, while soil treatment took place by placing soil in a Petri dish and using the part with the discharge as cover.

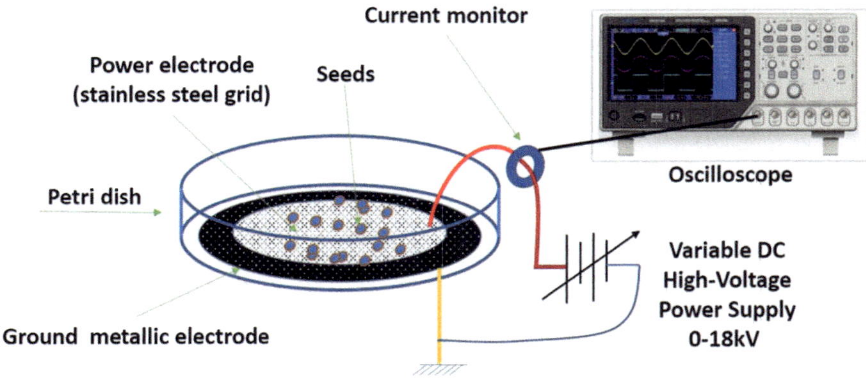

Fig. 1. Schematic representation of the experimental setup of sDBD for seed and soil treatment.

2.1 Seed Selection and Seedling Protocol

For the seed treatments we used two types of seeds: *Raphanus sativus* L. var. *longipinnatus*, which is Japanese radish, commonly known as daikon, very popular sprouts in Japan, and *Bassica oleranaceae* L. var. *italic*, which is broccoli sprouts, widely consumed around the world. Except for the different variety, the daikon seeds and broccoli seeds are different in size, with approximately 4–5 mm diameter in the case of daikon, and 2–3 mm in the case of broccoli. The seeds were commercially procured.

All the experiments were performed in triplicate. After sDBD treatments, the seeds were placed in vegetation pots with 4 ml of pure water per 100 seeds; each day the same amount of water was added. The average temperature during the experiments was 22 °C and the relative humidity varied between 55 and 70%.

The first count for the germination test was performed after the second day. The germination potential was calculated as being the number of germinated seeds in 3 days divided by the number of seeds times 100. The biometric measurements were performed during growth up to 7 days when the biochemical parameters were determined as well. The photosynthetic pigments were evaluated by spectrophotometric analysis of the extracts in acetone by recording the absorbance at 470, 662, and 645 nm with UV-1800 Spectrophotometer, according to the protocol [18].

2.2 Seed Analysis

The hydrophilicity of the seeds before and after plasma treatment was evaluated with the drop method. Because radish seeds are smaller than broccoli seeds, the latter were used to evaluate the water contact angle. A 1 μl of pure water was dropped on each seed and pictures were taken, then analyzed using free software imageJ with drop snake plugin to determine the water contact angle in each case [19]. For each condition 10 seeds were used, the values shown for discussions being statistically processed.

The morphology of the seed surface was studied as well for both types of seeds before and after sDBD exposure. For this, environmental scanning electron microscopy

(ESEM) was performed using FEI—Quanta 450. The seeds were analyzed without any preparation since ESEM allows such imaging for non-conductive samples. The seeds were directly mounted on stubs with adhesive carbon tape and imaged under high vacuum and using electrons accelerated at different voltages.

2.3 Soil Treatment and Analysis

The soil samples were collected from a field nearby Iasi, Romania, where the influence of no tillage system on soil properties is currently studied. The soil samples were collected in paper bags from a layer 0–10 cm in 3 repetitions. Following they were air dried for 72 h and then, using sieving apparatus from Eijelkamp, the aggregates with a diameter of 1–2 mm were separated and used for atmospheric plasma treatment.

After treatment the total and organic nitrogen fixed in the samples were determined using the Kjeldahl method based on the protocols provided by Gerhardt (based on the German standard method DIN 38409-28:1992-04): the samples were milled, weighted and put for digestion in Turbotherm system. The digested samples were cooled and then the water steam distillation was performed using Vapodest. The distillate was titrated until the color changed from green to violet. The total nitrogen was determined from the sample weight and titration volume. In the case of total nitrogen determination Devarda alloy was used for the extraction of inorganic nitrogen.

3 Results and Discussions

3.1 Seed Treatments

Using the configuration presented in Fig. 1, the seeds were treated for 30, 60, and 90 s, using 9, 10, and 11 kV, respectively. In these conditions, a bright surface DBD is ignited from the mesh up. Because the seeds are not very large, we can say that the treatment is relatively uniform.

The wettability of the seed surface is important for germination and growth. It has been already reported that low pressure plasma treatment can dramatically change the wettability of the seeds [7, 10]. Thus, we firstly analyzed the hydrophilicity of the seed surface. Figure 2 shows the results of water contact angle evaluation in the case of radish seeds. For all voltages there is a decrease of the water contact angle from about 114° to a minimum of 93°. It looks like the longer the exposure time, the lower the contact angle. However, we should note that there is no linearity between the decrease of contact angle with respect to increasing the discharge voltage. For the same exposure time, the decrease for 9 and 10 kV plasmas is higher than for 11 kV discharge, suggesting that there might be a threshold at which the changes rich a maximum without destruction. The most important decrease of the water contact angle of about 17% was determined for 10 kV, in which case we also performed measurements for 120 s (not shown here) and found that there is a saturation, the angle does not decrease over the accuracy range of its evaluation for 90 s.

The hydrophilicity change in the surface is usually related to the surface modification; there for we next studied the appearance of the seeds surface with and without

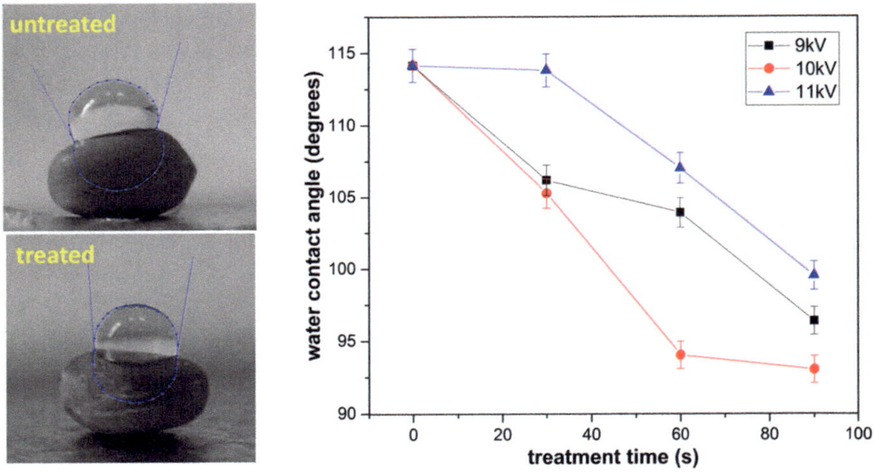

Fig. 2. Water contact angle change after plasma treatment of *Raphanus sativus* seeds.

plasma treatment using electron microscopy. The treated samples were exposed for 90 s to sDBD ignited with 10 kV HV signal. Figure 3 presents the appearance of the seeds at different magnifications and for different electron acceleration energies. Figure 4(b) and (c) show the appearance of radish seeds for the untreated and treated samples, respectively. A slight disintegration of the cell walls on the outer epidermis, irregular shaped agglomerations on the surface appear probably as a result of the integration with plasma species. Similar changes in the surface morphology have been recently reported for quinoa seeds processed in DBD plasma and at low pressure [20] and other species [21, 22]. Morphological changes in the seed coat structure can be also seen in the case of broccoli seeds both when imaging with 12.5 kV (Fig. 3(e) and (f)) or only the outmost surface using 1 kV (Fig. 3(g) and (h)), but the surface seems less affected than for the radish seeds.

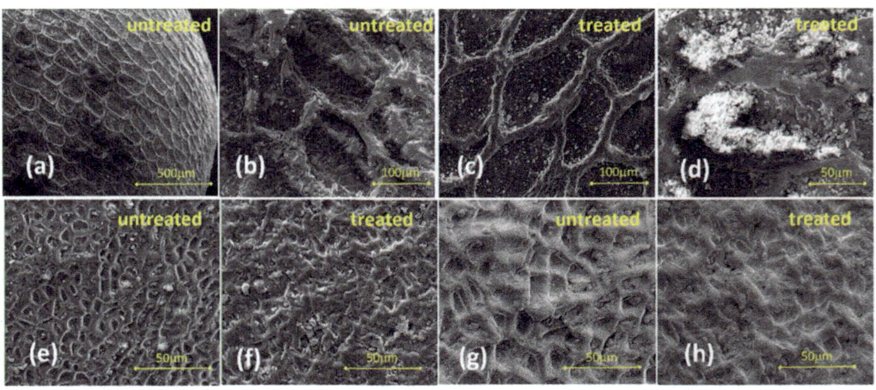

Fig. 3. ESEM images of (a)–(d) untreated and sDBD 10 kV 90 s treated *Raphanus sativus* seeds, and € to (h) *Brassica oleraceae*, respectively.

Regarding the viability of the seeds, Fig. 4 shows the results of the germination tests. A different response to atmospheric plasma treatment is found for the two species: in the case of the radish seeds, the viability decreases for all exposure conditions, while for broccoli it increases in some exposure circumstances. Although the majority of the recent reports discuss only the germination stimulation, we believe that the negative outcomes should also be taken in consideration to figure out the insights of the mechanisms by which the seeds are modified as a result of exposure to plasma. Up to our knowledge, the negative effects of plasma treatment were reported in the case of coffee seeds exposed to DBD [23]. Due to the complexity of the sDBD treatment, which is a combination of electric field, reactive species, UV radiation, and others, and the particularities of the seeds, it is very difficult to point out which cause lead to some specific effect due to the synergistic action of all factors. Studies shown that only the exposure to electric fields could influence the germination process, with stronger effects in the case of variable fields than for static ones [24, 25].

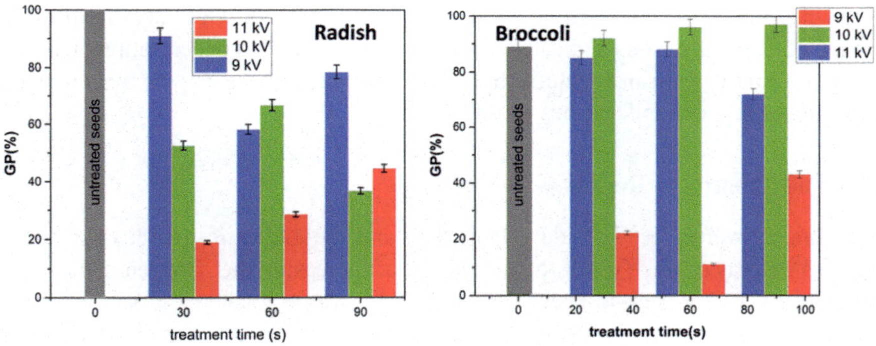

Fig. 4. Germination potential for (a) radish sprout seeds, and (b) broccoli sprouts seeds in different exposure conditions.

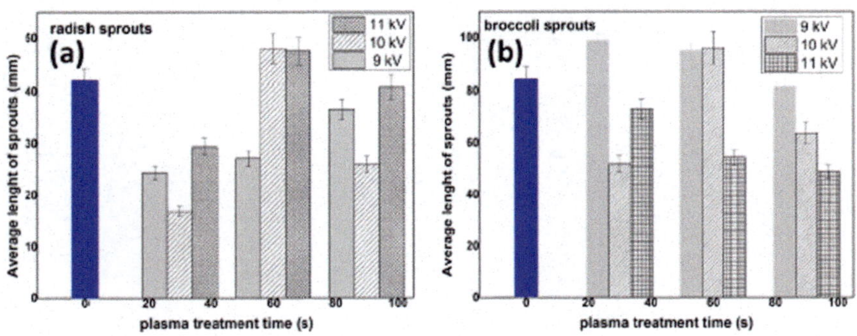

Fig. 5. Average length of sprouts measured in the 7th day for (a) radish, and (b) broccoli.

Although the germination shows an inhibition for some conditions, the biometric measurements indicate that plasma might enhance the growth of the sprouts. Figure 5 presents the average length determined for the viable plants in the same conditions as discussed above. Even if the germination potential was strongly affected for the radish seeds, some of the remaining ones, namely those treated for 60 s in plasma ignited in 10 and 11 kV, have longer length than control sprouts. For all other conditions the length is considerably smaller than untreated seeds, with up to 60%.

Broccoli sprouts have average length longer than control for the lower voltages and average treatment time (Fig. 5(b)), partially corresponding with the highest values of germination potential (Fig. 4(b)). 11 kV DBD seems to inhibit the development of the sprouts as well as their viability. It is interesting to notice that for 30 s treatment, the inhibition is stronger than for longer exposure, suggesting the existence of a defense mechanisms which initially blocks the metabolic pathways and following a longer synergistic impact of plasma components reopens them or facilitates by other means (e.g. changing cell morphology at seed surface, increasing hydrophilicity) the germination and growth mechanisms. However, the direct impact is not strong enough to affect the production and activity of chlorophyll pigments. Our measurements performed for all conditions indicated similar values of their concentration, within the error limit, so that we can conclude that the morphologic effects are the most noticeable rather than the biochemical ones in this case.

3.2 Soil Treatment Results

Soil samples were placed inside a Petri dish and the dielectric barrier electrode was placed on top as cover. To test the possibility of increasing the nitrogen content of the soil samples, 12 kV peak-to-peak signal at 10 kHz was used and treatment times of 10, 30, 60, 90, and 120 s for each sample. The experiment was performed in triplicate then each sample was divided in three parts which were tested to determine the total and organic nitrogen content as presented in the experimental part of the manuscript.

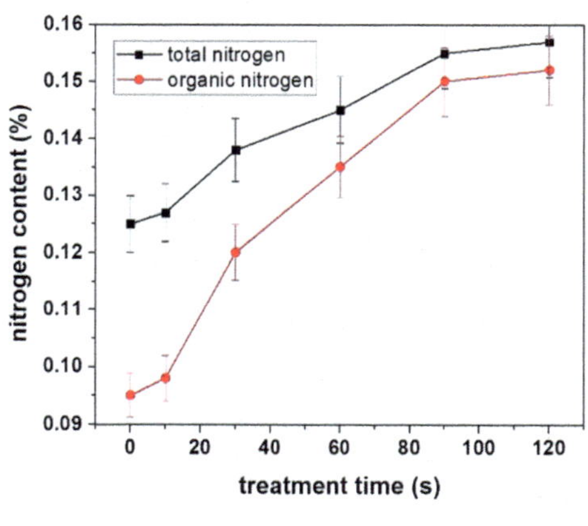

Fig. 6. Total and organic nitrogen content of untreated and plasma treated soil samples.

The results of nitrogen content of plasma treated and untreated soil samples are presented in Fig. 6. Both total and organic nitrogen are increasing with exposure time, in the first 60 s with about the same rate, then the total nitrogen content doesn't increase as much. From this information we can say that for short plasma exposure time, the nitrogen content of the soil increases due to an increase of the organic nitrogen content, possibly caused by the interaction of reactive nitrogen species (RNS) produced by the plasma and the organic part of the soil. For longer exposures, the reactive species, probably mainly oxygen related (ROS) among which ozone react with the inorganic nitrogen and produce volatile species, decreasing the total nitrogen content of exposed soil samples. These preliminary results are encouraging and show great premises of extending the application range in plasma agriculture to soil treatments. Of course, there are still many aspects to figure out that we shall address in another paper.

4 Conclusions

In this work we studied the effects of a sDBD atmospheric plasma on seeds and soil as agricultural applications. In the case of seeds, the ESEM microanalysis of their surface morphology and the measurements of water contact angle shown a modification of the seed surface with an increase of seed hydrophilicity in the case of radish seeds. The decrease of water contact angle might be one of the factors influencing the germination stimulation that was found in the case of broccoli seeds.

Tuning the exposure conditions can lead to both stimulation of germination and growth. We should also consider the negative impact plasma can have on plant development when seeds are irradiated, and more attention from the research community should be given not only to positive impact but on the inhibition effects, as to have a detailed image on the interaction and effects. SDBD can be easily used for seed treatment applications; the optimum conditions seem to strongly depend not only on plasma parameters but very much on the type of treated seeds.

Soil treatments using sDBD shown the possibility of using atmospheric discharges to increase the nitrogen content of soil. The results indicate that the increase might be due to the interactions of NRS with the organic compounds found in soil. Further investigations are necessary to understand the complex interactions and better control the mechanisms.

References

1. Kitazaki S, Sarinont T, Koga K, Hayashi N, Shiratani M (2014) Plasma induced long-therm growth enhancement of Raphanus sativus L. using combinatorial atmospheric air dielectric barrier discharge plasmas. Curr Appl Phys 14:S149–S153
2. Dhayal M, Lee SY, Park S-U (2006) Using low-pressure plasma for Carthamus tinctorium L. seed surface modification. Vacuum 80:499–506
3. Zahoranova Z, Henselova M, Hudecova D, Kalinakova B, Kovacik D, Medvecka V, Cernak M (2016) Effect of cold atmospheric pressure plasma on the wheat seedlings vigor

and on the inactivation of microorganisms on the seed surface. Plasma Chem Plasma Process 36:397–414
4. Ito, M., Ohta, T., Hori, M.: Plasma agriculture. J Korean Phys Soc **60**, 937–943 (2012)
5. Iseki, S., Ohta, T., Aomatsu, A., Ito, M., Kano, H., Higashijima, Y., Hori, M.: Rapid inactivation of *Penicillium digitatum* spores using high-density nonequilibrium atmospheric pressure plasma. Appl Phys Lett **96**, 153704 (2010)
6. Koga K, Thapanut S, Amano T, Seo H, Itagaki N, Hayashi N, Shiratani M (2016) Simple method of improving harvest by nonthermal air plasma irradiation of seeds of *Arabidopsis thaliana* (L.). Appl Phys Express 9:016201
7. Jiafeng, J., Xin, H., Ling, L., Jiangang, L., Hanliang, S., Qilai, X., Renhong, Y., Yuanhua, D.: Effect of cold plasma treatment on seed germination and growth of wheat. Plasma Sci Technol **16**, 54 (2014)
8. Jiang, J., Lu, Y., Li, J., He, X., Shao, H., Dong, Y.: Effect of seed treatment by cold plasma on the resistance of tomato to Ralstonia solanacearum (Bacterial Wilt). PLoS ONE **9**, e97753 (2014)
9. Ling, L., Jiangang, L., Minchong, S., Chunlei, Z., Yuanhua, D.: Cold plasma treatment enhances oilseed rape seed germination under drought stress. Sci Rep **5**, 13033 (2015)
10. Bormashenko, E., Grynyov, R., Bormashenko, Y., Drori, E.: Cold radiofrequency plasma treatment modifies wettability and germination speed of plant seeds. Sci Rep **2**, 741 (2012)
11. Donkor, O.N., Stojanovksa, L., Ginn, P., Ashton, J., Vasiljevic, T.: Germinated grains—sources of bioactive compounds. Food Chem **135**(3), 950–959 (2012)
12. Gan, R.Y., Lui, W.Y., Wu, K., Chan, C.L., Dai, S.H., Sui, Z.Q., Corke, H.: Bioactive compounds and biactivities of germinated edible seeds and sprouts: an updated review. Trends Food Sci Technol **59**, 1–14 (2017)
13. Nestle, M.: Broccoli sprouts in cancer prevention. Nutr Rev **56**(4), 127–130 (1998)
14. Eto, H., Ono, Y., Ogino, A., Nagatsu, M.: Low-temperature sterilization of wrapped materials using flexible sheet-type dielectric barrier discharge. Appl Phys Lett **93**, 221502 (2008)
15. Jayasena, D.D., Kim, H.J., Yong, H.I., Park, S., Kim, K., Choe, W., Jo, C.: Flexible thin-layer dielectric barrier discharge plasma treatment of pork butt and beef loin: effects on pathogen inactivation and meat-quality attributes. Food Microbiol **46**, 51–57 (2015)
16. Patil, B.S., Wang, Q., Hessel, V., Lang, J.: Plasma N_2-fixation: 1900–2014. Catal Today **256**, 49–66 (2015)
17. Stryczewska, H.D., Ebihara, K., Takayama, M., Gyoutoku, Y., Tachibana, M.: Non-thermal plasma-based technology for soil treatment. Plasma Process Polym **2**, 238–245 (2005)
18. Lichtenthaler, H.K., Wellburn, A.R.: Determinations of total carotenoids and chlorophylls a and b of leaf extracts in different solvents. Biol Soc Trans **11**, 591–592 (1985)
19. https://imagej.nih.gov/ij/index.html
20. Gomez-Ramirez, A., Lopez-Santos, C., Cantos, M., Garcia, J.L., Molina, R., Cotrino, J., Espinos, J.P., Gonzalez-Elipe, A.R.: Surface chemistry and germination improvement of Quinoa seeds subjected to plasma activation. Sci Rep **7**, 5924 (2017)
21. Bafoil, M., Jemmat, A., Martinez, Y., Merbahi, N., Eichwald, O., Dunand, C., Mohammed, Y.: Effects of low temperature plasmas and plasma activated waters on *Arabidopsis thaliana* germination and growth. PLoS ONE **13**(4), e0195512 (2017)
22. Guo, Q., Meng, Y., Qu, G., Wang, T., Yang, F., Liang, D., Hu, S.: Improvement of wheat seed vitality by dielectric barrier discharge plasma. Bioelectromagnetics **39**(2), 120–131 (2017)

23. Tounekti, T., Mujahid, Z.I., Khemira, H.: Non-thermal dielectric barrier discharge (DBD) plasma affects germination of coffee and grape seeds. AIP Conf Proc **1976**, 020029 (2018)
24. Sidaway, G.H.: influence of electrostatic fields on seed germination. Nature **211**, 303 (1966)
25. Pietruszewski S (2014) Electromagnetic fields, impact on seed germination and plant growth. Encycl Agrophys Encycl Earth Sci Ser 267–269

Colonoscopy Videos: Towards Automatic Assessing of the Bowels Cleansing Degree

M. Luca[1(✉)], A. Ciobanu[1], and V. Drug[2,3]

[1] Institute of Computer Science, Romanian Academy, Iaşi Branch, Bd. Carol I, Nr. 8, 700505 Iaşi, Romania
mihaela.luca@iit.academiaromana-is.ro
[2] University of Medicine and Pharmacy, "Gr. T. Popa", Iaşi, Romania
[3] Institute of Gastroenterology and Hepathology, "Sf. Spiridon", Emergency Hospital, Iaşi, Romania

Abstract. In the attempt to decrease the number of colon cancers deaths, colonoscopy is one of the main screening tests recommended by the American and European guidelines, as well as the updated Asia Pacific consensus statements, meant to early detect abnormal structures formed on colon surface. In order to obtain the best images, a very effective colon cleansing is necessary. Thus, polyps, diverticulitis, or any peculiar aspects of the intestinal membrane, might be observed. Subjective evaluation influenced by various cleansing degrees might conduct to different results, or even to omissions. Expert assessment variability is another factor influencing the diagnosis. We further describe special software, useful for an objective, semi-supervised, evaluation of bowel cleansing degree.

Keywords: Video colonoscopy · Image processing · Narrow band imaging · La*b* · RGB color spaces

1 Introduction

Trying to improve the statistics (colo-rectal cancer being the second leading cause of deaths in United States [1]), one of the main steps is to certify the high quality of the medical procedures. Different evaluation standards for bowel preparation degree exist in actual medical practice all over the world. Two of the main standards used in medical practice are: the Boston Bowel Preparation Scale (BBPS) [2] and, more recently the Chicago Bowel Preparation Scale (CBPS) [3]. They mainly rely on practitioner experience and subjective evaluation. In order to be more accurate, some automatic attempts in computing the degree of bowel cleansing using colonoscopy video frames have been described in the last decades [4, 5]. These studies used the RGB color space, while we have approached it using the La*b* color space [6, 7], which allows a better and simpler (two-dimensional) color localization [8], making easier to find related colors and to estimate the covered areas. In our previous studies we used a video colonoscopy record with 5100 selected frames.

A. R. Várkonyi-Kóczy (Ed.): INTER-ACADEMIA 2019, LNNS 101, pp. 288–297, 2020.
https://doi.org/10.1007/978-3-030-36841-8_28

We continued our study on 17 normal video colonoscopies with diverse pathologies and a colonoscopy with Narrow Band Imaging, in order to identify new peculiarities impeding or alleviating fast and reliable automatic evaluation of video colonoscpies.

2 From Video Colonoscopy to Image Processing

Video colonoscopies together with occult tests are standard analyses recommended worldwide to early detect colorectal cancer and to decrease mortality, according to the actual statistics [9–13]. The colonoscopy has the advantage of permitting local resection and further biopsy of the abnormal polyps or adenomas which are detected. The quality of the colonoscopy and the results of the medical procedure depend primarily on two factors: the expert skills and the proper bowls preparation.

2.1 Boston Bowel Preparation Scale

The video colonoscopy is a laborious procedure, sometimes necessitating sedation, mainly for old people or for patients who were previously submitted to abdomino-pelvic surgery, sometimes being even at risk due to this [14].

The endoscope with the video camera (or, for the new types, cameras [15]) passes through the normal segments of the colon having a different temporization on its way to the terminal ileum and backwards.

Usually, the left segment (rectum, sigmoid, descending colon) has to be browsed in 30% of the insertion phase, transverse colon 30% and right segment 40% (ascending colon, then cecum, towards the terminal ileum). Backwards, experts agree that the withdrawal phase has a different temporisation: 30% right segment, 30% for transverse colon and 40% for the left segment of the colon. The withdrawal phase should last at least 6 min.

Main colon segments cleansing has to be observed, international standards suggesting marks (or ratings) some of them being quite *linguistically flue*, without standardized definitions [2].

Thus, in Boston Bowel Preparation Scale (BBPS), which is a "10-point scale assessing bowel preparation after all cleansing manoeuvres are completed by the endoscopist" [2], the physician has have to note if the cleansing degree is "*poor*" *or* "*unsatisfactory*", "*fair*", "*good*" or "*excellent*". Some remarks have to be done: it was not clear in BBPC [2] if the quality linguistic ratings have to be given in the insertion phase or in the withdrawal phase as they differ. The former give more clues about the purgative substance quality, while the latter, after washing and suctioning the fluids on the covered surfaces, might be a value also referring to the polyps' detection rate.

A correspondence to a scale of four-point marks was established [2] (for "more objective" scoring, yet still upon the practitioner appreciation, with intra and inter-practitioner variability), for each colon segment, in the withdrawal phase: 0 is a mark for unprepared colon, covered with solid faecal materials impossible to be cleaned; 1 is a mark given for partial covered areas of the colon segments, with solid residues or opaque liquid; 2 is corresponding to minor residues staining, yet mucosa of the colon segment is visible; 3 when the entire colon segment mucosa is well seen, no residual staining, no fragments, no opaque liquid.

The marks are given for every segment (from 0 to 3) and then summed, thus the final score will be in the range 0–9.

Note: A Korean study states that the BBPS is usually inversely correlated with the colonoscope withdrawal time [16], which is directly proportional with adenoma detection rate [17].

2.2 Pitfalls in Image Acquisition

Trying to clean the covered bowels, the fluids are modifying the images, resulting in a specific video acquisition. This justifies the necessity for image preprocessing and for computer aided systems procedures.

A video colonoscopy might last up to 20 min upon its complexity, yet a duration of 13–17 min is more usual. We might have 40–50000 frames in a split video, and we have chosen a rate of 1/10 images (a satisfactory amount of information which implies an acceptable amount of computing).

The problem arises when we study these frames, a lot of them being blurred, covered with water, reflections and shadows, white speckles and an excessive amount of light.

In order to process the image content we obviously have to select good images, normal/abnormal frames, issue that was the subject of some previous publications [6, 18–21].

The following figure shows normal, good images among the blurred, non-informative ones, irrelevant for further evaluations and time consuming if we continue to maintain them in the image data base that we have to select. It was often proven that good results rely on good data sets. Discarding the ambiguous, irrelevant data conducts to unexpected improvement for the expected results.

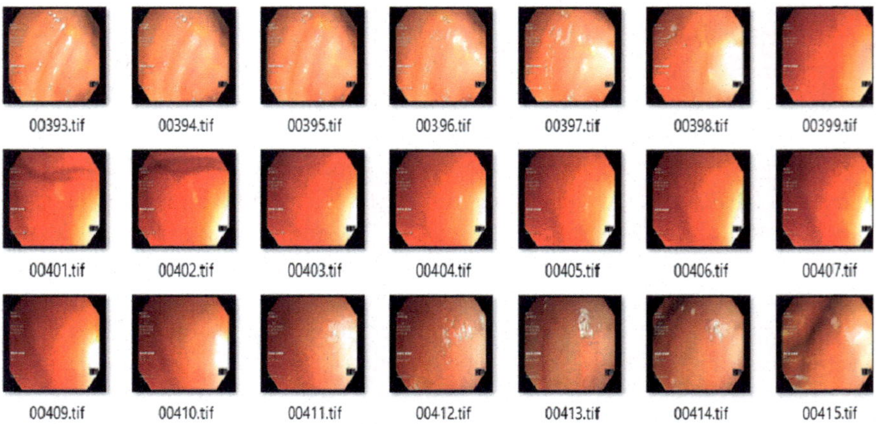

Fig. 1. Split video colonoscopy: good (valid) and (second raw) blurred (irrelevant) frames

Our method is a primary type of discarding useless images, based on sequential evaluation of image entropy on quarters of the colon picture situated inside. Entropy is well known in physics as a measure of disorder: still environment, "calm" background, is a sign of very low entropy, thus conducting to low quantity of information.

2.3 Informative Versus Non-informative Frames

In the following figure we illustrate the entropy variation on the images selected from the video colonoscopy. We selected a medium quality image (half is blurred, half contains some information about the stool covered areas). The picture given by the colonoscope is analyzed by quarters in a first stage, from upper left side in the hourly sense and a graphic is plotted, showing in different colors each value for each frame.

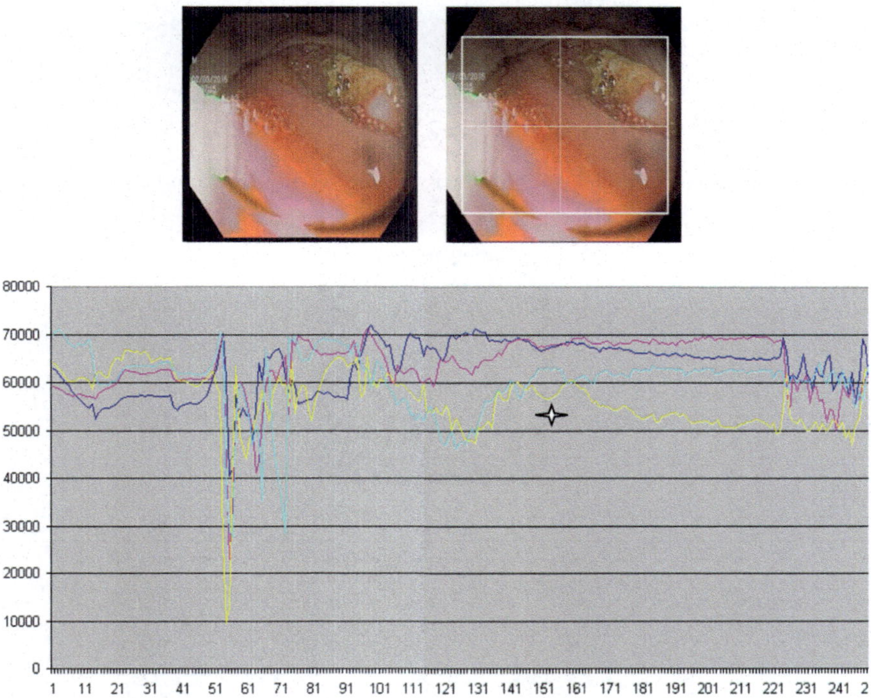

Fig. 2. Image 154/5000 video colonoscopy frames, fair to low quality of the image, characterized by medium entropy, computed for each of the four quadrants, starting from upper left magenta, upper right cyan, lower right blue, lower left yellow, anti-trigonometric sense.

On the Ox axes we have the number of the selected frames, from the original video colonoscopy, on the Oy axes we have the values corresponding to the entropies computed in Matlab. Very low entropies correspond to totally inutile images as observed in Figs. 3, and 4. Obviously, the selection procedure has to be further enhanced, as sometimes even a half of the image might contain an important informative clue.

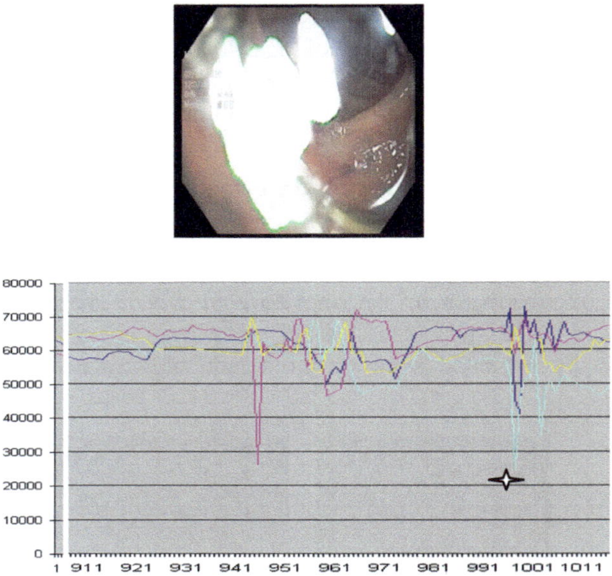

Fig. 3. Image 997/5000, very low entropy, poor quality, frame to be discarded

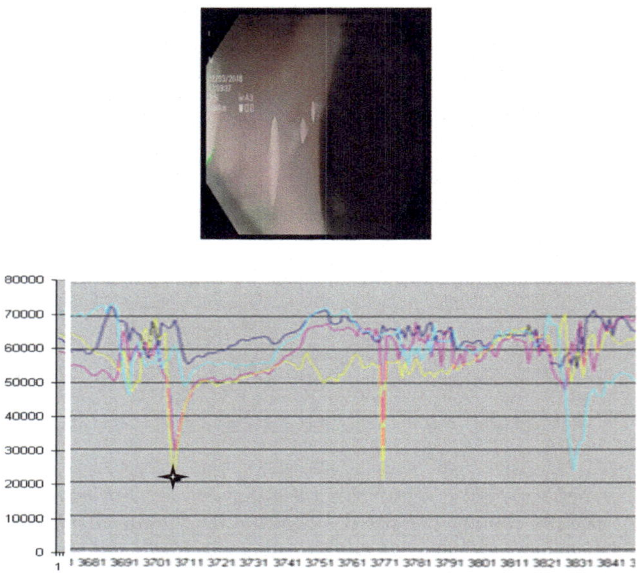

Fig. 4. Image 3704/5000, totally irrelevant, very low entropy, discarded frame

After this first step more than 10% of the frames have been discarded. Further the procedures differ depending on the purpose we aim to. In order to automatically compute the stool covered areas, which would be more objective than the endoscopist appreciation, subject of variability, the range of colors corresponding to the person (depending on the personal nutrition habits and bladder/liver health or disease) has to be established. Due to luminance it is more easy to work in the La*b* space than in RGB. If our purpose is to automatically detect polyps, adenomas or diverticula, another strategy has to be conceived [6, 31].

2.4 Using La*b* Color Space

La*b* color space is organized along a lightness axe (**L***), with perpendicular color planes arranged over a green–red (a*) axe and a blue–yellow (b*) axe, [6–8].

The results of La*b* color selection on video colonoscopy frames, are illustrated in Figs. 5 and 6: stool in red, and white light and speckles selection in blue.

Counting frame by frame for each colon segment, and reporting to the background surface we might have the objective values we search for.

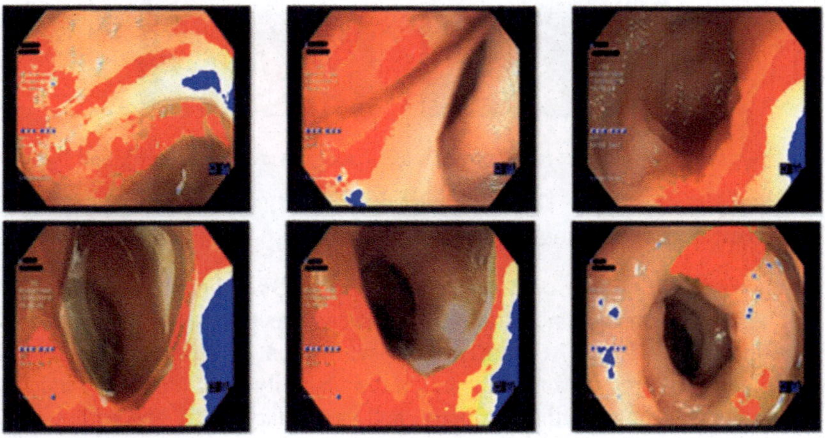

Fig. 5. Segmentation results in normal images, obtained by the La*b* method: red-stool presence, blue-white speckles areas

We will further describe the steps for the software computing of the covered areas. The software implementation in order to objectively compute a score on the Boston Bowel Preparation Scale (BBPS) consists of:

- the three main colon segments are marked by the endoscopist on the video sequence;
- the stool color range is identified for every colonoscopy by inspecting relevant frames;
- the colonoscopy video is parsed frame by frame;
- entropy is computed frame by frame;
- low entropy frames are discarded;
- the stool covered area is computed for each frame and a percent of covering is stored for each frame;

- stool covered area S is summed;
- S is compared to the uncovered surface;
- based on the BBPS and the stored percent of stool covered area, with a special aggregation method, a number is assigned to each segment of the colon;
- in the event of an extremely high ratio of stool presence in a frame, the corresponding image is stored as evidence of non-conformity to the standards, and its influence on the final score is increased by a reinforcement algorithm;
- *results*: aggregation is made, using an impact coefficient for the extreme cases (entirely covered frames);
- otherwise, each segment BBPS objectively evaluated is computed for the final evaluation.

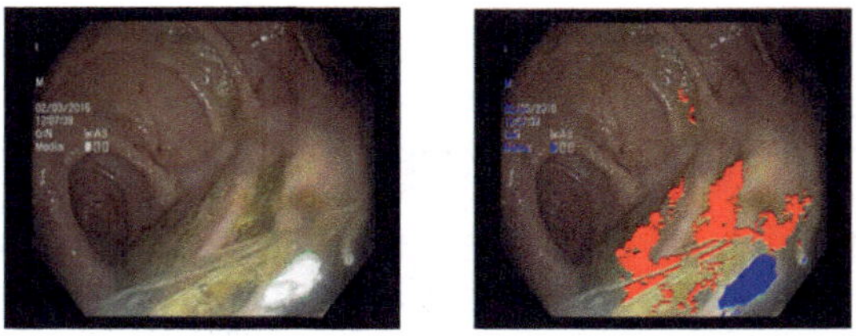

Fig. 6. Segmentation on a partly-blurred image: stool covered areas (red), and light (blue)

2.5 Narrow Band Imaging (NBI)

A new approach for colonoscopy consists in using NBI technique [22–30].

NBI is relying on light penetration properties, directly proportional to the wavelength. Blue light (415 nm) enhances the visualization of superficial mucosal capillaries while green light (540 nm) increases the visibility of submucosal and mucosal vessels" corresponding to the "primary and secondary light absorption peaks of hemoglobin, respectively" [15, 26, 30].

NBI completely modifies the range of colors in the intestinal observation, with clear advantages for evidentiating blood vessels and adenoma/polyp textures. Citing the OLYMPUS, NBI technology productor [30], "Capillaries in the superficial mucosa appear brown in 415 nm wavelength. The 540-nm wavelength penetrates slightly more deeply into the mucosa and submucosa and makes the deeper veins appear blue-green (cyan). Because most of the NBI light is absorbed by the blood vessels in the mucosa, the resulting images emphasize the blood vessels in sharp contrast with the nonvascular structures in the mucosa" [30].

We are questioning if this technique is also aleviating the automatic BBPS evaluation on video colonoscopy frames?

NBI sends blue and green light on the colon surface and for some reason, stool debris reflects the red wavelength, as well as some membrane components which also turn red under this light, This way, our method had difficulties in separating normal tissue color in NBI from the stool covered areas colors, even if the endoscopist slightly

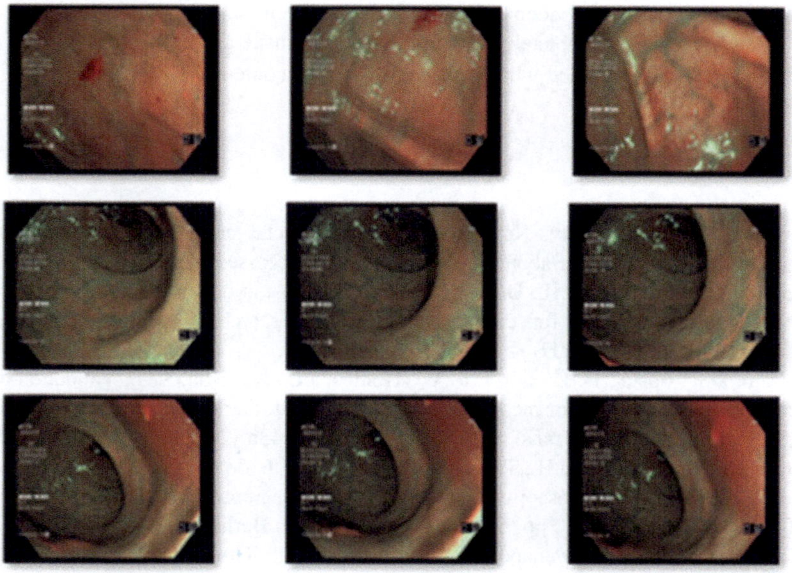

Fig. 7. Narrow Band Imaging (NBI) colonoscopy frames: colors are modified, blood vessels are dark green to brown, inner vessels are cyan, stool traces and some tissues appears in red

manage to explain the difference, and the results that we primarily obtained for software computing were not encouraging, thus the answer seams to be negative.

Yet, research in this field is in continuous progress, numerous scientific research teams concentrating on this challenge with a big impact on human's health.

An annual competition is taking place, MICCAI 2019, ENDOVIS, Endoscopic Vision Challenge [32].

More data bases are publically available as CVC Colon DB [33] and ASU MAYO DB [34] together with the previous one, providing data for the competition [32].

3 Conclusions

Previous attempts of automatic Boston Bowel Preparation Score computing have been made using RGB color space, characterized by difficulties in assigning a color 3D volume to be detected, corresponding for stool presence in colon. We developed a complementary method in the LAB color space, which offers an algorithm easy to compute, results being obtained faster. This procedure can be applied to the video recordings, saving time and facilitating the computer-assisted analysis of the cleansing aspects, relevant for colonoscopy. For the NBI technique the results are very promising for vessels and tissue/polyp texture identification, while the BBPS score is better obtained in the normal colonoscopy video frames.

Acknowledgement. All the video colonoscopy images were obtained with the written consent of the patients and were completely anonymised for the image processing.

No personal data is detained whatever upon the image content.

References

1. American Cancer Society, Key statistics for colorectal cancer. https://www.cancer.org/cancer/colon-rectal-cancer/about/key-statistics.html. Accessed June 2019
2. Lai, E.J., Calderwood, A.H., Doros, G., et al.: The Boston bowel preparation scale: a valid and reliable instrument for colonoscopy-oriented research. Gastrointest. Endosc. Clin. N. Am. **69**, 620–625 (2009)
3. Gerard, D.P., Foster, D.B., Raiser, M.W., Holden, J.L., Karrison, T.G.: Validation of a new bowel preparation scale for measuring colon cleansing for colonoscopy: The Chicago bowel preparation scale. Clin. Transl. Gastroenterol. **4**(12): e43n (2013)
4. Muthukudage, J.K., Oh, J.H., Tavanapong, W., Wong, J., de Groen, P.C.: Color based stool region detection in colonoscopy videos for quality measurements. In: Ho, Y.-S. (ed.) PSIVT 2011, Part I, LNCS 7087, pp. 61–72. Springer-Verlag, Berlin (2011)
5. Hwang, S., Oh, J., Tavanapong, W., Wong, J., de Groen, P.C.: Stool detection in colonoscopy videos. In: Proceedings of International Conference of the IEEE Eng. in Medicine and Biology Society (EMBC), Vancouver, British Columbia, Canada, pp. 3004–3007 (2008)
6. Ciobanu, A., Luca, M., Drug, V., Tulceanu, V.: Steps towards computer-assisted classification of colonoscopy video frames. In: 6th IEEE International Conference on E-health and Bioengineering—EHB Sinaia, Romania (2017)
7. Luca, M., Ciobanu, A., Drug, V.: LAB automatic evaluation of colon cleansing. In: ESGE Days 2019, ePP50, Abstract Volume, p. 145 (2019)
8. Ciobanu, A., Costin, M., Barbu, T.: Image categorization based on computationally economic lab colour features. In: Balas, V., Fodor, J., Várkonyi-Kóczy, A., Dombi, J., Jain, L. (eds.), Soft Computing Applications. Advances in Intelligent Systems and Computing, vol. 195, pp. 585–593. Springer, Berlin (2013)
9. World Health Organization, Fact Sheets: Cancer, Key Facts. http://www.who.int/news-room/fact-sheets/detail/cancer. Accessed June 2018
10. Colorectal Cancer Facts & Figures, 2017–2019, American Cancer Society. https://www.cancer.org/content/dam/cancer-org/research/cancer-facts-and-statistics/colorectal-cancer-facts-and-figures/colorectal-cancer-facts-and-figures-2017-2019.pdf (2019)
11. World Cancer Research Fund International: Colorectal cancer statistics (2015). https://www.wcrf.org/int/cancer-facts-figures/data-specific-cancers/colorectal-cancer-statistics. Accessed May 2019
12. Cancer statistics-specific cancers, statistics explained. http://ec.europa.eu/eurostat/statistics-explained/pdfscache/39738.pdf. Accessed June 2019
13. Eurostat, Statistics Explained. https://ec.europa.eu/eurostat/statistics-explained/index.php/Cancer_statistics. Accessed June 2019
14. Bun Kim, et al.: Quality of bowel preparation for colonoscopy in patients with a history of abdomino-pelvic surgery: retrospective cohort study. Yonsei Med. J. **60**(1), 73–78 (2019)
15. Ngu, W.S., Rees, C.: Can technology increase adenoma detection rate? Ther. Adv. Gastroenterol. **11**, 1–18 (2018)

16. Eun-Jin Kim, et al.: A Korean experience of the use of boston bowel preparation scale: a valid and reliable instrument for colonoscopy—oriented research. Saudi J. Gastroenterol. **20** (4), 219–224 (2014)
17. Douglas, K.R.: Colonoscopy withdrawal times and adenoma detection rates. Gastroenterol. Hepatol. (N Y) **3**(8), 609–610 (2003)
18. Manivannan, S., Wang, R., Trucco, E., Hood, A.: Automatic normal-abnormal video frame classification for colonoscopy. In: 2013 IEEE 10th International Symposium on Biomedical Imaging: From Nano to Macro, (ISBI), San Francisco, CA, USA, 7–11 (2013). https://pdfs.semanticscholar.org/696c/ef94b8656a86b01cda1c580ba586adef3265.pdf
19. Bashar, M.K., Kitasaka, T., Suenaga, Y., Mekada, Y., Mori, K.: Automatic detection of informative frames from wireless capsule endoscopy images. Med. Image Anal. **14**(3), 449–70 (2010). https://www.ncbi.nlm.nih.gov/pubmed/20137998. Accessed June 2018
20. Tajbakhsh, N., Chi, C., Sharma, H., Wu, Q., Gurudu, S.R., Liang, J.: Automatic assessment of image informativeness in colonoscopy. In: ABDI@MICCAI 2014, pp. 151–158 (2014)
21. Sanchez, F.J., Bernal, J., Sanchez-Montes, C., Rodriguez de Miguel, C., Fernandez-Esparrach, G.: Bright spot regions segmentation and classification for specular high-lights detection in colonoscopy videos. J. Mach. Vis. Appl. 1–20. http://refbase.cvc.uab.es/files/SBS2017.pdf. Accessed July 2018
22. Kuznetsov, K., Lambert, R., Rey, J.F.: Narrow-band imaging: potential and limitations. Endoscopy **38**, 76–81 (2006)
23. Gono, K., Obi, T., Yamaguchi, M., et al.: Appearance of enhanced tissue features in narrow-band endoscopic imaging. J. Biomed. Opt. **9**, 568–577 (2004)
24. Sano, Y., et al.: Narrow-band imaging (NBI) magnifying endoscopic classification of colorectal tumors proposed by the Japan NBI expert team. Rev. Dig. Endosc. **28**, 526–533 (2016)
25. Sano, Y., Kobayashi, M., Kozu, T., et al.: Development and clinical application of a narrow band imaging (NBI) system with builtin narrow-band RGB filters. Stom. Intest. **36**, 1283–1287 (2001)
26. Sano, Y.: NBI story. Early Colorectal Cancer **11**, 91–92 (2007)
27. Kaltenbach, T., Friedland, S., Soetikno, R.: A randomised tandem colonoscopy trial of narrow band imaging versus white light examination to compare neoplasia miss rates. Gut J. **57**, 1406–1412 (2008)
28. Nagorni A., Bjelakovic G., Petrovic B.: Narrow band imaging versus conventional white light colonoscopy for the detection of colorectal polyps. Cochrane Database Syst. Rev. 18 (2012)
29. Su M.Y., Hsu C.M., Ho Y.P., Chen P.C., Lin C.J., Chiu C.T.: Comparative study of conventional colonoscopy, chromoendoscopy, and narrow-band imaging systems in differential diagnosis of neoplastic and nonneoplastic colonic polyps. Am. J. Gastroenterol. **101**(12), 2711–2716. Accessed June 2019
30. OLYMPUS, Narrow Band Imaging (NBI): A new wave of diagnostic possibilities. https://www.olympus-europa.com/
31. Geetha, K., Rajan, C.: Automatic colorectal polyp detection in colonoscopy video frames. Asian Pac. J. Cancer Prev. **17**(11), 4869–4873 (2016)
32. MICCAI. https://endovis.grand-challenge.org/endoscopic_vision_challenge/. Accessed June 2019
33. CVC Colon DB. http://www.cvc.uab.es/CVC-Colon/index.php/databases/. Accessed June 2019
34. ASU Mayo DB. https://polyp.grand-challenge.org/site/polyp/asumayo/. Accessed June 2019

Modeling and Simulation

Numerical Investigation of Gas-Liquid Two-Phase Flow in a Stirred Tank

Nao Ishida[1], Al Abri Mohammed[1], Atsushi Sekimoto[1], Yasunori Okano[1](✉), Shinya Abe[2], and Kosuke Tanaka[2]

[1] Department of Materials Engineering Science, Osaka University, 1-3, Machikaneyama, Toyonaka, Osaka 560-8531, Japan
okano@cheng.es.osaka-u.ac.jp
[2] KANEKA, 1-8, Miyamae-Cho, Takasago-Cho, Takasago, Hyogo 676-8688, Japan

Abstract. Gas-liquid stirred tanks are widely used in various chemical engineering processes such as fermentation and pharmaceutical production. In the mixing process for fermentation, aerobic microbes produce high polymer compounds from oxygen and carbon sources which need to be supplied by agitation and aeration in a tank. It is essential to supply a sufficient and uniform amount of nutrients to provide high quality and quantity products, and hence, it is necessary to optimize the shape and operating conditions of the stirring tank to mix gas and materials in the liquid sufficiently and uniformly. However, it is difficult to obtain a guideline for the design of plant-scale tanks from laboratory-scale experiments because experimental measurements cannot provide the detailed distribution of materials. The present study aims to obtain a guideline for an optimized design by developing computational fluid dynamics (CFD) simulation of the gas-liquid two-phase flow in the stirred tank and investigating the mass transfer inside the stirred tank in detail. Applying k-ω shear stress transport (SST) model, we calculate the gas-liquid flow in the stirred tank and estimate overall gas hold-up in comparison with experimental data.

Keywords: Gas-liquid flow · Stirred tank · Gas hold-up · Numerical simulation

1 Introduction

Gas-liquid stirred tanks are widely used in chemical, biochemical and mineral processing industries to carry out a variety of operations, such as fermentation, pharmaceutical production, and wastewater treatment [15]. It is essential to provide aerobic microbes with a sufficient and uniform amount of nutritive compounds (source of carbon and energy, organic nitrogen and oxygen) for the improvement of the qualities and the production efficiency [3].

It is required to optimize the design and operating conditions of a stirred tank in order to perform a uniform mixing. There are some problems to optimize the stirred tank by experiments. Firstly, direct observation is difficult in the case of a high gas flow rate. Secondly, it is difficult to quantitatively evaluate the stirring performance and the

A. R. Várkonyi-Kóczy (Ed.): INTER-ACADEMIA 2019, LNNS 101, pp. 301–309, 2020.
https://doi.org/10.1007/978-3-030-36841-8_29

mass transfer of oxygen to the liquid phase. Thirdly, there is a large number of parameters (the shape of a stirred tank, the shape and the number of impellers and baffles, the gas flow rate and the agitation speed).

During the past few decades, hydrodynamics and transport phenomena in a stirred tank have been studied using computational fluid dynamics (CFD). A widely used approach to model gas-liquid two-phase flow is the Eulerian-Eulerian approach [3, 7, 15, 17, 18]. This approach considers both gas and liquid phase as a continuum, and the balance equations of mass and momentum are solved for each phase separately and weighted by the corresponding volume fraction [17]. Source terms in the balance equations describe the momentum exchange between the phases, and the distribution of bubble diameter is prescribed for the calculation of the momentum exchange term.

The bubbles injected in a stirred tank are non-uniformly distributed due to the bubble breakup and coalescence. Barigou *et al.* [1] measured the local bubble size distributions in the stirred tank by experiment. Hasan [4] investigated the bubble breakup mechanism using a Rushton turbine by experiments, and it is revealed that the possibility of the bubble breakup increases as the rotation speed of turbines is increased. The prediction of bubble size is a crucial factor to simulate the proper effect of aeration in a stirred tank. Ranganathan *et al.* numerically investigated a stirred tank using population balance model, which takes into account the effect of bubble breakup and coalescence [15]. However, the high-fidelity calculation of the bubble distribution requires a lot of computational resources [9], and hence, we need to model the effect of bubble size distribution using experimental data.

Also, since the flow in a stirred tank is often turbulent, the flow is analyzed using a turbulent flow model. Reynolds Averaged Navier–Stokes (RANS) turbulent model has been widely used in the literature to investigate gas-liquid two-phase flow in an aerated stirred tank [7, 15, 17]. The RANS model allows for fast computations of mean flow and dispersion patterns even in complex flow [5]. It is possible to predict a rough bubble profile, gas-liquid flow fields, and local gas hold-up distribution in an aerated stirred tank by combining the Eulerian-Eulerian approach with RANS turbulent model [7]. The most commonly used RANS model is the k-ε model for the simulation of the stirred tank [7, 9, 14, 15, 21]. By contrast, Singh *et al.* [20] has proposed that k-ω shear stress transport (SST) model can predict the velocity field, the turbulent kinetic energy and the turbulence energy dissipation rate more accurately than the k-ε model. Nevertheless, little work has been reported on the simulation of the stirred tank using k-ω SST model.

Furthermore, it is essential to investigate the distribution of oxygen concentration in the liquid phase in order to assess the performance of the stirred tank in biological processes. Kerdouss *et al.* [8] investigated the distribution of oxygen mass transfer coefficient in the liquid phase using a theoretical model, such as Higbie penetration model [6]. Also, Ranganathan *et al.* [15] investigated the concentration distribution of oxygen transferred from the gas phase to the liquid phase by solving the diffusion equation for oxygen concentration. However, optimization of the geometry and operating conditions of the stirred tank is a still challenging problem, and a well-established numerical model is required.

The present study aims to develop the numerical simulation of gas-liquid two-phase flows using the *k-ω* SST model in the aerated stirred tank which is used in the field of the chemical engineering, and the overall gas hold-up is validated with experimental results. This will pave a way of optimizing the geometry and the operating conditions of the stirred tank in order to achieve uniform oxygen supply in a stirred tank.

2 Numerical Method

2.1 Geometry of the Stirred Tank

The schematic of the numerical domain is shown in Fig. 1. It consists of an ellipse-bottom cylindrical shape stirred tank whose diameter and height are $D = 0.48$ m and $H = 1.02$ m. There are three impellers attached on the shaft, and the diameter of an impeller is $d = 0.15$ m and a shaft is $d_s = 0.04$ m. The impellers are Rushton turbines with six blades with 60-degree interval mounted, and the width w_t and height h_t of each blade is 0.0337 m and 0.0314 m, respectively. Three impellers are installed with intervals of $S_1 = 0.25$ m and $S_2 = 0.22$ m at the bottom of the shaft. The four baffles with the width $w_b = 0.048$ m are attached with 90-degree intervals near the side walls. The tank is filled with water with a height of $H_w = 0.82$ m, and the air is injected from a ring sparger near the bottom of the tank with a gas flow rate of 30 L/min. We consider three different agitation speeds of the Rushton turbines, $N = 300$ rpm, 450 rpm and 500 rpm.

2.2 Numerical Model

The Multiple Reference Frames (MRF) method is used to model the rotation of the impellers, in which the flow of the impellers regions is solved in the rotating reference frame, and the other is solved in the stationary reference frame [10]. The impeller Reynolds number is defined as $Re_d = Nd^2/\nu$, where ν is the kinematic viscosity. Generally, the gas-liquid flow is turbulent because of high $Re_d > 1000$, and actually in the present simulations, $Re \approx 113{,}000$ for the case of $N = 300$ rpm. In the present simulations, the turbulent flow is solved using one of RANS models, *k-ω* shear stress transport (SST) model.

2.3 Governing Equations

We have performed numerical simulations of gas-liquid two-phase flow using open-source software, OpenFOAM, multiphaseEulerFoam. This solver treats both phases as continua and performs the time-integration of the governing equations for each phase. The governing equations for each phase k (=liquid or gas) are the continuity,

$$\frac{\partial \alpha_k}{\partial t} + \nabla \cdot \left(\alpha_k \boldsymbol{u}_{\mathrm{R},k} \right) = 0 \tag{1}$$

and the momentum balance,

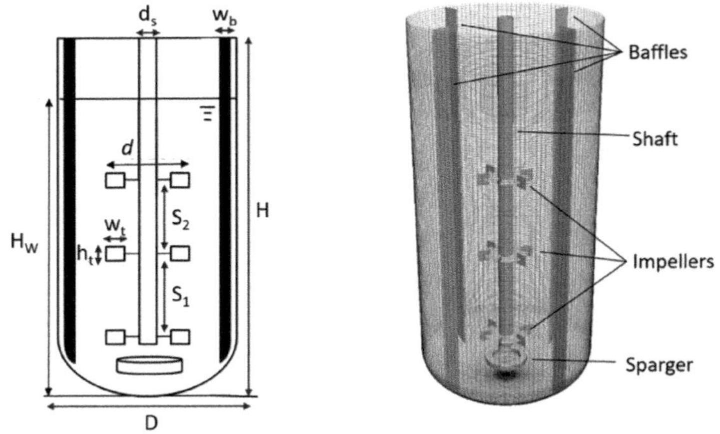

Fig. 1. (Left) the schematic of the stirred tank. (Right) the mesh of the computational domain

$$\frac{\partial}{\partial t}\left(\alpha_k \rho_k \boldsymbol{u}_{\mathrm{R},k}\right) + \nabla \cdot \left(\alpha_k \rho_k \boldsymbol{u}_{\mathrm{R},k} \boldsymbol{u}_{\mathrm{R},k}\right) = -\nabla p + \nabla \cdot \left(\alpha_k \left(\mu_k + \mu_{\mathrm{t},k}\right) \nabla \boldsymbol{u}_{\mathrm{R},k}\right) + \alpha_k \rho_k \boldsymbol{g} + \boldsymbol{F}_{\mathrm{I},k} + \boldsymbol{F}_{\mathrm{D},k} + \boldsymbol{F}_{\mathrm{M},k} + \boldsymbol{F}_{\mathrm{S},k} \tag{2}$$

where

$$\boldsymbol{u}_{\mathrm{R},k} = \boldsymbol{u}_k - \boldsymbol{\Omega} \times \boldsymbol{r} \tag{3}$$

α is the volume fraction, ρ is the density, t is the time, $\boldsymbol{u}$ is the absolute velocity, $\boldsymbol{u}_{\mathrm{R}}$ is the relative velocity, p is the pressure, μ is the molecular viscosity, μ_{t} is the turbulent viscosity, $\boldsymbol{g}$ is the gravity acceleration, $\boldsymbol{F}_{\mathrm{I}}$ are the inertial forces including Coriolis and centrifugal forces applied in the rotating reference frame in MRF model, $\boldsymbol{F}_{\mathrm{D}}$ is the drag force, $\boldsymbol{F}_{\mathrm{M}}$ is the virtual mass force and $\boldsymbol{F}_{\mathrm{S}}$ is the surface tension force. In the stationary reference frame, $\boldsymbol{\Omega} = 0$. The inertial forces $\boldsymbol{F}_{\mathrm{I}}$ are applied in the rotating reference frame and given by

$$\boldsymbol{F}_{\mathrm{I},k} = -2\alpha_k \rho_k \boldsymbol{\Omega} \times \boldsymbol{u}_{\mathrm{R},k} - \alpha_k \rho_k \boldsymbol{\Omega} \times (\boldsymbol{\Omega} \times \boldsymbol{r}) \tag{4}$$

The drag force $\boldsymbol{F}_{\mathrm{D}}$ represents the resistance opposed to the bubble motion in the fluid,

$$\boldsymbol{F}_{\mathrm{D},k} = \frac{3}{4} \alpha_{\mathrm{c}} \alpha_{\mathrm{d}} \rho_{\mathrm{c}} C_{\mathrm{D}} \frac{|\boldsymbol{u}_{\mathrm{d}} - \boldsymbol{u}_{\mathrm{c}}|(\boldsymbol{u}_{\mathrm{d}} - \boldsymbol{u}_{\mathrm{c}})}{d_{\mathrm{d}}} \tag{5}$$

where the subscripts c and d denote the continuous and dispersed phase [22]. d_d is the dispersed diameter of bubble which is assumed $d_d = 0.5$ mm. The drag coefficient C_D was computed using the Schiller and Naumann model [19]. In this model, the drag coefficient is a function of the dispersed Reynolds number Re_d according to

$$C_D = \begin{cases} \frac{24\left(1+0.15Re_d^{0.687}\right)}{Re_d}, & Re_d \leq 1000 \\ 0.44, & Re_d > 1000 \end{cases} \tag{6}$$

where the dispersed Reynolds number is defined as,

$$Re_d = \frac{\rho_c |\boldsymbol{u}_d - \boldsymbol{u}_c| d_d}{\mu_c} \tag{7}$$

The surface tension generates an additional pressure gradient at the gas-liquid interface, which is evaluated using the continuum surface force (CSF) model [2]. The surface tension force $\boldsymbol{F}_S$ is described by

$$\boldsymbol{F}_{S,k} = \sigma \kappa \nabla \alpha_k \tag{8}$$

where σ is the surface tension coefficient, and κ is the local surface curvature,

$$\kappa = -\nabla \cdot \left(\frac{\nabla \alpha}{|\nabla \alpha|} \right) \tag{9}$$

The virtual mass force $\boldsymbol{F}_M$ represents the force that accelerates the liquid surrounding the bubble. The virtual mass force is given by

$$\boldsymbol{F}_{M,k} = \alpha_d \rho_c C_{vm} \left(\frac{D_c \boldsymbol{u}_c}{Dt} + \frac{D_d \boldsymbol{u}_d}{Dt} \right) \tag{10}$$

where $C_{vm} = 0.5$ is the virtual mass coefficient between the two phases [16].

2.4 *K-ω* SST Model

The most popular turbulence models are two-equation eddy-viscosity models [11]. These models solve two transport equations, generally one for the turbulent kinetic energy and another one related to the turbulent length- (or time-) scale. Among the two-equation models, the k-ε model is the most widely used. However, flow analysis is inaccurate near the wall using the k-ε model [12]. The k-ω model is more accurate than the k-ε model in the near-wall layer, while it is sensitive to ω in the outer flow away from the walls [13]. The k-ω SST model combines the advantages of the k-ε model and the k-ω model. The k-ω SST model activates the k-ω model in the near-wall region and the k-ε model for the rest of the flow. Due to the low density and small spatial scales of the dispersed gas, it is sufficient to consider turbulence in the continuous liquid phase [18]. The transport equations for the turbulent kinetic energy k and the turbulent frequency ω (vorticity magnitude) in the liquid phase are, as in [13],

$$\frac{\partial(\alpha_L \rho_L k)}{\partial t} + \nabla \cdot (\alpha_L \rho_L \boldsymbol{u}_L k) = \alpha_L \left(\tilde{P}_k - \beta^* \rho_L k \omega\right) + \nabla \cdot \{\alpha_L (\mu_L + \sigma_k \mu_t) \nabla k\} \quad (11)$$

$$\frac{\partial(\alpha_L \rho_L \omega)}{\partial t} + \nabla \cdot (\alpha_L \rho_L \boldsymbol{u}_L \omega) = \alpha_L \left(\frac{\gamma \rho_L \tilde{P}_k}{\mu_t} - \beta \rho_L \omega^2\right) + \nabla \cdot \{\alpha_L (\mu_L + \sigma_k \mu_t) \nabla \omega\} + 2\alpha_L (1 - F_1) \frac{\rho_L \sigma_{\omega 2}}{\omega} \nabla k \cdot \nabla \omega \quad (12)$$

The production term $\tilde{P}_k$ has a limiter to prevent the building-up of turbulence in stagnant regions, which is defined as

$$\tilde{P}_k = \min(2\mu_t \nabla \boldsymbol{u}_L \cdot \boldsymbol{S}, 10\beta^* \rho_L k \omega) \quad (13)$$

where $\boldsymbol{S} = \frac{1}{2}\left(\nabla \boldsymbol{u}_L + (\nabla \boldsymbol{u}_L)^T\right)$ is the strain rate tensor. $\boldsymbol{F}_1$ denotes the blending function which approaches to zero asymptotically far away from the solid wall (*k*-ε model), and switches over to one side the boundary layer (*k*-ω model), which is defined as

$$F_1 = \tanh\left\{\left(\min\left(\max\left(\frac{\sqrt{k}}{\beta^* \omega y}, \frac{500\mu_L}{\rho_L \omega y^2}\right), \frac{4\rho_L \sigma_{\omega 2} k}{C_{k\omega} y^2}\right)\right)^4\right\} \quad (14)$$

with $\beta^* = 0.09$, $C_{k\omega} = \max\left(2\rho\sigma_{\omega 2} \frac{1}{\omega} \nabla k \nabla \omega, 10^{-10}\right)$ and y is the distance to the nearest wall. The other model constants $\chi = (\beta, \gamma, \sigma_k, \sigma_\omega)$ are computed by a blending form as

$$\chi = F_1 \chi_1 + (1 - F_1)\chi_2 \quad (15)$$

where the coefficients for *k*-ω model is $\chi_1 = (\beta_1, \gamma_1, \sigma_{k1}, \sigma_{\omega 1}) = (0.075, 5/9, 0.85, 0.1)$ and the ones for *k*-ε model is $\chi_2 = (\beta_2, \gamma_2, \sigma_{k2}, \sigma_{\omega 2}) = (0.0828, 0.44, 1.0, 0.856)$.

The turbulent eddy viscosity μ_t is defined as,

$$\mu_t = \frac{ak}{\max(a\omega, 2F_2 \boldsymbol{SS})} \quad (16)$$

where $a = 0.31$, and F_2 is the second blending function,

$$F_2 = \tanh\left\{\left(\max\left(\frac{2\sqrt{k}}{\beta^* \omega y}, \frac{500\mu_L}{\rho_L \omega y^2}\right)\right)^2\right\} \quad (17)$$

3 Results and Discussion

Figure 2 shows the time evolution of the overall gas hold-up rate, ε_G, which is the volume fraction of the gas in the liquid. At the initial state for 450 rpm and 500 rpm, the turbines are steadily rotating without aeration. The initial state for 300 rpm is the

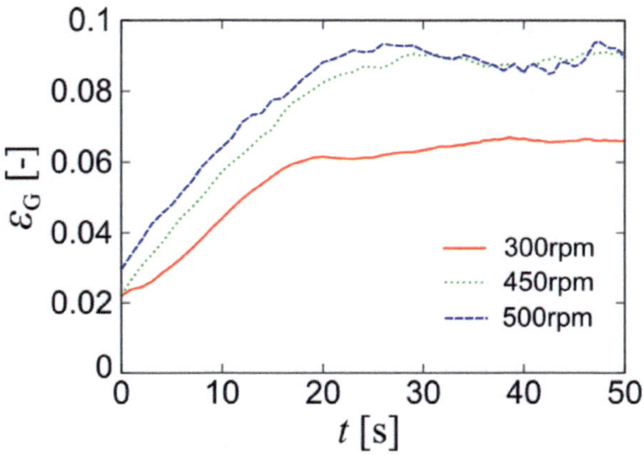

Fig. 2. The time evolution of the overall gas hold-up ε_G at each rotation speed, 300 rpm, 450 rpm, 500 rpm.

same as the initial state for 450 rpm. We discard the initial transient flow from the statistical analysis. There are some entrainments of air from the surface, which leads to a small ε_G up to 3%. The air is injected from the bottom sparger, assuming the bubble diameter being 0.5 mm at t = 0. The gas hold-up becomes roughly constant after 30 s.

The average values during the last 20 s (t = 30 − 50 s in Fig. 2) are compared with the experimental results as shown in Fig. 3. The result of $N = 450$ rpm ($Re_{\mathrm{d}} \approx 168,000$) was roughly consistent with the experimental measurement, the result of $N = 500$ rpm ($Re_{\mathrm{d}} \approx 188,000$) was underestimated, and the result of $N = 300$ rpm ($Re_{\mathrm{d}} \approx 113,000$) was overestimated. It is because the diameter of the bubble assumed in the simulations might be inappropriate for the calculation of $N = 300$ rpm and $N = 500$ rpm. The bubbles break up considerably when the agitation speed increases [4]. As the bubble diameter decreases, the rising velocity of the bubble decreases, then the total amount of gas in the liquid increases as Re increases. From the experiment, the mean bubble diameters are measured as $d_{\mathrm{d}} = 0.57$ mm, 0.5 mm and 0.42 mm at N = 300, 450 and 500 rpm respectively. Since the bubble diameter is assumed to be 0.5 mm in the present simulation, the gas hold-up estimated from the simulation is coincident with the experimental measurement only for the case $N = 450$ rpm.

4 Conclusion

We have successfully performed the numerical simulation of gas-liquid two-phase flows in an aerated stirred tank using OpenFOAM multiphase solver. Although the bubble diameter is assumed to be constant, the simulation shows the good agreement of

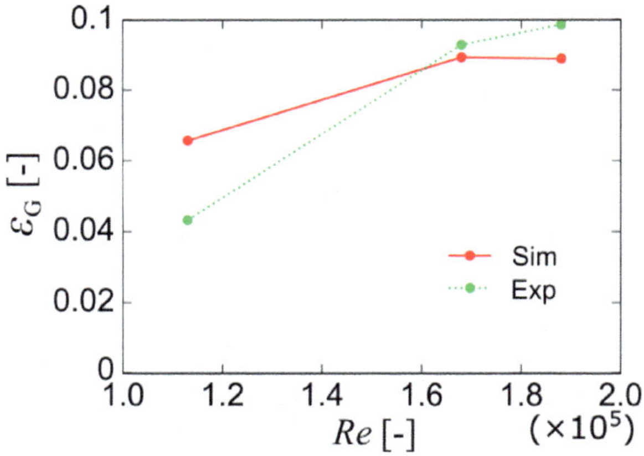

Fig. 3. The comparison of the overall gas hold-up ε_G of the present numerical results with experimental results.

the gas hold-up with the experimental results, while the estimation of the bubble diameter is not feasible. An accurate prediction of the mean bubble diameter from the simulated flow, such as local shear stress, and flow speed, would pave a way of more general gas-liquid stirred tank simulation.

Acknowledgements. The computation was carried out using the computational resources of Research Institute for Information Technology, Kyushu University.

References

1. Barigou, M.: Bubble-size distributions in a mechanically agitated gas-liquid contactor. Chem. Eng. Sci. **47**, 2009–2025 (1992)
2. Brackbill, J.U., Kothe, D.B., Zemach, C.: A continuum method for modeling surface tension. J. Comput. Phys. **100**, 335–354 (1992)
3. Galaction, A.-I., Cascaval, D., Oniscu, C., Turnea, M.: Prediction of oxygen mass transfer coefficients in stirred bioreactors for bacteria, yeasts and fungus broths. Biochem. Eng. J. **20**, 85–94 (2004)
4. Hasan, B.O.: Experimental study on the bubble breakage in a stirred tank. Part 1. Mechanism and effect of operating parameters. Int. J. Multiph. Flow **97**, 94–108 (2017)
5. Hertwig, D., Efthimiou, G., Bartzis, J., Leitl, B.: CFD-RANS model validation of turbulent flow in a semi-idealized urban canopy. J. Wind Eng. Ind. Aerodyn. **111**, 61–72 (2012)
6. Higbie, R.: The rate of absorption of a pure gas into a still liquid during short periods of exposure. Trans. Am. Inst. Chem. Eng. **35**, 365–389 (1935)
7. Kerdouss, F., Bannari, A., Proulx, P.: CD modeling of gas dispersion and bubble size in a double turbine stirred tank. Chem. Eng. Sci. **61**, 3313–3322 (2006)
8. Kerdouss, F., Bannari, A., Proulx, P., Bannari, R., Skrga, M., Labrecque, Y.: Two-phase mass transfer coefficient prediction in stirred vessel with a CFD method. Comput. Chem. Eng. **32**, 1943–1955 (2008)

9. Khopkar, A.R., Rammohan, A.R., Ranade, V.V., Dudukovic, M.P.: Gas-liquid flow generated by a Rushton turbine in stirred vessel: CARPT/CT measurements and CFD simulations. Chem. Eng. Sci. **60**, 2215–2229 (2005)
10. Lane, G.L., Schwarz, M.P., Evans, G.M.: Predicting gas-liquid flow in a mechanically stirred tank. Appl. Math. Model. **26**, 223–235 (2002)
11. Menter, F.: Zonal two-equation k-ω turbulence model for aerodynamic flows. AIAA Paper (1993)
12. Menter, F., Carregal, F., Esch, T., Konno, B.: The SST turbulence model with improved wall treatment for heat transfer predictions in gas turbines. IGTC2003-TC-059 (2003a)
13. Menter, F., Kuntz, M., Langtry, R.: Ten years of industrial experience with the SST turbulence model. Turbul. Heat Mass Transf. **4**, 625–632 (2003)
14. Pan, A., Xie, M., Li, C., Chu, J., Zhuang, Y.: CFD simulation of average and local gas-liquid flow properties in stirred tank reactors with multiple rushton impellers. J. Chem. Eng. Jpn **50**, 878–891 (2017)
15. Ranganathan, R., Sivaraman, S.: Investigations on hydrodynamics and mass transfer in gas-liquid stirred reactor using computational fluid dynamics. Chem. Eng. Sci. **66**, 3108–3124 (2011)
16. Rusche, H.: Computational fluid dynamics of dispersed two-phase flows at high phase fractions. Ph.D. Thesis, Imperial College, University of London, pp. 99–105 (2002)
17. Rzehak, R., Krepper, E.: CFD modeling of bubble-induced turbulence. Int. J. Multiph. Flow **55**, 138–155 (2013)
18. Rzehak, R., Kriebitzsch, S.: Multiphase CFD-simulation of bubbly pipe flow: a code comparison. Int. J. Multiph. Flow **68**, 135–152 (2015)
19. Schiller, L.A., Naumann, Z.: A drag coefficient correlation. Ver. Deutscher Ing. **77**, 318–320 (1935)
20. Singh, H., Fletcher, D.F., Nijidam, J.J.: An assessment of different turbulence models for predicting flow in a baffled tank stirred with a Rushton turbine. Chem. Eng. Sci. **66**, 5976–5988 (2011)
21. Wang, H., Jia, X., Wang, Z., Wen, J., Zhang, J.: CFD modeling of hydrodynamic characteristics of a gas-liquid two-phase stirred tank. Appl. Math. Model. **38**, 63–92 (2014)
22. Wardle, K.E., Weller, H.G.: Hybrid multiphase CFD solver for coupled dispersed/segregated flows in liquid-liquid extraction. Int. J. Chem. Eng. **2013**, 128–936 (2013)

Internal Temperature Estimation in Microwave Flow Reactors

Akiko Kitagawa(✉) and Kazuhiro Takeda

Shizuoka University, Johoku 3-5-1, Naka-Ku, Hamamatsu, Shizuoka 432-8561, Japan
tuinkuruakiko@gmail.com

Abstract. Microwave irradiation is a very effective tool in the field of synthesis because of its rapid heating, etc., based on its energy savings and improvement of selectivity as compared to conventional external heating. In particular, flow-type microwave devices in organic synthesis are suitable for difficult synthesis processes in that the synthesis can be performed under rapid heating and cooling and pressurized conditions. On the other hand, estimating the internal temperature profile during chemical synthesis is important for proper synthesis control. However, it is difficult to directly measure the internal temperature in the target device. This paper focuses on Fischer indole synthesis. A dynamic equation was established from the heat energy balance of the reaction tube. The internal temperature profile was estimated taking into account the correlation between the microwave absorption and temperature. This method could accurately estimate the temperature profile within a relative error of 3.4–6.3% under low power microwave conditions. By clarifying the internal temperature profile, it can be used for future control of organic synthesis.

Keywords: Microwave irradiation · Flow reactor · Internal temperature profile estimation

1 Introduction

The synthesis by microwave irradiation is advantageous in the field of synthesis because it has a specific heating mechanism [1]. For example, it is used as a heating tool for the synthesis of organic and inorganic materials [2, 3]. The heating characteristics of micro-irradiation, especially in organic synthesis, include faster reaction rates, higher purity, less byproducts and less solvent compared to conventional heating [4]. These features are also advantageous from the viewpoint of energy savings and improvement of selectivity.

Since the flow type microwave device for organic synthesis, which is the objective, can be synthesized under high pressure and high temperature conditions, highly concentrated and high yield products can be obtained even in difficult syntheses.

It is important to control various parameters such as quality, flow rate, concentration, temperature, etc., for a successful synthesis. Yokozawa et al. used an experimental

A. R. Várkonyi-Kóczy (Ed.): INTER-ACADEMIA 2019, LNNS 101, pp. 310–317, 2020.
https://doi.org/10.1007/978-3-030-36841-8_30

design to find the optimum conditions for a microwave heated flow reactor [5]. Although the microwave power, flow velocity, pressure, and reactant concentration can be easily measured and controlled in the target device, it is difficult to measure the internal temperature profile of the reactor. The reason why the measurement cannot be performed is that the microwave irradiation causes the metal to disturb the electromagnetic field, so that the thermocouple cannot be inserted into the cavity of the reactor. Moreover, since measurement by an optical fiber sensor is also a flow system, it cannot be inserted into the reactor due to the risk of disturbing the flow. Although numerical investigations were proposed [6, 7], they investigated a batch type of vessel not a flow type of one. Temperature profile of a flow tube with microwave irradiation was investigated using finite element method [8, 9]. In the tube, food material was heated, but there was with no reaction. In order to profile the internal temperature, we currently estimate the outlet temperature based on a dynamic energy balance equation [10].

In this paper, the correlation between the microwave absorption and temperature is made into a function, and by incorporating it into the dynamic energy balance equation, in addition to the exit estimation, it is possible to estimate the internal temperature profile.

2 Methods

2.1 Equation of Energy Balance

The flow type microwave heating apparatus is assumed to have the heat transfer shown in Fig. 1. Therefore, it was formulated by taking the heat balance of the reactor into account. The heat energy of the reaction tube is established from the heat transfer due to the flow rate of the substance, the heat input by the micro-irradiation, the heat dissipation accompanying it, and the reaction heat balance.

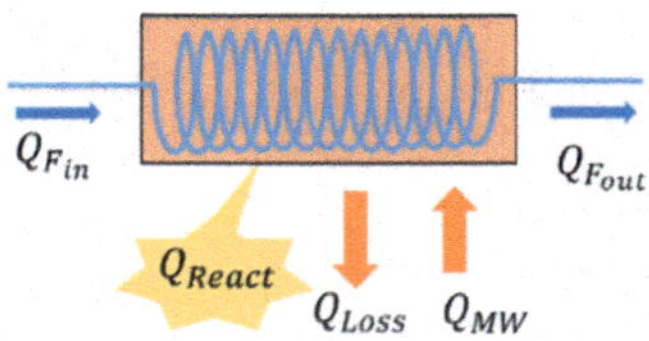

Fig. 1. Model diagram of energy balance formula in microwave heating flow type reactor.

The time series of the dynamic temperature of the reactor is given by solving the initial value problem of an ordinary differential equation as shown in (1).

$$C_{PA} \cdot V \cdot \frac{dT}{dt} = Q_{F_{in}} - Q_{F_{out}} + Q_{MW} - Q_{Loss} + Q_{React} \tag{1}$$

where C_{PA} [J/mL K] represents the specific heat of the whole reactor, and V [mL] represents the volume of the reactor. Furthermore, the heat transfer due to the fluid flow is given by the following Eq. (2)

$$Q_{F_{in}} - Q_{F_{out}} = \sum\nolimits_{i\ Component} F_{in,i} \cdot C_{p,i}(T_{in} - T_{out}) \tag{2}$$

where $F_{in,i}$ [kg s^{-1}] is the inlet flow rate of the component, T_{in} [K] is the inlet temperature, T_{out} [K] is the outlet temperature, $C_{p,i}$ [J kg K^{-1}] is the corresponding heat capacity. Also, Q_{MW} can be estimated by the microwave power and the fluid, and the tube's microwave absorption rate. However, the absorptivity of the microwaves varies depending on the substance and temperature, and specific values cannot be used because of the inherent parameters of complex substances such as the electromagnetic fields, permanent dipoles, and polarization. In addition, since there is no law that takes these into consideration, an approximation is made using the relationship of the microwave power as shown in (3).

$$Q_{MW} = a \cdot P_{MW} \tag{3}$$

where Q_{MW} [W] is the microwave power and a [-] is an experimental parameter of the overall absorption rate of the fluid and the tube. Also, since the tube in the cavity is not thermally insulated by the microwave radiation, the heat dissipation from the surface of the tube cannot be ignored. Q_{Loss} is given by (4).

$$Q_{Loss} = b \cdot \Delta T \tag{4}$$

where ΔT [K] is a parameter of the temperature difference between the surface of the tube and the ambient temperature around it. This parameter corresponds to the product of the surface area of the tube and the sum of the heat transfer coefficient of the layer between the fluid and the tube. Also, since they change with the reaction, the dissipation parameters are determined by experiment. Q_{React} is given by the enthalpy difference between the reactant and the product and the product mass (5).

$$Q_{React} = \left(\sum\nolimits_{Component\ i} H_{p,i} - \sum\nolimits_{Component\ i} H_{R,i}\right) \cdot n \tag{5}$$

where n [mol s^{-1}] is the amount of the product, $H_{p,i}$ [J mol^{-1}] is the specific enthalpy of the product, $H_{R,i}$ [J mol^{-1}] is the specific enthalpy of the reactant. Also, as the reaction proceeds, products are produced. n was referenced to Eq. (6) in [10].

Also, the concentration distribution in the conversion rate distribution and the residence time was based on the previously used method of [10]. Based on the previous equalions, the unknown parameters (a, b, C_{PA}) are determined by experiments described in the next chapter, and the other parameters can be converted from the physical property values, thus the heat balance can be established for the reaction tube using (1) It is now possible to estimate the internal temperature profile.

Estimation of temperature time change by a transient analysis. Assuming that the temperature of the entire reactor is uniform, the temperature time change is estimated by establishing an energy balance equation for the entire reactor. The outside air temperature was 298 K, T'_{out} was the current outlet temperature, and T_{out} was the outlet temperature after one unit time. Using Eqs. (1) to (5), solving for T_{out} gives (6).

$$T_{\text{out}} = \frac{C_{PA} \cdot V \cdot T'_{out} + C_P \cdot F \cdot T_{in} + a \cdot P_{MW} + b \cdot 25 + \left(\sum_{Componenti} H_{P,i} - \sum_{Componenti} H_{R,i}\right) \cdot \mathrm{n}}{C_{PA} \cdot V + C_P \cdot F + b} \tag{6}$$

In addition, the product mass n of the product is the amount of production in the entire reactor in unit time. The rate constant k is $5.0 \times 10^4 \times exp\left(-\frac{5.7\times10^4}{8.31\times T}\right)$ and the reaction order is 1.1.

Temperature profile inside the reactor during steady state. Since the radius of the reactor is negligibly small, to for account for the temperature distribution in the radial direction, it is divided into small sections in the flow direction, and the internal temperature profile is estimated by establishing an energy balance equation in each section. This estimates the outlet temperature in the first section, estimates the inlet temperature in the next section as having the same value as the outlet temperature in the previous section, and estimates the temperature profile by repeating this process.

If constant, the energy balance equation of each section is given as (7) using (1).

$$0 = Q_{F_{in}} - Q_{F_{out}} + Q_{MW} - Q_{Loss} + Q_{React} \tag{7}$$

In addition, (8) is given by organizing $\mathrm{T_{out}}$ using (2) to (5) and (7).

$$\mathrm{T_{out}} = \frac{C_P \cdot F \cdot T_{in} + \frac{a \cdot P_{MW}}{20} + \frac{b \cdot 25}{20} + \frac{\left(\sum_{Componenti} H_{P,i} - \sum_{Componenti} H_{R,i}\right) \cdot \mathrm{n}}{20}}{C_P \cdot F + \frac{b}{20}} \tag{8}$$

2.2 Target Reaction System

The reaction of interest is the Fischer indole synthesis, which is shown in Fig. 2. The reaction products of the target reaction were cyclohexanone and phenyl-hydrazine. Cyclohexanone was dissolved in 1.0 M acetic acid and phenyl-hydrazine was dissolved in 1.0 M acetonitrile. The solutions were mixed in a 1: 1 ratio. The mixture was heated by microwave irradiation. The reaction tube was pressurized at 2.5 MPa using a serpentine made of borosilicate glass. The reactor was enclosed in a TM110 rectangular resonant cavity. The ambient temperature was 298 K.

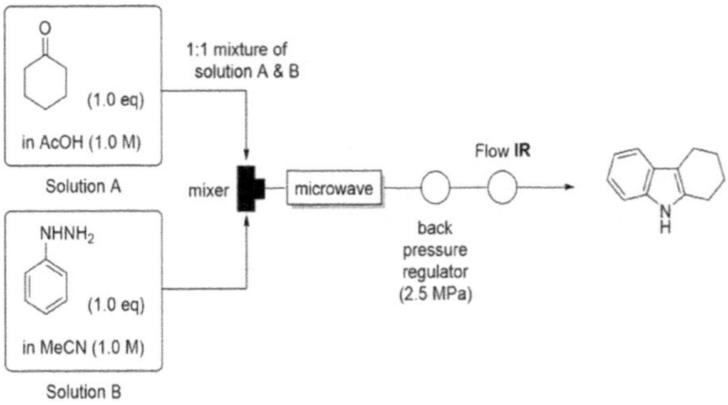

Fig. 2. Fisher indole synthesis

The standard enthalpy of formation of the reactants and products was estimated using the PM6 semi-empirical method in Gaussian 16. The heat of reaction was estimated to be 245.8 kJ/mol. The enthalpy values for the reactants and products are shown in Table 1.

Table 1. Estimated enthalpy values for products and reactants

	Standard enthalpies of formation [kJ/mol]
Cyclohexanone	29.1
Phenyl-hydrazine	201.7
n-phenyl-benzenamine	225.1

Results under flow conditions. In the reaction system, the solvent, the product and the density, viscosity and heat capacity of the reaction and the product were negligible since the solvent was sufficiently greater than the reaction and the product. The density, viscosity, and heat capacity of the mixture in the reactor were determined by the mixing rules shown in Table 2. Also, the radius of the reaction tube is 1.2 mm. Since the Reynolds number of the flow is 2228, it is laminar.

Table 2. Solvent component density, viscosity and heat capacity [10]

	Density [g/cm^3]	Viscosity [cp]	Heat capacity [J/(kg K)]
Acetic acid	1.05	1.1	1272
Acetonitrile	0.786	0.35	2152
Mixture	0.899	0.725	1712

Flow rate and microwave output condition of the target device. The prepared mixture was microwave irradiated under the device conditions shown in Table 3. The irradiation time continued to be measured until a steady state was reached. Moreover, in order to avoid measurement data bias, measurement was performed three times under each condition.

Table 3. Equipment conditions for flow rate and microwave output

	Flow rate [mL/min]	MW [W]
Set 1	1	10

3 Results and Discussion

3.1 Calculation of *a*, *b* and c_{PA}

In order to calculate the parameters of the microwave absorption coefficient *a*, the heat dissipation coefficient *b*, and the specific heat of the entire reactor C_{PA}, the method of [10] was used. As a result, it was estimated that $a = 0.82$ [-], $b = 0.069$ [J/s K], and C_{PA} was 10.78 [J/(kg K)].

3.2 Estimation of Internal Temperature Profile of Reaction Tube Versus Time

The internal temperature profile was determined based on the estimation method. The estimation of the temperature-time change at the non-stationary time and the estimation of the internal temperature profile at a steady-state were performed using the method shown in [10] and above, but the error of the measured value and estimated value of the outlet temperature ware high. Therefore, we made a function that considered the microwave absorption and temperature dependence, and estimated the internal temperature profile versus time. The result is shown in Figs. 3 and 4.

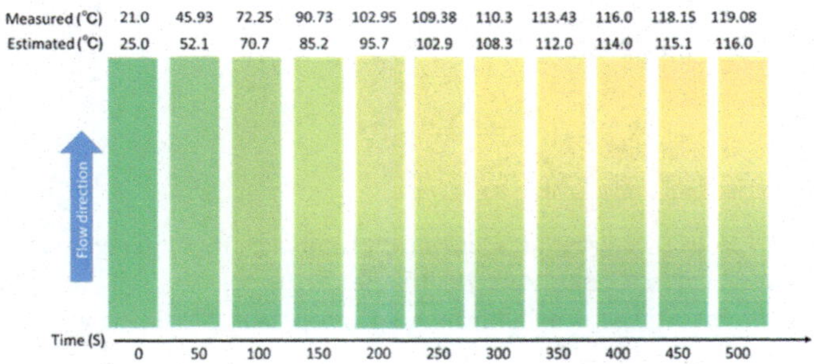

Fig. 3. Estimation of internal temperature profile of reaction tube versus time in Set 1.

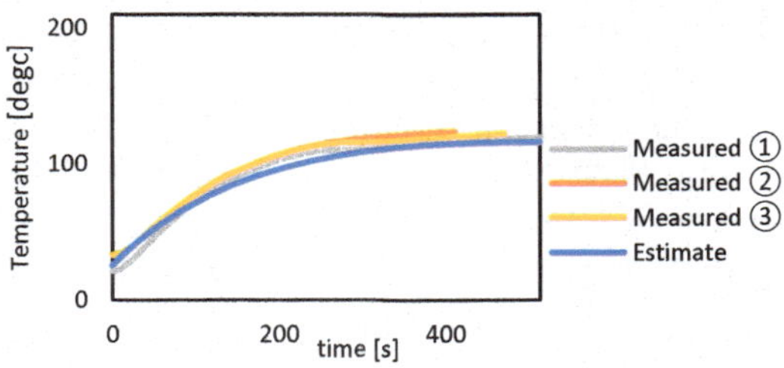

Fig. 4. Outlet temperature in internal temperature profile estimation and measured of reaction tube versus time in Set 1.

Figure 3 shows the result of estimating the internal temperature profile in Set 1. The value shown at the top in this figure is the measurement result of the first outlet temperature, and the value below is the outlet temperature of the estimated internal temperature profile. It can be seen that the internal temperature profile is stable after about 250 s and the temperature inside the reaction tube hardly changes. Figure 4 plots the outlet temperature values in Fig. 3 with time on the horizontal axis and temperature on the vertical axis. From this result, it is clear that the estimated value follows the measured value, and it is proved that the estimated value was accurately estimated.

Also, the estimated value of the outlet temperature at the end of the measurement time is 116 °C, and the measured value is within 119 to 121 °C. Therefore, the relative error range is within 3.4–6.3%. From the above, it became clear that the internal temperature profile was accurately estimated.

4 Conclusion

By establishing the heat balance for the reaction tube and considering the relationship between the temperature and microwave absorption coefficient, we were able to estimate the internal temperature profile during the microwave irradiation. In particular, the relative error could be estimated within 3.4–6.3% under the stated device conditions at this low flow rate and low microwave output. For future research, it is necessary to verify that the estimated internal temperature profile is realistically correct. In addition, it should be examined whether it is possible to estimate even with complex reaction systems and different flow rates. Under this equipment condition, there was a limit to temperature rise due to low output. Therefore, it is a challenge to create an accurate estimation model in a wide temperature range by conducting experiments under high power microwave conditions.

References

1. Kappe, C.O.: My twenty years in microwave chemistry: from kitchen ovens to microwaves that aren't microwaves. Chem. Rec. **19**, 15–39 (2019)
2. Fotiou, I., Baltopoulos, A., Vavouliotis, A., Kostopoulos, V.: Microwave curing of epoxy polymers reinforced with Carbon Nanotubes. J. Appl. Polym. Sci. **12**, 2754–2764 (2013)
3. Makul, N., Rattanadecho, P., Agrawal, D.K.: Application of microwave energy in cement and concrete. Renew. Energy. Rev **37**, 715–733 (2014)
4. Kappe, C.O., Stadler, C.A.: Microwave. Organ. Med. Chem. 1–409 (2005)
5. Yokozawa, S., Ohneda, N., Okamoto, T., Odajima, H.: RSC Adv. **5**, 10204–10210 (2015)
6. Navarro, M.C., Ortiz, A.D., Prieto, P., de la Hoz, A.: A spectral numerical model and an experimental investigation on radial microwave irradiation of water and ethanol in a cylindrical vessel. Appl. Math. Modeling **66**, 680–694 (2019)
7. Navarro, M.C., Burgos, J.: A spectral method for numerical modeling of radial microwave heating in cylindrical samples with temperature dependent dielectric properties. Appl. Math. Modeling **43**, 268–278 (2017)

8. Choi, W., Lee, S.H., Kim, C.T., Jun, S.: A finite element method based flow and heat transfer model of continuous flow microwave and ohmic combination heating for particulate foods. J. Food Engng. **149**, 159–170 (2015)
9. Tuta, S., Palazoglu, T.K.: Finite element method of continuous-flow microwave heating of fluid foods and experimental validation. J. Food Engng. **192**, 79–92 (2017)
10. Takeda, K., Yanagi, N., Nonaka, K., Mase, N.: Estimation of outlet temperature of a flow reactor heated by microwave irradiation. Chem. Rec. **19**, 140–145 (2019)

Risk Evaluation Model for Information Technology Services in Integrated Risk Assessment

Noriaki Matsumura[1,2(✉)], Masakatsu Nishigaki[1], and Takahiro Hasegawa[1]

[1] Shizuoka University, 3-5-1 Johoku, Naka-ku, Hamamatsu 432-8561, Japan
matsumura.noriaki.17@shizuoka.ac.jp
[2] Shinshu University, 3-1-1 Asahi, Matsumoto 390-8621, Japan

Abstract. A risk evaluation model for information technology (IT) services in integrated risk assessment is proposed in this paper. The model covers management systems for information security and IT services. The component-impact coefficient parameter is introduced to define the strength of the relation between assets and IT services. The concept of composition of relations and the weighted sum principle are applied to analyze and evaluate the risk of IT services. When we applied the model to IT services in operation, the risk evaluation was output as quantities that reflect the component-impact coefficient, and risk treatment prioritization was attained in the descending order of numerical values. The proposed model therefore improves the precision of risk evaluation, and application of the model allows more accurate risk evaluation than the conventional method.

Keywords: Integrated risk management · Risk assessment · Weighted sum · Information security management system · IT service management system

1 Introduction

With information technology (IT) becoming increasingly widespread and complex in society, improvement in the efficiency of large-scale risk management has become more critical. ISO/IEC 27013 provides guidance concerning integration of information security management systems (ISMS) and IT service management systems (ITSMS), specified by ISO/IEC 27001 and ISO/IEC 20000-1, respectively. ISO/IEC 27013 states that using a common approach to risk management for both systems is efficient [1]. The term risk management here mainly involves risk assessment and risk treatment. Studies exist on comparison of risk management in ISO management systems [2], integration of quality management system and ITSMS [3], and ISO 31000-based integrated risk management process assessment model for IT organizations [4]. However, explicit integrated risk assessment methods for ISMS and ITSMS have not been investigated yet. We therefore proposed an integrated risk assessment method for these management systems based on the "composition of relations" concept in Ref. [5]. The relation between risk and assets and the relation between assets and IT services are composed

A. R. Várkonyi-Kóczy (Ed.): INTER-ACADEMIA 2019, LNNS 101, pp. 318–325, 2020.
https://doi.org/10.1007/978-3-030-36841-8_31

for integration. Generally, when an organization operates both ISMS and ITSMS, workload relating to management operation will increase. By using of this integration method at Shizuoka University, the work time could be reduced by approximately 24% compared to that by individually conducting ISMS and ITSMS risk assessments, thereby demonstrating improvement in efficiency. A simplified risk evaluation was used on the basis of the practically necessary depth of risk assessment. If there are multiple "High" risk IT services as evaluated by the method, the risk treatment priority must be determined by assessing other information. This paper proposes a model that provides more accurate risk evaluation to solve this issue by introducing parameters in integrated risk assessment.

The contributions of this paper to the risk assessment field are as follows:

- To explain a novel integrated risk assessment method for both ISMS and ITSMS described in Ref. [5], which can reduce workloads.
- To propose a model which can improve the precision of risk evaluation for IT services in the integration method.

2 Integrated Risk Assessment

This section provides an outline of the integrated risk assessment method that is described in detail in Ref. [5].

2.1 Assumptions

IT services are provided by combining various assets as service components. An effective operation of ISMS and ITSMS considers ITSMS-related organizations and activities that exist in ISMS as ITSMS infrastructure. ITSMS is further built by appropriately managing processes necessary for various services on top of the infrastructure. Therefore, integrated risk assessment discussed in this section assumes that IT services comprise assets within the scope of ISMS, and that risk assessment of ISMS also incorporates the approach of ITSMS risk assessment.

2.2 Concepts

The concept of integrated risk assessment in ISMS and ITSMS is explained. Figure 1 compares the concepts of conventional and integrated risk assessment. Assets are denoted as a_i ($i = 1, \ldots, n$), IT services as s_j ($j = 1, \ldots, m$), and risk as r_{ik} ($i = 1, \ldots, n$; $k = 1, \ldots, \ell$). The number of assets is n, the number of IT services is m, and ℓ is the number of risks related to asset a_i. It is assumed that an organization owns and operates both a set of assets A and a set of IT services S. Here, A and S are defined by

$$A = \{a_1, a_2, \ldots, a_n\}, \tag{1}$$

$$S = \{s_1, s_2, \ldots, s_m\} \tag{2}$$

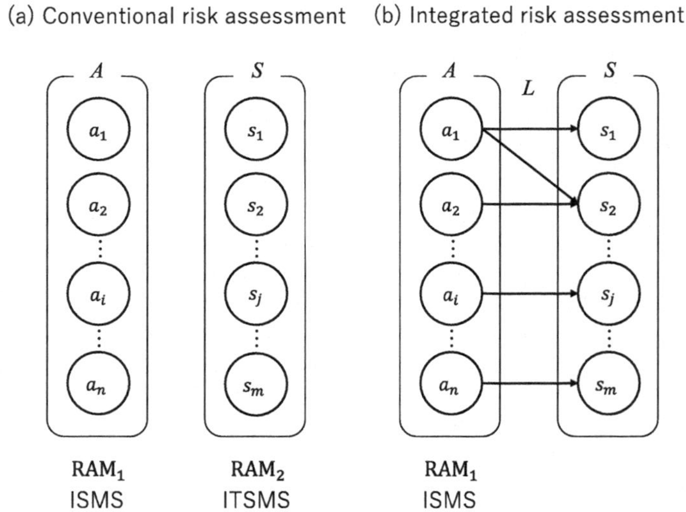

Fig. 1. Comparison of conventional and integrated risk assessment concepts.

Conventional risk assessment applies a risk assessment method (RAM_1) to assets a_i in ISMS and a different method (RAM_2) to IT services s_j in ITSMS. Because RAM_1 and RAM_2 are risk assessment methods focusing on assets and IT services, respectively, there is no intrinsic consistency. A further review may be necessary to confirm consistency based on the individual results (Fig. 1(a)).

In integrated risk assessment, a risk assessment method for assets (RAM_1) is initially employed as in the conventional procedure. The risk assessment method for IT services (RAM_2) is not applied, but instead relation L is defined to link the set of assets A and set of IT services S (Fig. 1(b)).

RAM_1 in integrated risk assessment is a risk assessment based on the consequence/probability matrix technique [6]. Here, assets are identified, threats and vulnerabilities are clarified, and then risks are identified. The set of risks R identified for a set of assets A is

$$R = \{r_{1k}, r_{2k}, \ldots, r_{nk}\} \tag{3}$$

This step defines the relation L' between the set of risks R and the set of assets A. The relation L between the set of assets A and the set of IT services S is defined in integrated risk assessment. The composition of relations L' and L, $L' \circ L$, is then obtained. Figure 2 shows a schematic of relations in integrated risk assessment. The composition of relations $L' \circ L$ is denoted as

$$L' \circ L = \{(r_{1k}, s_1), (r_{1k}, s_2), (r_{2k}, s_2), \ldots, (r_{ik}, s_j), \ldots, (r_{nk}, s_m)\} \tag{4}$$

The risk of IT services is defined by the risk of assets closely related to IT services.

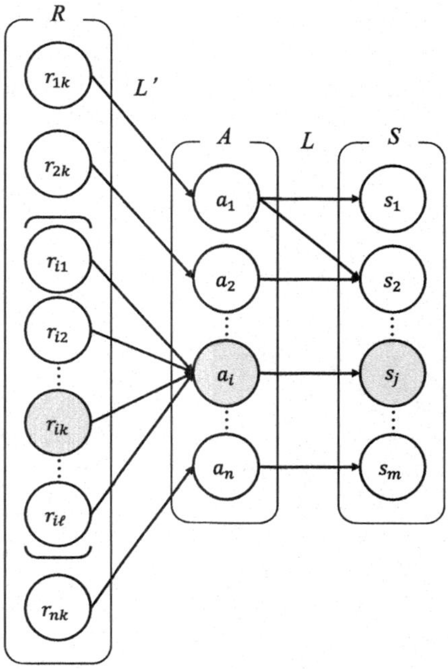

Fig. 2. Schematic of relations in integrated risk assessment.

3 Risk Evaluation Model for IT Services

ISO31000 [7] defines risk assessment as the "overall process of risk identification, risk analysis and risk evaluation". This section overviews the conventional risk evaluation method that we presented in Ref. [5] and then discusses the proposed model.

3.1 Conventional Method

The procedure of risk evaluation for IT services in the conventional method is as follows.

- The risk evaluation of an asset is determined by the highest risk related to the asset.
- The risk evaluation of an IT service is determined by the highest risk among assets related to the IT service by the relation L.

For example, if r_{ik}, which is one of the risks related to asset a_i, exceeds the risk criterion and is evaluated as "High" risk, then asset a_i carries a "High" risk evaluation (Fig. 2). IT service s_j is related to asset a_i through the relation L; hence, its risk evaluation is "High".

3.2 Proposed Model

A risk evaluation model for IT services incorporating parameters in integrated risk assessment is proposed. The component-impact coefficient (w_{ij}) parameter denotes the strength of the relation between assets and IT services. Figure 3 shows a schematic of the proposed model.

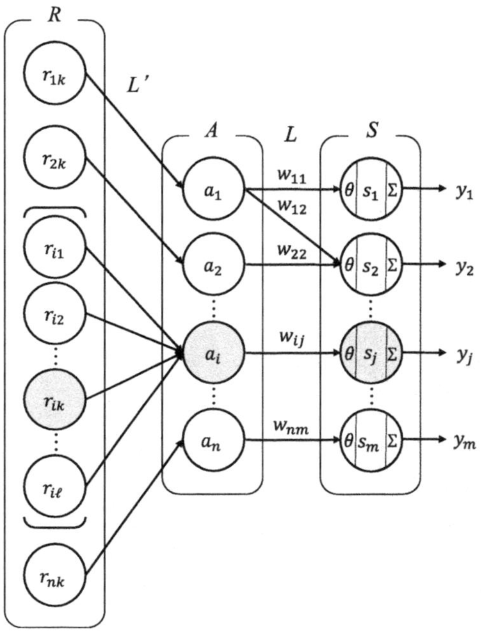

Fig. 3. Schematic of the proposed risk evaluation model.

The individual risk related to IT services obtained from the composition of relations in integrated risk assessment is denoted by x_{jh} ($j = 1, \ldots, m$; $h = 1, \ldots, p$) in this explanation. Here, m is the number of IT services, and p is the number of risks related to IT service s_j. The proposed model calculates values of x_{jh} and then conducts risk evaluations based on them. First, we assume that the risk values r_{ik} of asset a_i are available from the consequence/probability matrix technique conducted within RAM_1, and these are taken as input. Next, as there may be variations in the strength of the relation L, parameter w_{ij} is multiplied by r_{ik} to obtain x_{jh}. Here, w_{ij} is the strength of the relation between asset a_i and IT service s_j. The equation form is

$$x_{jh} = w_{ij} r_{ik} \tag{5}$$

Multiple risks are related to IT services as a result of applying composition of relations in integrated risk assessment; thus, the simple sum of x_{jh} is the risk value of the IT services as a whole. Taking y_j as the total risk of IT service s_j, the following holds:

$$y_j = \sum_{h=1}^{p} x_{jh} \tag{6}$$

Thus, the weighted sum of risks is derived. However, this simple weighted sum cannot be evaluated using the risk criterion in RAM_1, so the following risk evaluation model based on a weighted sum considering the risk criterion in RAM_1 as a threshold is adopted.

As risk evaluation, the risk management process in ISO31000 compares the risk criterion established in advance with the calculated risk values. After the risk evaluation, risk treatment is then conducted. Integrated risk assessment conducts RAM_1 only; thus, using the risk criterion of RAM_1 for risk evaluation of IT services is desirable. Therefore, the proposed model first evaluates x_{jh} with the risk criterion of RAM_1 and then sums x_{jh} that exceeds the risk criterion to obtain the risk of an IT service, y_j. This process in the proposed model is expressed as follows:

$$y_j = \sum_{h=1}^{p} x_{jh} C_h \tag{7}$$

$$C_h = \begin{cases} 0, x_{jh} \leq \theta \\ 1, x_{jh} > \theta \end{cases} \tag{8}$$

Here, θ is the value of risk criterion in RAM_1, C_h is the result of comparing x_{jh} with θ, and $C_h = 1$ indicates that x_{jh} exceeds the risk criterion. The IT service with the highest y_j obtained from Eqs. (7) and (8) must be treated with the highest priority, and the order of priority is determined by the descending order of y_j. If $y_j = 0$, the risk is judged to be tolerable and no treatment is necessary.

4 Results and Discussion

Evaluation experiments were conducted by applying the proposed model to two IT services operated by Shizuoka University, namely, "e-mail" and "global IP address assignment". In this section, the results of the conventional risk evaluation method and the proposed model are compared in order to demonstrate the effectiveness of the proposed model in solving issues found in the conventional method.

w_{ij}, which defines the strength of the relation L, and the risk criterion θ are parameters in the proposed model. Risk evaluation according to the situation of the organization can be conducted by adjusting these parameters. In our experiments, w_{ij} was set to 1 for a "typical relation" and 1.2 for a "strong relation," where the service level agreement (SLA) of the related IT service cannot be satisfied if there is a problem with an asset. The risk criterion θ in RAM_1 was 24 for "risk regarding confidentiality and/or integrity" and 16 when "availability is at risk". Availability was emphasized in particular because of the public nature of the university and its obligations to society through continuity of research and education. Consequently, the availability standard was set approximately 30% higher than the standards for confidentiality and integrity.

Table 1 presents the results of the conventional risk evaluation method and the proposed model applied to the two IT services at Shizuoka University. Both IT services exceed the risk criterion and are evaluated as "High" risk in the conventional method. The risk evaluation by the conventional method shows that risk treatment is necessary for both IT services, but which IT service needs higher treatment priority cannot be judged immediately. When there are multiple "High" risk IT services as evaluated by the conventional method, the risk treatment priority must be determined by assessing other information such as the number of related assets and risks.

In the proposed model, the risk evaluation result is calculated as a number from Eqs. (7) and (8). The results for both IT services exceed the risk criterion, implying that risk treatment is necessary. The number for "global IP address assignment" is higher, indicating a higher risk treatment priority. An assessment with other information is not necessary. Thus, the proposed model can solve the issue in the conventional method. The effect of parameter w_{ij}, defining the strength of the relation L, appeared in the proposed model. When this parameter was not considered, the risk value of one of the risks related to availability of the asset, "external DNS server," which is related to the IT service "global IP address assignment", was 16 and did not exceed the risk criterion. However, the person in charge of ISMS and ITSMS set the weight of the relation between the asset "external DNS server" and "global IP address assignment" to 1.2 in the current experiment; hence, the risk value became 19.2, exceeding the risk criterion, and the need for risk treatment was triggered. Setting parameters allows risk evaluation with higher precision, which in turn more accurately reflects the situation of the organization.

Table 1. Results of risk evaluation.

IT service	Risk evaluation	
	Conventional method	Proposed model
E-mail	High	32
Global IP address assignment	High	57.6

The results of the experiment showed that risk evaluation that considers the parameter w_{ij} can be conducted without issues using the proposed model. Moreover, a more precise risk evaluation compared to the conventional method is possible. The proposed model can solve the issue found in the conventional method.

5 Conclusions

A risk evaluation model for IT services in integrated risk assessment of ISMS and ITSMS was proposed in this paper. The proposed model introduced the component-impact coefficient parameter w_{ij} and applied the concept of composition of relations and the weighted sum principle. An evaluation experiment was conducted by applying the proposed model to two IT services at Shizuoka University, namely, "e-mail" and "global IP address assignment". The result indicated that risk evaluation with parameter w_{ij} taken into consideration could be implemented without a problem in the proposed model. Furthermore, a more precise risk evaluation compared with the conventional method could be attained.

The parameter in the proposed model can be changed according to the situation of the organization, and the component-impact coefficient w_{ij} in the proposed model can be assigned various values to further improve precision. However, only two values were used in this experiment in light of the depth of the practical risk assessment from the operation of ISMS and ITSMS. How to determine parameters is an issue for future analysis, because accumulation of risk assessment data may enable derivation of appropriate parameters from the data. Furthermore, it is conceivable to set parameters for the relationship between risk and assets. We would like to treat that as the subject of our analysis from now on.

ISO management system standards are periodically reviewed. Requirements for ITSMS were revised in 2018. However, ISO/IEC 27013, which is the guideline on integration of ISMS and ITSMS, has not been updated. This paper is based on the version of ISO/IEC 27013 prior to the update; nevertheless, we intend to modify the proposed model to the new version after updates in the future.

References

1. ISO/IEC 27013:2015: Information technology—Security techniques—Guidance on the integrated implementation of ISO/IEC 27001 and ISO/IEC 20000-1, pp. 8–9. International Organization for Standardization, Geneva (2015)
2. Barafort, B., Mesquida, A.L., Mas, A.: Integrating risk management in IT settings from ISO standards and management systems perspectives. Comput. Standard. Interfaces **54**, 176–185 (2017)
3. Mesquida, A.L., Mas, A.: Integrating IT service management requirements into the organizational management system. Comput. Standard. Interfaces **37**, 80–91 (2015)
4. Barafort, B., Mesquida, A.L., Mas, A.: ISO 31000-based integrated risk management process assessment model for IT organizations. J. Softw.: Evolut. Process **31**(1), e1984 (2019)
5. Matsumura, N., Hasegawa, T.: Integration of the risk assessment for an information security management system and that for an IT service management system using composition of relations. IPSJ (Inf. Process. Soc. Japan) J. **60**(1), 250–259 (2019) (in Japanese)
6. IEC/ISO 31010:2009: Risk management—Risk assessment techniques, pp. 82–86. International Organization for Standardization, Geneva (2009)
7. ISO 31000:2009: Risk management—Principles and guidelines, International Organization for Standardization, Geneva (2009)

A Model of Induced Motion of Inclusions in Inhomogeneously Stressed Crystals

Oleksandr P. Kulyk[1(✉)], Leonid A. Bulavin[2], Stella F. Skoromnaya[3], and Victor I. Tkachenko[1,3]

[1] School of Physics and Energy, V.N Karazin Kharkiv National University, 4 Svobody Sq, Kharkiv 61022, Ukraine
kulykop@gmail.com

[2] Faculty of Physics, Taras Shevchenko National University of Kyiv, Prospekt Akademika Glushkova 4, Kiev 03127, Ukraine

[3] National Science Center "Kharkiv Institute of Physics and Technology" of the National Academy of Sciences of Ukraine, 1 Akademichna St, Kharkiv 61108, Ukraine

Abstract. A physical model of liquid inclusion motion in an inhomogeneously stressed crystal is stated. The model is based on the phenomenon of induced transitions of atoms of the matrix into the solution and back to the matrix. The dependence of the speed of inclusion motion on its size is obtained and it describes the experimental results with high accuracy. Numerical estimates of the inclusion's characteristic parameters correspond to tabulated data and results obtained by other authors. The proposed model of induced inclusion motion in a crystal with an inhomogeneous dislocation distribution can be applied to the crystals with inhomogeneity of another nature.

Keywords: Crystallization · Solution · Inhomogeneously stressed crystal · Liquid inclusions · Induced motion

1 Introduction

During the last few decades, studies of crystallization process from solution became rather topical in connection with the tasks have arisen in biology (biophysical growth of proteins, cell and tumor growth, polymerization of the cytoskeleton in the immune system) [1] and chemical industry (purification by crystallization and producing crystalline drugs, sugar, salt, fertilizer, and detergent) [2]. However, despite the progress achieved in the study of crystallization from solution, the mechanism of formation and motion of 3D defects, such as liquid inclusions, is still difficult to understand [3]. Therefore, to investigate the behavior of such defects (their motion and shape transformation caused by different effects) is relevant and interesting for various reasons.

On the one hand, the study of inclusion motion gives information about the kinetics of dissolution–growth processes of the matrix substance and the physical characteristics of these processes [4–6].

On the other hand, inclusions are present in many natural minerals and deposits of geological origin [7]. Some deposits are considered as natural medium for long-term

A. R. Várkonyi-Kóczy (Ed.): INTER-ACADEMIA 2019, LNNS 101, pp. 326–339, 2020.
https://doi.org/10.1007/978-3-030-36841-8_32

storage and disposal of radioactive waste from the nuclear power industry. And crystalline halide formations are one of the preferable geological sediments in terms of safety for the environment [8]. But over time, in the vicinity of radioactive waste, buried in a crystalline halide formation, temperature increases and areas of mechanical stress appear [9]. The gradients of temperature, mechanical stress, and radiation defects arising under these conditions can cause the motion of liquid inclusions within the formation, i.e. in salt medium. If the inclusion flow increases, they can concentrate eventually near the heat generating containers with waste and facilitate the destroying of storage walls [10].

Currently, the phenomenological theory developed by Geguzin and Krivoglaz [4], Anthony and Cline [5] is used to analyze the motion of inclusions in a crystal. It describes the matrix–inclusion interaction in the stationary mode when the fluxes of dissolution, diffusion and crystallization are steady-state. However, in order to predict the dynamics of liquid inclusion motion it is necessary to develop an acceptable physical model that describes the time dependences of inclusion's basic parameters.

2 Analysis of Approaches

It was experimentally found that macroscopic inclusions in a crystal move as a whole with nearly unchanged shape due to the chemical potential difference $\Delta\mu$ between the front and back inclusion surfaces [4]. The chemical potential difference can be caused by the existence of different types of stress in a crystal such as a nonuniform distribution of dislocations density [11], radiation damage [12], or an uneven temperature distribution [6]. In this case, the characteristic dimension of inhomogeneity in the crystal considerably exceeds the characteristic size of inclusions.

From the physics point of view, the motion of inclusions is determined by Le Chatelier principle [13], according to which the processes relieving external or internal stress in a crystal are stimulated. As experiments show, the motion of inclusions may occur spontaneously and it is accompanied by reduction of stress in the crystal. So, the motion of inclusions can be considered as the way of stress relaxation in crystals.

However, for inclusion motion there should be a mechanism of mass transfer of the matrix substance between the front and back inclusion interfaces. For example, in the inclusion with saturated solution the mass transfer occurs by dissolving the matrix substance on the front surface, its diffusion (bulk or surface) through the inclusion solution, and crystallization on its back surface.

The theoretical description of inclusion motion is based on classical works on the theory of crystal growth [14–17], taking into account that in a closed volume of inclusion the processes of dissolution, diffusion, and crystallization of the matrix substance are simultaneous.

The previously developed phenomenological theory [4] is based on the equations of motion and conservation of matter, with regard to the elementary processes occurring at inclusion interfaces. This theory allows us to describe the main characteristics of inclusions' motion: their velocity and velocity dependence on the inclusion's size. In view of the slowness of these processes, the proposed theory is stationary, i.e. it does not describe the dependence of inclusion's characteristic parameters on time. In many cases, the experimental results are well described by the proposed theoretical model. However, because of model's stationarity the dynamics (time dependence) of the parameters of occurring processes cannot be studied. Therefore, in order to describe the dynamics of processes this work introduces a model based on A. Einstein's principle of detailed balance [18].

This model makes it possible to describe the time dependence of the number of matrix atoms deposited on the inclusion interface or transferred to the solution and to determine the inclusion velocity dependence on time as well. These characteristics can be easily transformed into the experimental dependences of inclusions' velocities on their sizes or into other characteristics observed in experiments.

3 Results and Discussion

3.1 Formulation of the Problem

To describe the motion of a macroscopic inclusion in a crystal, we consider inclusion shape as a rectangular parallelepiped $a \times a \times c$ ($c \geq a$) (to simplify further calculations, two sides are the same) with its longer side c parallel to the direction of the gradient of dislocation density $\rho(z)$ (axis z in Fig. 1). Thus, there is a difference in dislocation density $\Delta\rho$ on opposite faces of the parallelepiped with area $a \times a$: $\Delta\rho = (d\rho/dz)c$, which we consider to be small, i.e. we assume that inequality $\Delta\rho/\rho = (d\rho/dz)c/\rho << 1$ is correct.

On the other hand, we assume that the dislocation density is so great that the distance between dislocations is small compared with the size of an inclusion in the direction of crystal inhomogeneity. This condition is equivalent to the inequality $c(\rho_{av})^{1/2} >> 1$, where

$$\rho_{av} = L^{-1} \int\limits_0^L \rho(x)dx \tag{1}$$

is the average density of dislocations in matrix along the length $L >> c$. Therefore the motion of a macroscopic inclusion along the direction of dislocations inhomogeneous distribution is considered at following conditions:

$$\left(\sqrt{\rho_{av}}\right)^{-1} << c << \left(\frac{d\rho}{\rho dz}\right)^{-1}. \tag{2}$$

A schematic diagram of the inclusion in a crystal is shown in Fig. 1.

We consider that the dislocation density varies linearly along the direction of dislocation inhomogeneity:

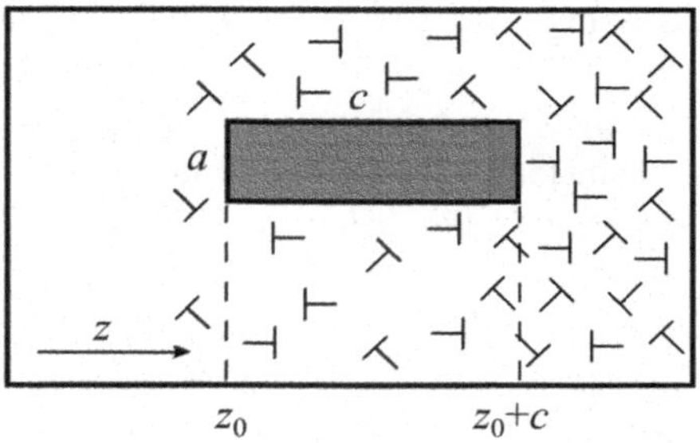

Fig. 1. Inclusion of dimensions $a \times a \times c$ ($c \geq a$) in a crystalline matrix with a nonuniform density of dislocations.

$$\rho(z) \approx \rho(z_0) + \left.\frac{d\rho}{dz}\right|_{z=z_0} (z - z_0). \tag{3}$$

Inclusion, by definition, contains a solvent that facilitates the stress relaxation in a crystal due to enhanced diffusion transfer of matrix material in the solvent. And the experimental fact that there is no significant change in the inclusion shape indicates that the processes of atom transitions from the matrix to the solution and vice versa occur on the opposite inclusion surfaces – the front and back ones respectively.

As mentioned above, such processes are observed in crystals with nonuniform distribution of various stresses, particularly, with inhomogeneous dislocation distribution. Thus, we consider in this paper that the dislocation inhomogeneity causes the motion of inclusions.

The processes of matrix atoms transition into the solution and back to the matrix are probabilistic in nature and can be described using the model of a two-level system.

3.2 Two-Level Model of Induced Transitions of Matrix Atoms into the Solvent and Back

Inclusion is represented as a rectangular parallelepiped filled with a solvent, the back and front faces of which have coordinates $z = z_0$ and $z = z_0 + c$ respectively. We consider that the dislocation density on the inclusion back face is $\rho_1 = \rho(z_0)$, and on its front face is $\rho_2 = \rho(z_0 + c)$ ($\rho_2 > \rho_1$), and they are constant. Since axis Oz is parallel to the direction of dislocation density increasing, the distribution of dislocation density is assigned as a function linearly depended on coordinate z:

$$\rho(z) = \rho_1(1 + \beta z), \tag{4}$$

where $\beta = \Delta\rho/\rho_1 c$, $\Delta\rho = \rho_2 - \rho_1$.

Similarly to the crystal growth from a supersaturated solution or its dissolution into an unsaturated solution [19], the growth and dissolution of crystal occur on the front and back inclusion faces.

High density of dislocations on the inclusion face stimulates the matrix atoms transfer into solution and leads to the dissolution of this surface and supersaturation of solution near the opposite inclusion face (with low dislocation density), which results in its growth. Thus, an inclusion as a unit moves to the area with greater dislocation density, i.e. along the direction of the dislocation density gradient.

To discuss the processes within an inclusion moving along the gradient of dislocation density in the crystal (axis Oz), we use the following statistical model.

Let consider the elementary processes of matrix atom transitions into the solution and back on the front inclusion surface on the example of a two-level system, whose scheme is shown in Fig. 2.

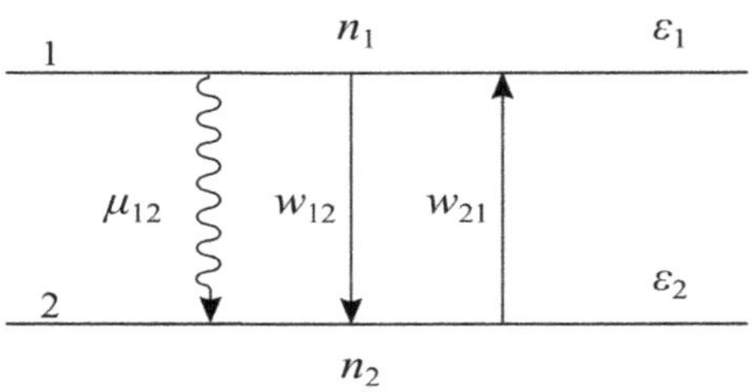

Fig. 2. Scheme of a two-level system with population n_1 and n_2, energies ε_1 and ε_2, and probability of transitions: μ_{12} - spontaneous, w_{12} and w_{21} - induced.

In Fig. 2, arrows show the direction of different atom transitions from one level to another: spontaneous ones with probability μ_{12}, induced ones with probability w_{12} and w_{21}; level 1 corresponds to the matrix atoms in the solvent near the surface and level 2 corresponds to those on the surface of matrix.

In a crystal with inhomogeneous dislocation distribution there is no inclusions motion when a difference in dislocation density is less than critical [11]. In this case, the difference between free energies of atom in the solvent and on the matrix surface is positive [20], i.e. $\varepsilon_1 > \varepsilon_2$.

Spontaneous transitions reduce the free energy of an atom in solution. Therefore, atoms spontaneously transfer from level 1 to level 2. The probability of spontaneous transitions μ_{12} defines the spontaneous transition probability of an atom per second. Spontaneous transitions from the upper level to the lower one ($\varepsilon_1 \rightarrow \varepsilon_2$) are described by the following equations:

$$\begin{aligned} \frac{dn_1(t)}{dt} &= -\mu_{12} n_1(t), \\ \frac{dn_2(t)}{dt} &= \mu_{12} n_1(t), \end{aligned} \tag{5}$$

where $n_1(t)$ and $n_2(t)$ are the population of level 1 and level 2, respectively, as a function of time t.

Since the inclusion motion is observed only when the dislocation density difference on the front and back inclusion faces is greater than the critical one [11], we assume that the probability of induced transition depends on the existing difference in dislocation density on the front and back inclusion faces. Then the probability of induced transitions can be written in the form:

$$w_{im} = \alpha_{im}(\rho(z_0 + c) - \rho(z_0)) = \frac{\alpha_{im}\beta\rho_1}{a^2} W = w^*_{im} N, \tag{6}$$

where the indices *i*, *m* have values 1, 2, ($i \neq m$), α_{im} are constant values, *W* is an inclusion volume, $w^*{}_{im} = \alpha_{im}\beta\rho_1 W_0/a^2$ is a proportionality coefficient, W_0 is a characteristic volume of the solvent molecule, and *N* is the number of solvent atoms (molecules) in the inclusion volume.

It should be noted that the number of solvent atoms within the inclusion is a time-varying quantity due to the chemical reactions of decomposition (front face of inclusion) or reduction (back face of inclusion) of solvent molecules. For example, these processes may occur as reactions [21]:

$$2\left\{\begin{array}{c} Na \\ K[11] \end{array}\right\} Cl + 2H_2O \underset{\text{Backface}}{\overset{\text{Frontface}}{\rightleftarrows}} 2\left\{\begin{array}{c} Na \\ K[11] \end{array}\right\} OH + H_2 + Cl_2. \tag{7}$$

We denote the number of solvent atoms at the initial instant as $N(t)|_{t=0} \equiv N_0$ and assume that it is sufficiently large, i.e. $N_0 >> n_{10}$, n_{20}, where $n_{10} = n_1(t)|_{t=0}$, $n_{20} = n_2(t)|_{t=0}$.

Taking into account the induced processes caused by dislocation density inhomogeneity, the form of Eq. (5) has changed to the next:

$$\begin{aligned} \frac{dn_1(t)}{dt} &= -\left(\mu_{12} + w^*_{12}N\right)n_1 + w^*_{21}Nn_2, \\ \frac{dn_2(t)}{dt} &= \left(\mu_{12} + w^*_{12}N\right)n_1 - w^*_{21}Nn_2. \end{aligned} \tag{8}$$

Equation (8) should be supplemented with the equation describing the number change of solvent atoms *N*(*t*), which provide induced processes in the two-level system:

$$\frac{dN(t)}{dt} = -\left(\mu_{12} + w^*_{12}N(t)\right)n_1(t) + w^*_{21}N(t)n_2(t). \tag{9}$$

The condition that there is a statistical equilibrium in the two-level system allows applying the Einstein relation to it: $\mu_{12} = w^*{}_{12} = w^*{}_{21}$ [18, 20]. In the case when induced processes dominate over spontaneous ones ($N_0 >> n_{10}$, n_{20}), this relation significantly simplifies the initial system of Eqs. (8), (9):

$$\begin{aligned}\frac{dn_1(t)}{dt} &= \mu_{12}(n_2(t) - n_1(t))N(t),\\ \frac{dn_2(t)}{dt} &= -\mu_{12}(n_2(t) - n_1(t))N(t),\\ \frac{dN(t)}{dt} &= -\mu_{12}(n_2(t) - n_1(t))N(t).\end{aligned} \tag{10}$$

Dislocation density difference on the front and back inclusion faces is less than critical. In this case the solution of equation system (10) is found using its integrals:

$$\begin{aligned} n_1(t) + n_2(t) &= n_{10} + n_{20} = C_1,\\ N(t) + n_1(t) &= N_0 + n_{10} = C_2,\end{aligned} \tag{11}$$

where C_1, C_2 are constants corresponding to the sums of different population of levels at initial instant.

Using (11), it is easy to obtain solution of equations system (10):

$$\begin{aligned} N(t) &= \frac{N_{eq}}{1 + \frac{n_{10}-n_{20}}{2N_0} exp(-2\mu_{12}N_{eq}t)},\\ n_1(t) &= \frac{\frac{n_{10}+n_{20}}{2} + \frac{n_{10}-n_{20}}{2} exp(-2\mu_{12}N_{eq}t)}{1 + \frac{n_{10}-n_{20}}{2N_0} exp(-2\mu_{12}N_{eq}t)},\\ n_2(t) &= n_{10} + n_{20} - \frac{\frac{n_{10}+n_{20}}{2} + \frac{n_{10}-n_{20}}{2} exp(-2\mu_{12}N_{eq}t)}{1 + \frac{n_{10}-n_{20}}{2N_0} exp(-2\mu_{12}N_{eq}t)},\end{aligned} \tag{12}$$

where $N_{eq} = N_0 + (n_{10}-n_{20})/2$ is an equilibrium number of solvent atoms.

In particular case $n_{10} = 0$ and $N_0 >> n_{20}$ the dependence of population of levels 1 and 2 on time is determined by expressions:

$$\begin{aligned} n_1(t) &= \frac{1}{2} n_{20}(1 - exp(-\gamma t)),\\ n_2(t) &= \frac{1}{2} n_{20}(1 + exp(-\gamma t)),\end{aligned} \tag{13}$$

where $\gamma \approx 2\mu_{12}N_0$.

This solution shows that over time, half of the atoms located on the matrix surface pass into the solvent and half of them remain in the matrix. At that, equilibrium in the system sets up when the number of atoms on the matrix surface and in the solvent is the same and equal to half of atoms of the matrix surface.

After the "half" equilibrium has established, due to diffusion the matrix atoms in the solvent move away from the surface and the system comes again to non-equilibrium state, when half of matrix atoms are at the lower level and the number of matrix atoms at the upper level is negligible, i.e. it can be equated to zero.

Under such conditions, the statistical model of induced transitions of matrix atoms into the solution and back (13) works again. As a result, the number of matrix atoms in solution is $(n_{20}/2 + n_{20}/4)$.

At the next step, diffusion of matrix atoms in the solution results in the problem when the number of matrix atoms in solution is zero and the number of matrix atoms on the front inclusion face is $n_{20}/4$. In this case, after establishing the equilibrium according to (13), the number of matrix atoms in solution is $(n_{20}/2 + n_{20}/4 + n_{20}/8)$.

At multiple repetition of the process establishing equilibrium, subject to enhanced diffusion, the number of matrix atoms appeared in solution is $n_{20}/2 + n_{20}/4 + n_{20}/8 + \ldots = n_{20}$, i.e. all matrix atoms located on the front inclusion face pass into solution.

At the back inclusion face, where the solution is supersaturated, there is deposition of matrix atoms and, as a result, this face displaces along the gradient of dislocation density.

The process of deposition of matrix atoms on the back inclusion face is opposite to the process of dissolution of the front inclusion face. It is described by system of Eq. (10), in which level 1 and level 2 are reversed: level 1 corresponds to atoms on the matrix surface and level 2 corresponds to atoms in the solvent.

It should be noted that there are no processes of induced transitions of matrix atoms into solution and back to surface on the side inclusion faces, where there is no gradient of dislocation density. Therefore, the side inclusion faces do not displace across the gradient of dislocation distribution.

Dislocation density difference on the front and back inclusion faces is greater than critical. As known from experiments, inclusions are able to move only if their longitudinal size exceeds a certain critical value, i.e. the inclusion velocity is a threshold quantity [11]. In this case, it is convenient to use a frame of reference related to inclusion. Then the matrix velocity, as well as the velocity of dislocations, can be written as $U(c) = V(c) - V_0$, $V_0 \geq V(c)$, where $V(c) = V[b_0 + (c - c_0)\theta(c - c_0)]$ is the velocity of inclusion with longitudinal size c, $V_0 = V(c_0)$ is the velocity of the inclusion having a threshold size c_0, for which $U = 0$. In the definition of inclusion velocity we used a unit function $\theta(x)$, which satisfies $\theta(x < 0) = 0$; $\theta(x \geq 0) = 1$.

Thus, in the chosen frame of reference, depending on the inclusion longitudinal size, the matrix with dislocations either is at rest or moves in the negative direction of axis Oz.

As a result of the transition to a new reference frame, in which the matrix moves with velocity U, the unknown quantities n_1, n_2, N, μ_{12} in Eq. (10) become dependent on coordinate z and Eq. (10) are converted into the next form:

$$\begin{aligned}
&\frac{\partial n_1}{\partial t} + U\frac{\partial n_1}{\partial z} = \mu_{12}(z)(n_2 - n_1)N, \\
&\frac{\partial n_2}{\partial t} + U\frac{\partial n_2}{\partial z} = -\mu_{12}(z)(n_2 - n_1)N, \\
&\frac{\partial N}{\partial t} + U\frac{\partial N}{\partial z} = -\mu_{12}(z)(n_2 - n_1)N.
\end{aligned} \tag{14}$$

We find the solution to Eq. (14) for the simplest case, when the probability of induced transitions μ_{12} can be represented as a linear function of coordinate z:

$$\mu_{12}(z) = \mu_{12}(0)(1 + \lambda z). \tag{15}$$

The transition in Eq. (14) to a variable $\xi(z, t) = t + z^2\lambda U^{-1}/2$ converts them to the form:

$$\begin{aligned}\frac{\partial n_1}{\partial \xi} &= \mu(n_2 - n_1)N,\\ \frac{\partial n_2}{\partial \xi} &= -\mu(n_2 - n_1)N,\\ \frac{\partial N}{\partial \xi} &= -\mu(n_2 - n_1)N,\end{aligned} \tag{16}$$

where $\mu = \mu_{12}(0)$ is used to simplify the equations.

The type of equation system (16) completely coincides with the type of equation system (10), the only difference is that the argument of system (16) is a function $\xi(z,t)$. Therefore, the solutions of system (16) coincide with solutions of (13), whose argument t should be replaced by the argument $\xi(z, t)$ and μ_{12} replaced by μ.

The inclusion velocity when dislocation density difference on the front and back inclusion faces exceeds a critical value. To determine the inclusion velocity V we set a plane of constant phase that is defined by expression:

$$\xi(z, t) = C, \tag{17}$$

where C is constant.

From Eq. (17) it follows that for a given time $t = t_S$ and $C = C_S$ it is possible to determine coordinate of the plane $z = z_S$.

If we fix only $C = C_S$, then over time the coordinate of the plane z_S changes, i.e. the plane of constant phase moves with velocity $V_S = dz/dt$. We define this velocity evaluating the total time derivative of Eq. (17):

$$V_S = -\frac{U}{z\lambda}, \tag{18}$$

Expression (18) is valid for $z \neq 0$.

If $z = c$, velocity V_S determines the velocity of the phase plane corresponding to the inclusion front face.

On the basis of energy considerations, velocity of the plane of constant phase V_S can be related to the velocity of matrix motion U.

Flux of dislocations with density $\rho(z)$ creates a "wind" that moves with velocity $U(c)$. Pressure force of the dislocation wind on inclusion is equal to the difference of pressure forces on the front and back faces with area $S = a^2$:

$$F_D = C_w \frac{1}{2}(\rho_2 - \rho_1) m_D S U^2 = C_w \frac{1}{2} \beta \rho_1 m_D W U^2. \quad (19)$$

where $m_D = (\rho_m b^2/4\pi)\ln(L/b)$ is effective mass of a unit dislocation length, the characteristic value of which is $m_D \approx 10^{-15}$ g/cm [22], C_w is dimensionless resistance coefficient, b is the Burgers vector magnitude, whose characteristic value has the order of several Å, $L \sim (10 \div 10^4)b$ is a characteristic dislocation length.

If an inclusion moves with velocity V' in the direction of the force (19), then consequently, the pressure force acting on it decreases:

$$F'_D = C_w \frac{1}{2} \beta \rho_1 m_D W (U - V')^2. \quad (20)$$

Power produced by this force is equal to:

$$P_D = C_w \frac{1}{2} \beta \rho_1 m_D W (U - V')^2 V'. \quad (21)$$

From (21) it follows that the maximum power is reached at $V' = U/3$ [23] and equal to:

$$P_D^{\max} = \frac{2}{27} C_w \beta \rho_1 m_D W U^3. \quad (22)$$

The maximum power (22) is taken on the basis of the principle of least action: stress in the crystal should be reduced in a minimum time. The minimum time of stress relaxation in a crystal corresponds to the maximum power of inclusion motion.

On the other hand, the rate of work doing by resistance force F_R at the inclusion motion through a distance $z(t + \Delta t) - z(t)$ for a short time interval Δt is determined by expression:

$$P_R = F_R \lim_{\Delta t \to 0} \left(\frac{z(t + \Delta t) - z(t)}{\Delta t} \right) = F_R \frac{dz}{dt} = F_R V_S, \quad (23)$$

where $F_R = k(U - V') = 2kU/3$ is the resistance force caused by nonuniform distribution of dislocation density, k is a proportionality coefficient.

From equality $P_D = P_R$ we determine inclusion velocity V, considering coordinate z equal to the coordinate of the inclusion front face $z = c$:

$$V = V_0 - \frac{\delta}{c}, \quad (24)$$

where $\delta = 9\, k/C_w \lambda \beta \rho_1 m_D W$.

Comparing dependences of the dislocation density inhomogeneity $\rho(z)$ and induced transitions $\mu_{12}(z)$ on coordinate z, one can assume their similarity and consider that $\lambda = \beta$.

Firstly, this similarity follows from the principle called "Occam's Razor" [24].

Secondly, this similarity is based on the fact that inhomogeneity of the dislocation density β and inhomogeneity of the induced transitions probability can be associated with inhomogeneous distribution of the chemical potential of atoms in a crystal $\Delta\mu_k$ [11, 25]: $\Delta\mu_k = \varepsilon\omega_k\Delta\rho$, where ε is energy of a unit length dislocation, ω_k is an atomic volume in solid phase.

On this basis, we reduce the number of independent constants in (24) by one that simplifies the interpretation of experimental results.

We use (24) to describe the experimental dependences of the inclusions velocity on their sizes [11].

In Fig. 3 solid lines show the graphs constructed using analytical expression (24). The experimental data are shown there with circles. The points of curve 1, marked with white circles O, have an abscissa that corresponds to the inclusion longitudinal size at a scale: a division is $c \cdot 10^3$ cm. The points of a straight line 2, marked with black circles •, have an abscissa that corresponds to the inclusion inverse longitudinal size at a scale: a division is $10 \cdot (c \cdot 10^3)^{-1}$ cm^{-1}.

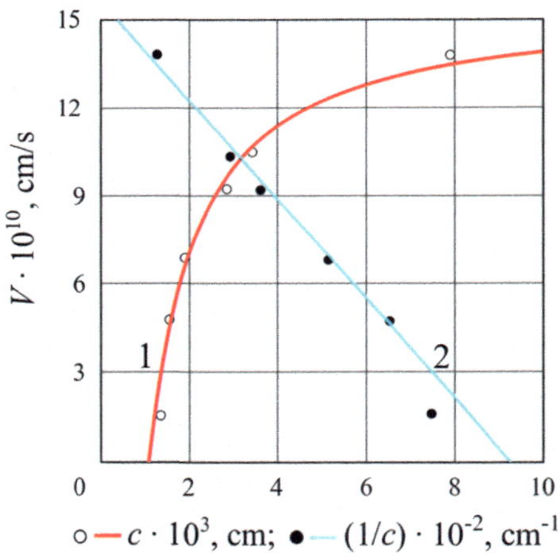

Fig. 3. Dependences of inclusion velocity on its longitudinal size (○) or on the inverse longitudinal size (•)

Comparison of theoretical calculations with experimental data indicates an adequate description of the experiment by the proposed theoretical model.

Calculations show that the deviation of theoretical dependences from experimental data, without taking into account one outlier on each curve, does not exceed 5%. This result confirms the applicability of the proposed theoretical model for describing the motion of inclusions in a non-uniformly stressed crystal.

Superimposition of dependence (24) on the experimental points [11] and its optimization gives the following values of constants: $V_0 = 1.558 \cdot 10^{-9}$ cm/s, $\delta = 1.683 \cdot 10^{-12}$ cm^2/s, $c_0 = 1.08 \cdot 10^{-3}$ cm.

Expression (24) corresponds to the inclusion motion in the diffusion mode [11]. Therefore, constants V_0 and δ have the following form:

$$V_0 = 1.558 \cdot 10^{-9} = \varepsilon\omega_k^2 \frac{c_i D_i}{\omega_i kT} \frac{\partial\rho}{\partial z} = \varepsilon\omega_k^2 \frac{c_i D_i}{\omega_i kT} \frac{\Delta\rho}{c_0} \quad \text{(cm/s)}, \tag{25}$$

$$\delta = 1.683 \cdot 10^{-12} = \frac{\omega_k c_i D_i}{\omega_i kT} \Delta\mu^* \quad (\text{cm}^2/\text{s}), \tag{26}$$

where c_i, ω_i, D_i are concentration of atoms in the inclusion volume, the atomic volume and diffusion coefficient in the inclusion, respectively, k is the Boltzmann constant, T is temperature, $\Delta\mu^*$ is a threshold jump of the chemical potential.

Assuming in the expression (26) $kT \approx 4.1 \cdot 10^{-14}$ erg, $\omega_k \approx \omega_i \approx 3 \cdot 10^{-23}$ sm^3, $c_i \approx 0.3$, and $D_i \approx 5 \cdot 10^{-5}$ sm^2/s, we determine the magnitude of a threshold kinetic jump of the chemical potential $\Delta\mu^* \approx 4.6 \cdot 10^{-21}$ erg.

If the energy of a unit length dislocation is set as $\varepsilon \approx 10^{-4}$ erg/cm [11], then from expression (25) we obtain the value of the difference in dislocations density on the front and back inclusion faces $\Delta\rho \equiv 1.5 \cdot 10^6$ cm^{-2}.

The data obtained on the basis of the model of induced transitions of the matrix atoms into the solution and back quantitatively correspond to the data given in [11].

4 Conclusion

Basing on the phenomenon of induced transitions of matrix atoms into solution and back, a physical model of the liquid inclusion motion in a non-uniformly stressed crystal is stated. From the energy principle, an analytical dependence of the inclusion velocity on its size is obtained. Threshold nature of the inclusion motion is shown. It is shown that the inclusion moves due to dissolution of the front face caused by the process of induced transitions of matrix atoms into the solution. On the back face the process is opposite: matrix atoms are induced to transfer from solution to the matrix. The rates of both processes are the same. It is shown that the analytical expression for inclusion velocity describes the experimental results with a sufficiently high accuracy. Numerical values of the characteristic parameters of inclusions correspond to tabular data and results obtained by other authors. The proposed model of induced motion of inclusions in a crystal with a nonuniform dislocation distribution can be applied to crystals with inhomogeneity of another nature.

Acknowledgements. A part of this research is based on the Cooperative Research Projects of Research Center for Biomedical Engineering, Japan: "Radiation-induced motion of liquid inclusions in crystals" (No 2076) and "Formation of a quasiperiodic relief on vicinal surfaces" (No 2077), adopted as the 2019 Cooperative Research at Research Institute of Electronics, Shizuoka University.

References

1. McPherson, A., Kuznetsov, Y.G.: Mechanisms, kinetics, impurities and defects: consequences in macromolecular crystallization. Acta Crystallographica F **70**(4), 384–403 (2014)
2. Lorentz, H., Beckmann, W.: Purification by crystallization. In: Beckmann, W. (ed.) Crystallization: Basics Concepts and Industrial Applications, pp. 129–149. Wiley-VCH Verlag GmbH & Co. KGaA, Weinheim (2013)
3. Bobo, E., Lefez, B., Caumon, M.-C., Petit, S., Coquerel, G.: Evidence of two types of fluid inclusions in single crystals. CrystEngComm **18**(28), 5287–5295 (2016)
4. Geguzin, Ya.E., Krivoglaz, M.A.: Migration of macroscopic inclusions in solids. Translated from Russian, Studies in Soviet Sciences, Consultants Bureau, New York (1973)
5. Anthony, T.R., Cline, H.E.: Thermal migration of liquid droplets through solids. J. Appl. Phys. **42**(9), 3380–3387 (1971)
6. Kruzhanov, V.S., Podshyvalova, O.V.: Defining the kinetic coefficients of dissolution using the experimental data on the motion of liquid inclusions in the crystals. Sov. Crystallography **35**(6), 1534–1539 (1990) [in Russian Определение кинетических коэффициентов растворения из опытов по движению жидких включений в кристаллах]
7. Boudreau, A.: Hydromagmatic processes and platinum-group element deposits in layered intrusions. Cambridge University Press, Cambridge (2019)
8. Popp, T., Kern, H.: Ultrasonic wave velocities, gas permeability and porosity in natural and granular rock salt. Phys. Chem. Earth **23**(3), 373–378 (1998)
9. Nikitin, A.N., Pocheptsova, O.A., Matthies, S.: Aspects of the thermal and transport properties of crystalline salt in designing radioactive waste storages in halogen formations. Crystallogr. Rep. **55**(3), 450–457 (2010)
10. Nikitin, A.N., Kulakovsky, A.L., Rodkin, M.V., Yurchenko, O.A., Ivankina, T.I., Vasin, R. N.: About some mechanisms of rock permeability in connection with geoecological safety of highly-radioactive waste storages. Geophys. Res. **6**, 85–95 (2006) [IPE of RAS, Moscow, in Russian О некоторых механизмах проницаемости горных пород в связи с геоэкологической безопасностью хранилищ ВАО]
11. Dzyuba, A.S.: Motion of a macroscopic inclusion in the field of structural inhomogeneity in a crystal. Sov. Solid State Physics **19**(1), 78–82 (1977) [in Russian Движение макроскопического включения в поле структурной неоднородности в кристалле]
12. Kruzhanov, V.S., Podshyvalova, O.V.: Motion of liquid inclusions in the crystal caused by radiation defects. Sov. Solid State Phys. **32**(2), 373–378 (1990) [in Russian Движение жидких включений в кристалле, обусловленное радиационными дефектами]
13. Landau, L.D., Lifshitz, E.M.: Statistical physics. Volume 5 of Course of theoretical physics. 3rd edn. Pergamon Press (1980)
14. Volmer, M.: Kinetik der Phasenbildung. Steinkopf Verlag, Dresden und Leipzig (1939)
15. Buckley, H.E.: Crystal growth. Wiley, New York (1954)
16. Burton, W.K., Cabrera, N., Frank, F.C.: The growth of crystals and the equilibrium structure of their surfaces, Philos. Trans. R. Soc. London, A **243** (866), 299–358 (1951)
17. Chernov, A.A.: The spiral growth of crystals. Sov. Phys. Usp. **4**(1), 116–148 (1961)
18. Einstein, A.: Zur Quantentheorie der Strahlung. Mitteilungen der Physikalischen Gesellschaft Zürich **18** (1916)
19. Chernov, A.A.: Modern crystallography III. Crystal growth with contributions by Givargizov, E.I., Bagdasarov, K.S., Kuznetsov, V.A., Demianets, L.N., Lobachev, A.N. Springer-verlag, Berlin, Heidelberg, New York, Tokyo (1984)

20. Kuklin, V.M.: On the emission processes in nonequilibrium media. The Journal of Kharkov National University (Physical series "Nuclei, Particles, Fields") **933**(4/48), 4–27 (2010) [in Russian О процессах излучения в неравновесных средах]
21. Chemical Equations, http://chemequations.com/en/
22. Hirth, J., Lothe, J.: Theory of Dislocations. 2nd edn. John Wiley & Sons, Inc. (1982)
23. Betz, A.: Die Windmühlen im Lichte neuerer Forschung. Die Naturwissenschaften **15**(46), 905–914 (1927)
24. Gauch, H.G.: Scientific Method in Practice. Cambridge University Press, Cambridge (2003)
25. Dzyuba, A.S., Gordeeva, T.F.: The effect of dislocations on the inclusion shape in a crystal. Sov. Solid State Phys. **17**(1), 307–309 (1975) [in Russian Влияние дислокаций на форму включения в кристалле]

The Dialogic Advance of Educational Practices

Valerie A. Wilkinson(✉)

Faculty of Informatics, Shizuoka University, Hamamatsu, Shizuoka 432-8011, Japan
vwilk@inf.shizuoka.ac.jp

Abstract. The fourth year of our collaboration and third iteration of our experimental research is composed of a structured Control class of 62 students and an Experimental class of 51 students. The start-up preparation before the first class forced close examination of the practices and strategies in structuring the two groups. The one group is conceived of as a Project Based Learning (PBL) simulation; the other group is Task-Based language learning (TBLL). The Project Team simulation promotes development of layered relationships and valorizes leadership in teams. The TBLL group has professor, technical assistant, and researcher with orderly routines. The university asks us as educators to work to promote diversity, innovation and creativity. Consistent high quality group development is also desired. Are these goals in conflict?

Keywords: General systems theory (GST) · Game theory (GT) · Integrated dynamic systems

1 Introduction

The university is a player in the game of knowledge creation on the international stage. University cooperates with industry and society to manage progressive, orderly development, if possible. Tensions arise and priorities are challenged with issues of funding, allocation of resources, along with knowledgeable development. In order to explore how the tensions are articulated in the communication processes that are integral to the co-production of knowledge, we make a case study of collaborative research in developing Academic English for juniors in electrical/electronic (EE) engineering. The Professor of Electrical/Electronic Engineering is the right person to transmit the engineering mindset to upperclassmen preparing to move on to career or further study. The humanist English teacher is the principal investigator (PI) who brings a sturdy confidence in the value of human communication as it applies to development of relationships, teamwork, cooperation, and a positive atmosphere. The funding is administered through Faculty of Informatics, to be used in educating Faculty of Engineering students. Navigation of the delicate and complicated negotiations across language and cultural differences makes for a deeply nuanced educational environment.

At the same time the collaboration of these two educators requires contributions of time and creativity for preparation, execution and evaluation. In the first place, the class for both Control and Experiment is structured as a workshop rather than a lecture. For

A. R. Várkonyi-Kóczy (Ed.): INTER-ACADEMIA 2019, LNNS 101, pp. 340–344, 2020.
https://doi.org/10.1007/978-3-030-36841-8_33

this, we need enough space and adjustable seating for the groups and teams to do their work. The Moodle education software platform, which we have used for both groups since the first year of collaboration, can be used to manage a significant amount of record keeping. Nevertheless, constant monitoring and maintenance is necessary for the Experiment group to succeed in the goals of self-organization, self-correction, and emergence, which are the hall marks of authentic living systems. It takes a delicate touch to create a class environment in which the students can develop without invasive supervision by the professor or PI. In the other class, the TBLL Control group is given more routine work and requires less monitoring. PI wonders if this is more natural and less stressful. Hopefully testing will indicate which of the class organizations is more effective.

At the same time, the bigger picture of four years of collaboration and the three-year research project have implications for development in the larger game of international university collaboration and integration throughout the world. It is quite rare to get the kind of opportunity that our team got to this point. Where do we go from here?

The academy is the university, the select realm of higher study where a select few can go to develop themselves. The university ought to be The Knowledge-creating Company [2]. Takeuchi and Nonaka's book by that name found creativity and innovation in a "middle-up-down" organizational structure, in which the middle management, authorized by executive management, created new products which upper management could then disseminate and promote. The structure of each lab with a professor, assistant professors, doctoral, master, and undergraduates, replicates the "middle-up-down" structure. Research is an essential part of the identity of the university.

As educators we face bigger questions of "sustainability" and "corporate social responsibility" which the university can field at the same time that it provides "requisite variety" to the community as a repository of learning, wisdom, and cutting-edge research.

2 Background of Research

Begin with an anecdote: Takefusa Kubo signed with FC Real Madrid on 14 June for the 2019–20 year. He is just 18 years old. What kind of life vaulted this young man to the center stage in the world of sport? What kind of network sent this news to PI? What madness possesses one to put an item hot-off-the-press in a serious article?

It is a question of quality. The world of sport includes Olympic sports as well as international team sports. The background stories catch the imagination. What brought Kubo as 10-year old Japanese child to Barcelona, or who? Was it something inside him or was his parents. Step by step PI traces the trail back to her own quest: "What is good ground?" It is the question of environment added to background. Throughout the world attention is paid to the winners. Who gets into a good school? Who gets the awards?

It is the quest for quality, for the cultivation of quality, the necessary conditions for quality; perhaps this is sufficient to hold the stage for a moment. The *International Conference on Global Research and Education, Inter-Academia*, began in 2002 with a main goal of providing a well-articulated international forum to review, stimulate, and understand recent trends in both fundamental and applied research. This 18th year has brought PI to Hungary to cap 4 years of involvement with this academic community.

When co-PI, EE Professor Chandler and PI began our collaboration, we wanted to teach English well. It has been a daunting job in Japan. However, there is one thing that keeps the fire going; there is a chance that we will find the key to activating the "growth mindset" in students! Another factor is important as well; in 1986 PI became a GST theorist. The path to Ludwig von Bertalanffy, the founder of GST, came through Gregory Bateson.

"What is the difference between the physical world of pleroma, where forces and impacts provide sufficient basis of explanation, and the *creatura*, where nothing can be understood until *differences* and *distinctions* are invoked?" [1] As a teacher, one wants to believe that students are being taught! Or at least students are provided with a context in which they can learn.

The reason that PI learned of Takefusa Kubo is that a friend passed along the story **because** of the context, the connection with communication. A mere tale about a soccer player would have no impact. A tale about a family who gave their son the space to develop as an international class player was the right kind of story. There is also the element of chance, of time, place, and opportunity. In 40 years in Japan and 24 years at Shizuoka University, PI had not had an opportunity to collaborate with a scientist in the task of teaching English. Then we got the chance to do this project.

3 Interface Engineering: The Human Sphere

I have a set of tapes of Gregory Bateson having discussions at recorded at Esalen before the publication of *Mind and Nature*. One of the tapes is labeled "Interfaces." Another is labeled "It used to matter." The great divide in the life of PI has been digital vs analog information. There is a still greater divide in the history of human thought: concrete vs abstract.

It was Heraclitus (c.535-c.445 BCE) who said, "You can't step into the same river twice." A look into the PreSocratic philosphers suggests Heraclitus and Parmenides were articulating the dialectic of becoming and being. Matters of belief can be urgent. The dialectic tension was important. And no. I am not going to go into the ancient texts here. This is simply to point out that the cognitive history of the human race has been developing over eons.

In "It used to matter" [3] Gregory spent a long time unspooling the story of the Pythagoreans, who held the secret of $\sqrt{2}$ only for the initiates. It is probably the first number known to be "irrational."

The field in which co-PI and PI meet is an interface environment in which the delicate mechanisms of student development, group work, and the professor-student relationship are being tested and refined.

4 The Micro and Macro Management Over Time

Quite a large number of individual interactions occur in the training of the executive staff and the upperclassmen staff, to delineate their roles in the guidance of the teams. The "work-shop" conference simulation is an environment in which self-organized teams can display emergence, as postulated by GST. We had to keep two things in mind at all times. The Control and the Experimental have the same curriculum. Tests and quizzes will be the same and executed in the same way. The way we have distinguished the two groups from each other it with PBL and TBLL framing.

4.1 Micro-managing

We had to make foundational decisions about roles, duties, and staffing from the start. We broke both classes groups as evenly as possible. We asked the PBL Experimental Teams to create a team name, choose a team leader, and create an icon. The TBLL control group does not require these levels of interaction and definition. The control group really is a class "as usual." For example, we advised them to take turns taking leadership. It is clear, however, that control group students are accustomed to group work and scheduling. Our control group is neither "tabula rasa" nor "lecture class."

4.2 The Bigger Picture

The character of the group of participants altered significantly from the 1st to the 3rd iteration. In 2017, for the Experimental class, we had a volunteer older student to play the role of the coach. He was not perfectly reliable. Some retired Japanese business men were kind to join the simulated conference. The Control group did not have these embellishments. In 2018, the assignment for both Experiment and Control was to find an International Robot Contest. The student coach, whose supervising professor joined us as a colleague, really enjoyed translating for the professor. We developed good instruments for evaluation, but the results were to difficult to compile. Finally, in 2019, the student coach did not bring his professor, but he was motivated. A graduate student from India, with a much higher level of language and nuanced academic experience joined us, and an administration staff member from Logistics requested to be included. Finally, an alumnus of Faculty of Engineering joined joined the group of retired business men. These additions racked up the Experiment simulation energy considerably.

5 Conclusion

In this age of promoting diversity, sustainable science, and corporate social responsibility, the scientific university can create a forum for discussing the aims of scientific education vis-a-vis the client students. Does the education environment permit them to follow their talent and inspiration or does it guide them to choose roles that are considered useful by society and industry. Ideally, the society served by the university will find what they need at an institution which provides the requisite variety to accommodate all comers.

It has not been easy to collaborate for four years. It is a tight fit to use the room which we have been using. It is over-supplied with technical equipment which we cannot use. The screen and projector make it possible to utilize the Moodle. The overhead projector is very quick and easy to use. Each desk is equipped with a desk and four desks are set around a hub, with many cables. We are able to borrow carousel panels from the library and supply our own magnets. One feels that the designers of the room put everything in there, but it is overkill. There are often power-failures and other problems in this Computer Assisted Language Lab room (CALL).

We are very glad to have enjoyed grant support for our research project. Subjectively and practically the learning environment has improved. One thinks of the enthusiastic participation by "stake holders," the volunteers who want to be there. True, we have not received much assistance from Informatic and Engineering staff to promote our work. There is a layered level of resistance against change!

There are no surprises here. What is a 2nd Order Cyberneticist to do? Recognize the ancient dialectic. It is restated in words attributed to Jean-Baptiste Alphonse Karr, "plus ça change, plus c'est la même chose!" The more things change, the more they stay the same.

This is the moment, at the end of the paper, to remember! Who looks at a class of 110, 170, 400 students as a defeat? Some one who feels that it cannot be managed! Perhaps there is someone who has done amazing things in fixed-seat class theaters? Of course, one 90-min class each week is not recommended for contact with students, effective practice, and language development. One way to approach the 90 min is to break it up into segments and to divide the class into groups. It is also important to find ways to get students to stand up and move around.

The Inter-Academia Community is a forward-looking enterprise, moving into a future world in which young scientists and scholars are welcomed to shine their light. The Asia Bridge Program began in 2015, bringing a large number of highly qualified English speaking students from around Asia. Good communication among students is highly to be desired. The idea of legitimate peripheral participation is not new. Team work and leadership are known to bring value. We end on a note that is buoyed up by the success of our research, yet aware that our success has not established the practice. Rather, we must keep working at it. That is, in itself, no bad thing. And on this note PI concludes.

References

1. Bateson, G.: Mind and Nature: A Necessary Unity, p. 7 (1979)
2. Takeuchi, H., Nonaka, I.: The Knowledge Creating Company: How Japanese Companies Create the Dynamics of Innovation (1995)
3. Bateson, G.: "It Used to Matter," *Big Sur Tapes*. Taped live discussion, Gregory Bateson with students: Esalen, California. Side one (1980)

Prediction of Combine Harvester Performance Using Hybrid Machine Learning Modeling and Response Surface Methodology

Tarahom Mesri Gundoshmian[1], Sina Ardabili[2], Amir Mosavi[3,4](✉), and Annamária R. Várkonyi-Kóczy[3,5]

[1] Department of Biosystem Engineering, University of Mohaghegh Ardabili, Ardabil, Iran
[2] Institute of Advanced Studies Koszeg, Koszeg, Hungary
[3] Kalman Kando Faculty of Electrical Engineering, Obuda University, Budapest, Hungary
amir.mosavi@kvk.uni-obuda.hu
[4] School of the Built Environment, Oxford Brookes University, Oxford OX3 0BP, UK
[5] Department of Mathematics and Informatics, J. Selye University, Komarno, Slovakia

Abstract. Automated controlling the harvesting systems can significantly increase the efficiency of the agricultural practices and prevent food wastes. Modeling and improvement of the combine harvester can increase the overall performance. Machine learning methods provide the opportunity of advanced modeling for accurate prediction of the highest performance of the machine. In this study, the modeling of combine harvesting id performed using radial basis function (RBF) and the hybrid machine learning method of adaptive neuro-fuzzy inference system (ANFIS) to predict various variables of the combine harvester for the optimal performance. Response surface methodology (RSM) is also used to optimize the models. The comparative study shows that the ANFIS method outperforms the RBF method.

Keywords: Combine harvester · Hybrid machine learning · ANFIS · Response surface methodology (RSM) · Artificial intelligence in agriculture · Radial basis function (RBF)

1 Introduction

A large part of agricultural products is destroyed for various reasons during production, consumption and also harvesting stage that is spent a lot of costs, energy, and hard work to produce them and also imposed pressure on the environment. The amount of waste of agricultural products (both quantitatively and qualitatively) has a major amount in Iran and is causing huge damages in the agricultural sector. Efforts to reduce wastes of agricultural products is more important and less costly compared to efforts to increase production [1]. Identification of the effective parameters in harvesting is the first and important step to control and reduce these types of wastes. These parameters

A. R. Várkonyi-Kóczy (Ed.): INTER-ACADEMIA 2019, LNNS 101, pp. 345–360, 2020.
https://doi.org/10.1007/978-3-030-36841-8_34

are the time of harvesting, type of harvesting (mechanized or manual), the correct settings of the harvesting machines, transmission to the target market and etc. According to the unique significance of machines in agricultural production systems as a power source, evaluation of mechanisms and performance of equipment is the inevitable priority of management in agricultural units. Harvester combines have a specific position in the harvest of agricultural crops [2] due to their sensitive mechanisms and processes applied to the Strategic Grain Products [3]. Adjustment and optimization of the internal components of this machine are very important [4]. Factors affecting on improper functioning of combines or any other machine can be achieved using kinematic and dynamic analysis. These factors are placed in three groups include geometric parameters, working conditions, and product properties that inappropriate level of each of them led to reducing the performance of the combined [1]. Each component of combines will affect the flow and movement of the product based on the geometry and properties of the product. Geometric changes affect the machine performance that prediction and modeling of these changes will be an effective proceeding in machine design and new components production [5]. The present research studies the effect of three parameters include clearance of threshing unit, fan speed and openness of sieves on product damage, loss and the amount of non-grain materials for conventional combines. Based on a report in a study by Spengler et al. [6], optimization of the threshing unit decreased 4 to 6% of total machine loss in Germany in 1986. Harvesting by harvester combine is a complex nonlinear process [7] that is affected by a wide range of data [8]. Many mathematical models are used for modeling relationships between inputs and outputs of a process, but classical logic requires precise definitions of relationships of mathematical models to describe phenomena [9]. The use of artificial intelligence methods such as artificial neural networks, fuzzy logic, etc., to develop predictive models possible very complex data to be compared and analyzed [10]. One of these methods is the fuzzy method. Fuzzy method has a lot of benefits that one of this benefits is to process the improper data. Fuzzy method provides linguistic labels for modeling complex systems [11]. Artificial neural network is another processing method. In general, The artificial neural network method is used in a wide area such as mathematics, engineering, medicine, economics, the environment and agriculture [12]. The advantage of both systems is focused on ANFIS structure. ANFIS is widely used in the study of complex systems modeling, control, or estimate parameters. A combined learning algorithm is the basis of ANFIS to identify parameters using the Sugeno Fuzzy Inference System. This system is combination of least squares and back propagation error methods to train membership functions to compete with total training data to achieve the best output [11].

Harvester combines a wide range of products in various environmental conditions [4]. Harvester combines have variable rate Feature that enables combines to harvest in a wide range of conditions [7]. Harvester combines have five general duties: Harvesting and feeding, threshing, separating, cleaning, and loading of product. The threshing unit separates about 60 to 90% of threshed seeds from clusters [13]. Separating unit is after the threshing unit and separates grain from different parts of the product. Cleaning of seeds from impurities refers to the final stage of the separation process. Cleaning is performed by both mechanical and aerodynamic simultaneously. Finally, cleaned seeds poured into transporters of the clean grain and are transferred to the tank by grain elevators. The simulating and modeling of the desired system (combine in this study) give better judge about the performance of various sectors [14, 15]. The threshing process is one of the essential processes in harvesting stage that is evaluated by factors include thresher efficiency, cleaner efficiency, damaged seeds and chopped stalks [1]. System modeling using conventional mathematical tools such as deferential relations is not suitable in systems with unclear behavior and not well-defined systems [16]. Despite the classical systems, intelligent control systems do not need to know the mathematical models of behavior system. Now a days, intelligent systems and soft computing-based systems are used in all scientific fields [17]. Therefore, using the prediction methods such as soft computing and intelligent methods, that are increased recently, help evaluate the desired systems [18]. These methods have various types that most popular of them are fuzzy methods and artificial neural networks [19]. There are several studies on modeling, studying or optimizing the performance of combine harvesters. Some of them are as follow that studying them can help to define the aim of present study and the novelty of the work:

Craessaerts et al. [4, 20] studied genetic-based methodology for input selection to identify the cleaning process on a combine harvester in two parts, first for selection the input variables for identification of the sieve losses and second for identification of MOG content in a grain bin. Maertens and De Baerdemaeker [21] prepared a dynamic separation model to avoid using the non-linear, complex and uncertain relations. Zhao et al. [22] presented an indirect grain separation loss monitoring method to show the grain loss in separation unit based on the analysis of the relationship between grain separation loss and grain separation flux in the area under the concave. Mirza Zadeh et al. [23] employed a multi-layered perceptron method of ANN to predict the grain separation of the combine harvester. They used the parameters of feed rate, stem height, the rotational revolution of thresher, and clearance ratio as the independent variables of the test. Results showed a correlation coefficient of 0.9. Maertens et al. [24, 25] made an analytical approach in two parts to maintain the grain flow model for a combine harvester. The first part for model designing and the second part for analysis and application of the model. Miu and Kutzbach [26, 27] did a two-part study on modeling the threshing and separation units of the combine harvester. These studies were developed based on mathematical modeling. They found a good correlation between predicted and experimental data. Miu [28] did a study on designing an optimized threshing process using a genetic algorithm. It was formulated a multi-objective genetic algorithm to optimize the functional parameters of threshing units. It was obtained that this method can be adapted to other threshing units in various crops. Miu and Kutzbach [29] simulated the threshing and separation process in the threshing unit of a combine harvester. It was developed, two models. One model describes the percentage of unthreshed grains, and the other model quantifies the cumulative percentages of separated MOG, separable MOG, and unfragmented MOG. In a study by Ryszard and Jachimczyk [30] it was developed the mathematical model of grain separation in a straw walker. Based on results, kinematic parameters of walkers significantly effected on the quality of separation. In a study by Bulgakov et al. developed a mathematical model for the renewal of the combine harvester fleet on the basis of integral equations for the fleet of combine harvesters. In a two part-study by Craessaerts et al. it was considered the identification of the cleaning process on combine harvesters. The first part for developing a fuzzy model for prediction of the material other than grain (MOG) content in the grain bin and the second part for prediction of the sieve losses.

As is clear from previous studies, there are a limited number of studies with soft computing methods on modeling and studying combine harvester performances. While the soft computing methods can reduce the complexity of system and process and can increase the precision. The present study tries ANFIS and RBF models as the models that were not applied to modeling the combine harvester to determine and evaluation of this method on combine modeling. Moreover, the RSM method is used to optimize the product loss, MOG content, and broken seeds, simultaneously. However, so far, an optimization study with this dependent parameters has not been performed with the RSM method, and this is the main novelty of present work.

2 Material and Method

Measuring and collecting the required data was conducted from 1055i john deer combine in Agricultural Research Station of Ardabil Province of Iran. The specifications of cleaning and threshing units were:

Threshing drum (TD) with a diameter of 610 mm and the length of 1080 include 8 blades and rotational speed of 410 to 1160 rpm.

- Modular and adjustable and slider type concave.
- Sieves by the area of 1.2 m^2.
- The fan speed of 440 to 1060 rpm include 5 mechanical adjustable vanes.
- Thresher drum and concave distance of 10 mm in input and 3 mm in output.

Experiments were performed in form of the factorial test include 3 repetitions in three levels of variables. The independent variables of the test were:

- The distance of TD and concave (A)
- Fan speed (B)
- Sieves openness (C)
- The dependent variables were:
- The number of broken seeds inside the tank (BS)
- Product loss (PL)
- Material other than grain (MOG).

2.1 ANN Modeling

Modeling with machine learning methods have been studied in a vast number of studies covering a wide range of applications [31–49]. Among the machine learning method the hybrid and ensemble are reported to outperform other conventional machine learning methods [50–78]. This has been the reason that in this study the hybrid machine learning method of ANFIS have been selected to model along with RBF.

In the present study, the RBF of ANN was employed to develop a model among dependent and independent variables. An RBF network is a three-layered feed-forward network with the structure which is presented in Fig. 1.

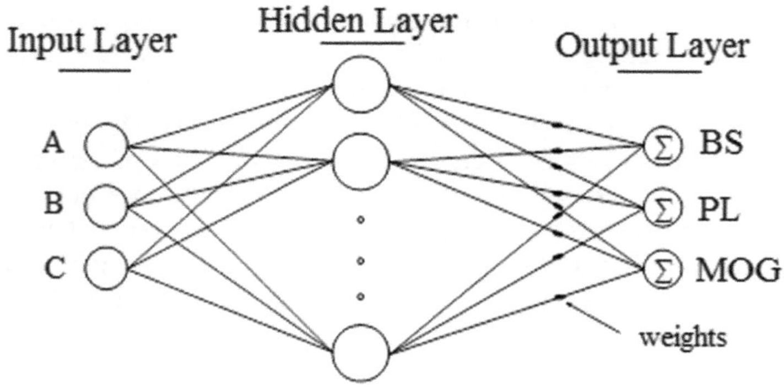

Fig. 1. The structure of RBF network

Inputs directly enter to the hidden layer and after multiplying to related weights enter to a summation box and generate the outputs. The input layer is composed of neurons that are a divider of input signals to hidden layer neurons. In the hidden layer of the neurons are executed a non-linear mapping of the input space into hidden layer space based on Eq. 1 that are dimensionally equal [19, 65, 66]. In the present study, the Gaussian function was employed to define the neurons as local receivers.

$$\phi_i(x) = \exp\left[-\frac{||x - c_i||^2}{2\sigma^2}\right]_{;\, i=1,2,\ldots,p} \quad (1)$$

Where c_i and σ_i are the center and width of the ith hidden layer, respectively. The output of the network is a linear mapping of $\emptyset(x)$ to Y. this mapping is calculated as Eq. 2 and presents the output of the network.

$$Y = \sum_{i=1}^{r} w2_{ik}\phi_i(x)_{;k=1,2,\ldots,m} \quad (2)$$

Where k is the node of the output layer, W_{ik} is the weight of connectors of hidden layer to the output layer and $\phi_i(x)$ is the response of hidden layer for nodes of the output layer. Infect, the RBF network realizes the following operations to train the network:

- Uncontrolled learning algorithm to train the centers and width of the main functions
- Assignment of weights related to the connections between hidden and output layers.

2.2 Network Training

The experimental data of target combine harvester were employed to train the network. The trained network has to be able to connect the inputs and outputs to be able to predict and model the behavior of the system. Therefore, the factors of A, B, and C were considered as the input variables (independent variables) and the factors of BS,

PL and MOG were considered as the output variables (dependent variables) of the network. These factors were selected based on adjustments at the disposal of the operator and the effect of the factors on system performance. 70% of data were selected as training data, and 30% of them were selected as testing data. The aim of the training process is to reduce the errors of the target and output values of the network. Mean absolute error (MSE) was used to compare the target and network's output values. The training process was started with five neurons in the hidden layer, and in each training step, 5 neurons were added to the number of previous neurons at the next run. Adding neurons to hidden layer continues as long as reducing the errors and taking a constant trend of errors.

2.3 ANFIS Modeling

Adaptive neuro-fuzzy inference system (ANFIS) is an approach to model the nonlinear complex problems that use Sugeno model with fuzzy inputs and rules to prepare a strong predicting tool [67]. ANFIS is a class of adaptive feed-forward network that has 5 layers (Fig. 2).

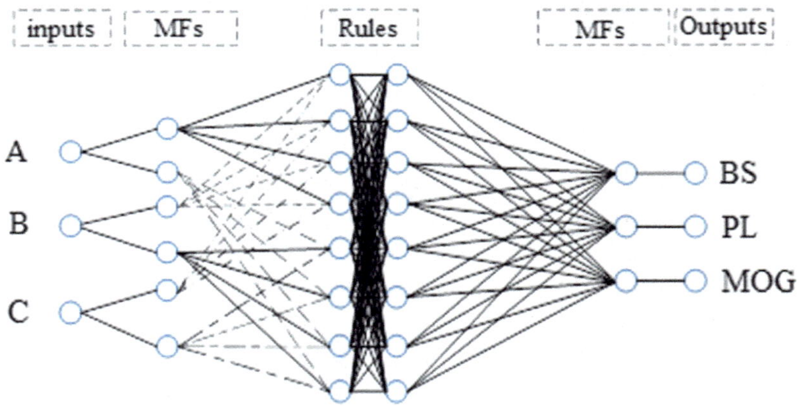

Fig. 2. ANFIS structure

This system generates fuzzy rules based on input and output data (i.e. training data). A simple rule using Sugeno fuzzy model is as follow:

$$\text{If x is } A_i \text{ and y is } B_i \text{ then } z = f(x; y)$$

Where A and B are fuzzy categories and z = f (x; y) is usually a polynomial function [68]. In this study, training and testing data for developing ANFIS were the same data related to developing the RBF model. ANFIS was developed by MATLAB 2012a software. The used algorithm was a combined algorithm and was selected to change the initial membership functions. In order to determine the best training network, the ANFIS method was developed based on different types of membership functions. The trim type function with a linear method using three membership functions prepared the best response to network modeling based on comparing parameters.

Comparing parameters was the root mean square error (RMSE), correlation coefficient (r), and mean absolute error (MAE) to compare the target and output values of networks.

$$RMSE = \sqrt{\frac{1}{N}\sum_{i=1}^{N}(A-P)^2} \tag{3}$$

$$R = \left(1 - \left(\frac{\sum_{i=1}^{n}(A-P)^2}{\sum_{i=1}^{n}A_i^2}\right)\right)^{1/2} \tag{4}$$

$$MAE = \frac{\sum_{i=1}^{N}|A-P|}{N} \tag{5}$$

Where A is related to actual values and P is related to predicted values and N is the number of data. Using this parameter helps to choose the best structure and best network and provides the possibility of understanding the proximity of model.

3 Results

The present study was performed on a combine harvester and data were recorded experimentally. Figure 3 presents the initial results of recorded data and their Interactions. Based on Fig. 3a, increasing A decreases the BS and MOG but decreases PL and then increases. This says there is an optimum value for A. Based on Fig. 3b, increasing B decreases MOG and about two other parameters, it decreases and then increases them. Based on Fig. 3c, the relation of C and the BS and MOG are opposite. The openness of sieve up to a certain range decreases the PL and after that range, increases it. Studying the milestone points takes the optimum range of each variable.

According to Fig. 4(a) by considering the fixed value of other parameters, opening hot water tap, increases the growing hall temperature. Figure 4(b) shows the variation of growing hall temperature by opening and closing air dampers when other parameters are fixed. Accordingly, if the rate of opening and closing of circulation and fresh air dampers be equal, respectively, the hall temperature almost will be fixed. In Fig. 4(c) by reducing the temperature of the water and by increasing the ambient temperature during the day, the growing hall temperature has undergone a constant trend.

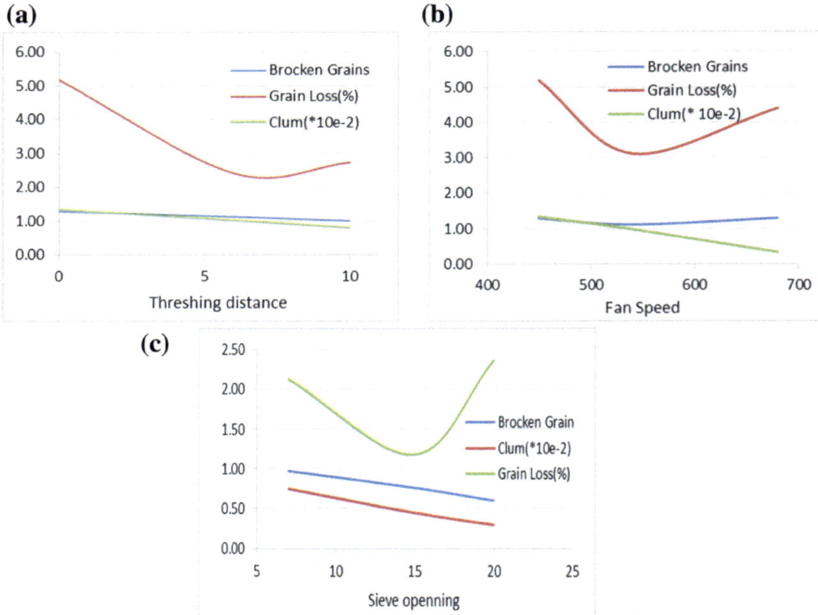

Fig. 3. Interaction of input variables on output variables (a) Effect of threshing distance on BS, PL, MOG (b) Effect of Fan speed on BS, PL, MOG (c) Effect of Sieve opening on BS, PL, MOG

3.1 Modeling Results

Training stage

Studying and presenting a model based on RBF and ANFIS methods were considered as the main aim. One of the essential stages of preparing a precise model is the training stage.

Training process of RBF and ANFIS models were performed and the results were extracted. These results help us to choose the best model to enter the testing process. Prediction process was performed by the ANFIS and RBF networks. To perform modeling operations, BS, PL, and MOG were considered as the independent variable (output of network) and A, B and C were considered as independent variables (inputs of the network). In order to train target networks, test data were employed to develop the network. This stage was performed to create a precise network for the test stage. Results of training for RBF and ANFIS methods were presented in Tables 1 and 2, respectively:

Table 1. Results of training RBF network

No. neurons	Output	r	MSE
8	BS	0.84	1.028 e-2
	PL	0.9	200
	MOG	0.95	0.1
10	BS	0.9	6.87 e-3
	PL	0.94	129.19
	MOG	0.96	7.81 e-2
12	BS	0.98	1.09 e-3
	PL	0.99	7.918
	MOG	0.97	5.5 e-2
14	BS	0.981	1.06 e-3
	PL	0.99	7.918
	MOG	0.987	5.5 e-2

Table 2. Results of training ANFIS network

Structure	Output	Membership function	MSE	Oprim. method
3 3 3	BS	*Tri*	*1.2396 e-5*	*Hybrid*
	PL		*6.9 e-4*	
	MOG		*5.907 e-5*	
3 3 3	BS	Trap	1.4209 e-5	Hybrid
	PL		7.5 e-4	
	MOG		6.74 e-5	
3 3 3	BS	Gbell	4.6544 e-5	Hybrid
	PL		27.3 e-4	
	MOG		18.7 e-5	
3 3 3	BS	Gauss	3.3099 e-5	Hybrid
	PL		17.4 e-4	
	MOG		9.38 e-5	

Based on results of Tables 1 and 2 the network with 14 neurons in the hidden layer for RBF network with lowest MSE (1.06 e-3, 7.918 and 5.5 e-2 for BS, PL and MOG, respectively) and the network with Tri. Membership function for ANFIS with lowest MSE (1.2396 e-5, 6.9 e-4, and 5.907 e-5, for BS, PL and MOG, respectively) were selected to be employed in the test stage.

Testing stage

The test data were imported to the selected network in the training stage. The output of networks was compared with target data and the comparison results were presented in Table 3.

Table 3. Results of comparison parameters

		RMSE	R	MAE
ANFIS	BS	1.24 e-5	0.999	8.814 e-6
	PL	1.746 e-3	0.999	1.165 e-3
	MOG	2.96 e-4	0.999	7.95 e-5
RBF	BS	0.0401	0.9775	0.02
	PL	5.66	0.986	2.48
	MOG	0.2414	0.9753	0.083

Based on report of Table 3, ANFIS with RMSE of 1.24 e-5, 1.746 e-3 and 2.96 e-4 and r value of 0.999, 0.999 and 0.999, for BS, PL and MOG, respectively, presented the best result compared to RBF network due to its low value of RMSE and high value of r compared to RBF network. Therefore, ANFIS was selected as the best prediction model in the present study.

3.2 Optimization

The optimization process was performed by response surface methodology (RSM). The RSM method is a statistical method to find the relationships between several explanatory variables (input variables) and one or more response variables (output variable (s)). The main idea of RSM is to use a sequence of designed experiments to obtain an optimal response [69].

Figure 4 indicates the optimum values of PL, MOG and BS and system behavior against the variations of A, B, and C by RSM using Design-Expert software. Optimization was performed to minimize the PL, BS, and MOG values. A quadratic model with r value of 0.9841, 0.9 and 0.87, respectively for BS, MOG, and PL were employed to model the process. Based on results, it is clear, because all three outputs are optimized simultaneously, therefore the optimal condition occurs in one point which three parameters have their minimize value. i.e., 0.66, 18.25 and 1.37, respectively for BS, MOG and PL which are accrued on 7.35, 598.1 and 16.82, of A, B, and C, respectively.

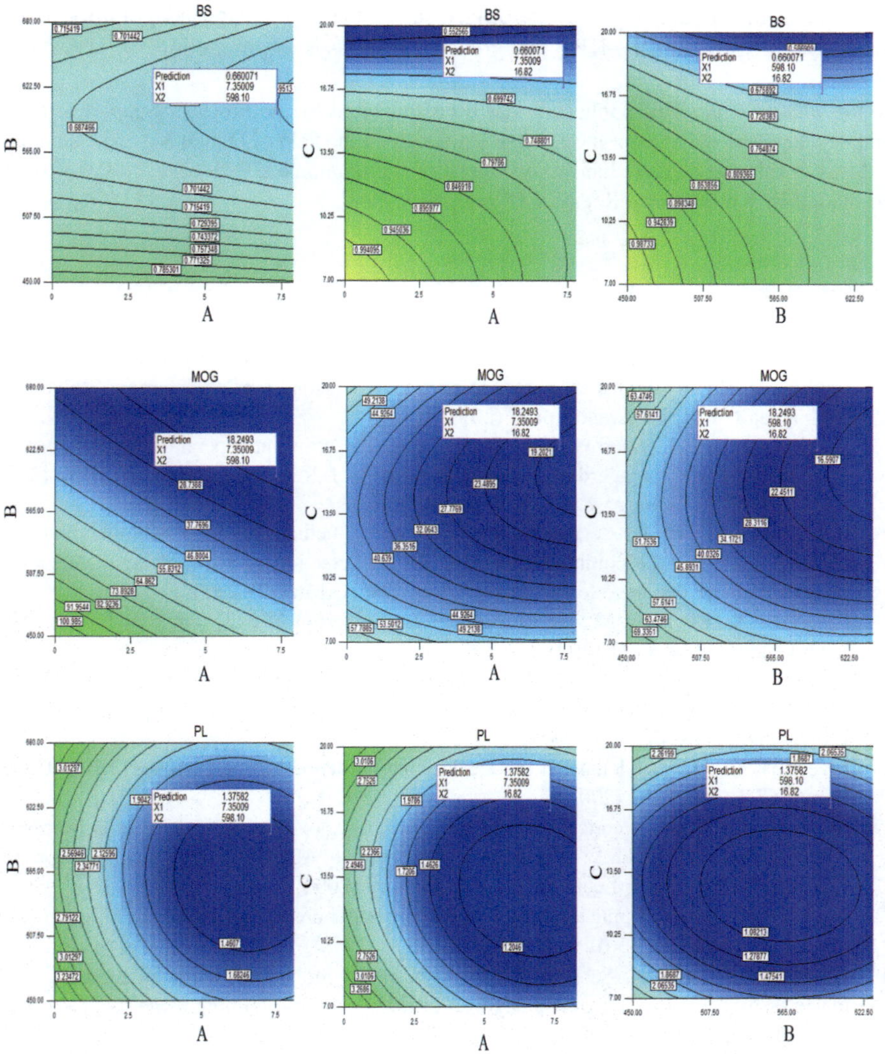

Fig. 4. Results of the optimization process

4 Conclusion

In the present study, the performance factors of combine harvester including the BS, PL, and MOG was modeled based on three factors, A, B and C, using MLP and RBF of artificial neural networks. By the statistical analysis and checking the functional parameters, using the correlation test, the relationship between these parameters was significant at 5% probability level. After modeling and using the obtained results, it was observed that with an increasing number of neurons in the hidden layer, results would be better and better. The best result and the highest correlation value was obtained in 20 neurons in the hidden layer. Regarding results, due to the high adaptability and low error, we can say that the RBF has great value for system modeling.

Acknowledgments. This publication has been supported by the Project: "Support of research and development activities of the J. Selye University in the field of Digital Slovakia and creative industry" of the Research & Innovation Operational Programme (ITMS code: NFP313010T504) co-funded by the European Regional Development Fund.

References

1. Khorram, T.: The interaction effect of seive openess and thresher clearence on threshing quality, in Thesis of Master scienc. Department of Biosystem engineering. University of Mohaghegh Ardabili. Ardabil, Iran (2013)
2. Singh, A., Garg, I., Sharma, V.: Effect of different crop and operational parameters of a combine on grain damage during paddy harvesting. J. Res. **38**(3 and 4), 241–252 (2001)
3. FAO, FAOSTAT data base. FAO, Rome (2008)
4. Craessaerts, G., et al.: A genetic input selection methodology for identification of the cleaning process on a combine harvester, Part I: Selection of relevant input variables for identification of the sieve losses. Biosys. Eng. **98**(2), 166–175 (2007)
5. Maertens, K., Reyniers, M., De Baerdemaeker, J.: Design of a Dynamic Grain Flow Model for a Combine Harvester (2001)
6. Spengler, A., Mehne, S., Feiffer, A.: Combine harvesting at large scale enterprises in Europe. In: Electronic Proceedings of the International Conference on Crop Harvesting and Processing, Louisville, Ky (2003)
7. Maertens, K., De Baerdemaeker, J.: Design of a virtual combine harvester. Math. Comput. Simul. **65**(1–2), 49–57 (2004)
8. Maertens, K.: Data-driven techniques for the on-the-go evaluation of separation processes in combine harvesters. Ph.D. Thesis. Department of Agro-Engineering and Economics, Katholieke Universiteit Leuven, Leuven, Belgium (2004)
9. Naderloo, L., et al.: Application of ANFIS to predict crop yield based on different energy inputs. Measurement **45**(6), 1406–1413 (2012)
10. Gautam, R., Panigrahi, S., Franzen, D.: Neural network optimisation of remotely sensed maize leaf nitrogen with a genetic algorithm and linear programming using five performance parameters. Biosys. Eng. **95**(3), 359–370 (2006)
11. Soyguder, S., Alli, H.: An expert system for the humidity and temperature control in HVAC systems using ANFIS and optimization with fuzzy modeling approach. Energy Build. **41**(8), 814–822 (2009)
12. Safa, M., Samarasinghe, S.: Determination and modelling of energy consumption in wheat production using neural networks:"A case study in Canterbury province, New Zealand". Energy **36**(8), 5140–5147 (2011)
13. Mansouri raad, D.: Tractors and Agricultural Machinery, vol. 2. Publication of Bo-ali sina university. Hamadan, Iran (1993)
14. Faizollahzadeh_Ardabili, S.: Simulation and Comparison of Control System in Mushroom Growing Rooms Environment. Thesis of Master science. Department of Mechanic of Agricultural Machinery Engineering. University of Tabriz. Tabriz, Iran (2014)
15. Faizollahzadeh_Ardabili, et al.: Modeling and comparison of fuzzy and on/off controller in a mushroom growing hall. Measurement **90**, 127–134 (2016)

16. Jang, J.-S.R.: ANFIS: adaptive-network-based fuzzy inference system. Syst. Man Cybern. IEEE Trans. **23**(3), 665–685 (1993)
17. Faizollahzadeh_Ardabili, Mahmoudi, A., Mesri Gundoshmian, T.: Modeling and simulation controlling system of HVAC using fuzzy and predictive (radial basis function, RBF) controllers. J. Build. Eng. **6**, 301–308 (2016)
18. Soyguder, S.: Intelligent system based on wavelet decomposition and neural network for predicting of fan speed for energy saving in HVAC system. Energy Build. **43**(4), 814–822 (2011)
19. Chen, X.-T., Zhang, L.-H.: High-quality voice conversion system based on GMM statistical parameters and RBF neural network. J. China Universities Posts Telecommun. **21**(5), 68–75 (2014)
20. Craessaerts, G., et al.: A genetic input selection methodology for identification of the cleaning process on a combine harvester, Part II: Selection of relevant input variables for identification of material other than grain (MOG) content in the grain bin. Biosys. Eng. **98** (3), 297–303 (2007)
21. Maertens, K., De Baerdemaeker, J.: Design of a virtual combine harvester. Math. Comput. Simul. **65**(1), 49–57 (2004)
22. Zhao, Z., et al.: Grain separation loss monitoring system in combine harvester. Comput. Electron. Agric. **76**(2), 183–188 (2011)
23. Mirzazadeh, A., et al.: Intelligent modeling of material separation in combine harvester's thresher by ANN. Int. J. Agric. Crop Sci. **4**(23), 1767–1777 (2012)
24. Maertens, K., et al.: PH—power and machinery: an analytical grain flow model for a combine harvester, Part I: design of the model. J. Agric. Eng. Res. **79**(1), 55–63 (2001)
25. Maertens, K., et al.: PA—precision agriculture: an analytical grain flow model for a combine harvester, part ii: analysis and application of the model. J. Agric. Eng. Res. **79**(2), 187–193 (2001)
26. Miu, P.I., Kutzbach, H.-D.: Modeling and simulation of grain threshing and separation in threshing units—Part I. Comput. Electron. Agric. **60**(1), 96–104 (2008)
27. Miu, P.I., Kutzbach, H.-D.: Modeling and simulation of grain threshing and separation in axial threshing units: Part II. Application to tangential feeding. Comput. Electron. Agric. **60** (1), 105–109 (2008)
28. Peter, I.M.: Optimal Design Threshing Units Based on a Genetic Algorithm. I. Algorithm. ASAE
29. Miu, P.I., Kutzbach, H.D.: Simulation of threshing and separation processes in threshing units. Agrartechnische Forschung **6**, 1–7 (2000)
30. Myhan, R., Jachimczyk, E.: Grain separation in a straw walker unit of a combine harvester: Process model. Biosys. Eng. **145**, 93–107 (2016)
31. Mosavi, A., Edalatifar, M.: A Hybrid Neuro-Fuzzy Algorithm for Prediction of Reference Evapotranspiration, in Lecture Notes in Networks and Systems, pp. 235–243. Springer (2019)
32. Mosavi, A., Lopez, A., Varkonyi-Koczy, A.R.: Industrial applications of big data: state of the art survey, D. Luca, L. Sirghi, and C. Costin, Editors, pp. 225–232. Springer (2018)
33. Mosavi, A., Ozturk, P., Chau, K.W.: Flood prediction using machine learning models: literature review. Water (Switzerland) **10**(11) (2018)
34. Mosavi, A., Rabczuk, T.: Learning and intelligent optimization for material design innovation. In: Kvasov, D.E. et al. (eds.), pp. 358–363. Springer (2017)
35. Mosavi, A., Rabczuk, T., Varkonyi-Koczy, A.R.: Reviewing the novel machine learning tools for materials design. In: Luca, D., Sirghi, L., Costin, C. (eds.), pp. 50–58. Springer (2018)

36. Mosavi, A., et al.: State of the art of machine learning models in energy systems, a systematic review. Energies **12**(7) (2019)
37. Mosavi, A., et al.: Prediction of multi-inputs bubble column reactor using a novel hybrid model of computational fluid dynamics and machine learning. Eng. Appl. Comput. Fluid Mech. **13**(1), 482–492 (2019)
38. Mosavi, A., Varkonyi-Koczy, A.R.: Integration of machine learning and optimization for robot learning. In: Jablonski, R., Szewczyk, R. (eds.), pp. 349–355. Springer (2017)
39. Nosratabadi, S., et al.: Sustainable business models: a review. Sustainability (Switzerland) **11**(6) (2019)
40. Qasem, S.N., et al.: Estimating daily dew point temperature using machine learning algorithms. Water (Switzerland) **11**(3) (2019)
41. Rezakazemi, M., Mosavi, A., Shirazian, S.: ANFIS pattern for molecular membranes separation optimization. J. Mol. Liq. **274**, 470–476 (2019)
42. Riahi-Madvar, H., et al.: Comparative analysis of soft computing techniques RBF, MLP, and ANFIS with MLR and MNLR for predicting grade-control scour hole geometry. Eng. Appl. Comput. Fluid Mech. **13**(1), 529–550 (2019)
43. Shabani, S., Samadianfard, S., Taghi Sattari, M., Shamshirband, S., Mosavi, A., Kmet, T., Várkonyi-Kóczy, A.R.: Modeling Daily Pan Evaporation in Humid Climates Using Gaussian Process Regression. Preprints (2019), 2019070351. https://doi.org/10.20944/preprints201907.0351.v1
44. Shamshirband, S., Hadipoor, M., Baghban, A., Mosavi, A., Bukor J., Varkonyi-Koczy, A. R.: Developing an ANFIS-PSO Model to predict mercury emissions in Combustion Flue Gases. Preprints (2019), 2019070165. https://doi.org/10.20944/preprints201907.0165.v1
45. Shamshirband, S., et al.: Ensemble models with uncertainty analysis for multi-day ahead forecasting of chlorophyll a concentration in coastal waters. Eng. Appl. Comput. Fluid Mech. **13**(1), 91–101 (2019)
46. Shamshirband, S., Mosavi, A., Rabczuk, T.: Particle swarm optimization model to predict scour depth around bridge pier. arXiv preprint arXiv:1906.08863 (2019)
47. Taherei Ghazvinei, P., et al.: Sugarcane growth prediction based on meteorological parameters using extreme learning machine and artificial neural network. Eng. Appl. Comput. Fluid Mech. **12**(1), 738–749 (2018)
48. Torabi, M., et al.: A Hybrid clustering and classification technique for forecasting short-term energy consumption. Environ. Prog. Sustain. Energy **38**(1), 66–76 (2019)
49. Torabi, M., et al.: A hybrid machine learning approach for daily prediction of solar radiation. In: Lecture Notes in Networks and Systems, pp. 266–274. Springer (2019)
50. Aram, F., et al.: Design and validation of a computational program for analysing mental maps: aram mental map analyzer. Sustainability (Switzerland). **11**(14) (2019)
51. Asadi, E., et al.: Groundwater Quality Assessment for Drinking and Agricultural Purposes in Tabriz Aquifer, Iran (2019)
52. Asghar, M.Z., Subhan, F., Imran, M., Kundi, F.M., Shamshirband, S., Mosavi, A., Csiba, P., Várkonyi-Kóczy, A.R.: Performance Evaluation of Supervised Machine Learning Techniques for Efficient Detection of Emotions from Online Content. Pre-prints 2019, 2019080019. https://doi.org/10.20944/preprints201908.0019.v1
53. Bemani, A., Baghban, A., Shamshirband, S., Mosavi, A., Csiba, P., Várkonyi-Kóczy, A.R.: Applying ANN, ANFIS, and LSSVM Models for Estimation of Acid Sol-vent Solubility in Supercritical CO2. Preprints (2019), 2019060055. https://doi.org/10.20944/preprints201906.0055.v2
54. Choubin, B., et al.: Snow avalanche hazard prediction using machine learning methods. J. Hydrol., 577 (2019)

55. Choubin, B., et al.: An ensemble prediction of flood susceptibility using multivariate discriminant analysis, classification and regression trees, and support vector machines. Sci. Total Environ. **651**, 2087–2096 (2019)
56. Dehghani, M., et al.: Prediction of hydropower generation using Grey wolf optimization adaptive neuro-fuzzy inference system. Energies **12**(2) (2019)
57. Dineva, A., et al.: Review of soft computing models in design and control of rotating electrical machines. Energies **12**(6) (2019)
58. Dineva, A., et al.: Multi-Label Classification for Fault Diagnosis of Rotating Electrical Machines (2019)
59. Farzaneh-Gord, M., et al.: Numerical simulation of pressure pulsation effects of a snubber in a CNG station for increasing measurement accuracy. Eng. Appl. Comput. Fluid Mech. **13**(1), 642–663 (2019)
60. Ghalandari, M., et al.: Investigation of submerged structures' flexibility on sloshing frequency using a boundary element method and finite element analysis. Eng. Appl. Comput. Fluid Mech. **13**(1), 519–528 (2019)
61. Ghalandari, M., et al.: Flutter speed estimation using presented differential quadrature method formulation. Eng. Appl. Comput. Fluid Mech. **13**(1), 804–810 (2019)
62. Karballaeezadeh, N., et al.: Prediction of remaining service life of pavement using an optimized support vector machine (case study of Semnan-Firuzkuh road). Eng. Appl. Comput. Fluid Mech. **13**(1), 188–198 (2019)
63. Menad, N.A., et al.: Modeling temperature dependency of oil - water relative permeability in thermal enhanced oil recovery processes using group method of data handling and gene expression programming. Eng. Appl. Comput. Fluid Mech. **13**(1), 724–743 (2019)
64. Mohammadzadeh, S., et al.: Prediction of compression index of fine-grained soils using a gene expression programming model. Infrastructures **4**(2), 26 (2019)
65. Wen, X.-L., Wang, H.-T., Wang, H.: Prediction model of flow boiling heat transfer for R407C inside horizontal smooth tubes based on RBF neural network. Procedia Eng. **31**, 233–239 (2012)
66. Jiang, H., et al.: Intelligent optimization models based on hard-ridge penalty and RBF for forecasting global solar radiation. Energy Convers. Manag. **95**, 42–58 (2015)
67. Riverol, C., Di Sanctis, C.: Improving adaptive-network-based fuzzy inference systems (ANFIS): a practical approach. Asian J. Inf. Technol. **4**(12), 1208–1212 (2005)
68. Chaabene, M., Ammar, M.B.: Neuro-fuzzy dynamic model with Kalman filter to forecast irradiance and temperature for solar energy systems. Renew. Energy **33**(7), 1435–1443 (2008)
69. Ardabili, S.F., et al.: A novel enhanced exergy method in analysing HVAC system using soft computing approaches: a case study on mushroom growing hall. J. Build. Eng. (2017)
70. Ardabili, S., Mosavi, A., Mahmoudi, Mesri Gundoshmian, T., Nosratabadi, S., Varkonyi-Koczy, A.: Modelling temperature variation of mushroom growing hall using artificial neural networks, Preprints (2019)
71. Mesri Gundoshmian, T., Ardabili, S., Mosavi, A., Varkonyi-Koczy, A.: Prediction of combine harvester performance using hybrid machine learning modeling and re-sponse surface methodology, Preprints (2019)
72. Ardabili, S., Mosavi, A., Varkonyi-Koczy, A., Systematic review of deep learning and machine learning models in biofuels research, Preprints (2019)
73. Ardabili, S., Mosavi, A., Varkonyi-Koczy, A.: Advances in machine learning model-ing reviewing hybrid and ensemble methods, Preprints (2019)
74. Ardabili, S., Mosavi, A., Varkonyi-Koczy, A.: Building Energy information: demand and consumption prediction with Machine Learning models for sustainable and smart cities, Preprints (2019)

75. Ardabili, S., Mosavi, A., Dehghani, M., Varkonyi-Koczy, A.: Deep learning and machine learning in hydrological processes climate change and earth systems a systematic review, Preprints (2019)
76. Mohammadzadeh, D., Karballaeezadeh, N., Mohemmi, M., Mosavi, A., Varkonyi-Koczy, A.: Urban Train Soil-Structure Interaction Modeling and Analysis, Preprints (2019)
77. Mosavi, A., Ardabili, S., Varkonyi-Koczy, A., List of deep learning models, Preprints (2019)
78. Nosratabadi, S., Mosavi, A., Keivani, R., Ardabili, S., Aram, F.: State of the art survey of deep learning and machine learning models for smart cities and urban sustainability, Preprints (2019)

Urban Train Soil-Structure Interaction Modeling and Analysis

Danial Mohammadzadeh[1,2], Nader Karballaeezadeh[3], Morteza Mohemmi[4], Amir Mosavi[5,6(✉)], and Annamária R. Várkonyi-Kóczy[5,7]

[1] Department of Civil Engineering, Ferdowsi University of Mashhad, Mashhad, Iran
[2] Department of Elite Relations with Industries, Khorasan Construction Engineering Organization, Mashhad, Iran
[3] Faculty of Civil Engineering, Shahrood University of Technology, Shahrood, Iran
[4] Department of Civil Engineering, Iran University of Science and Technology, Tehran, Iran
[5] Institute of Automation, Kalman Kando Faculty of Electrical Engineering, Obuda University, Budapest 1034, Hungary
amir.mosavi@kvk.uni-obuda.hu
[6] School of the Built Environment, Oxford Brookes University, Oxford OX3 0BP, UK
[7] Department of Mathematics and Informatics, J. Selye University, Komarno, Slovakia

Abstract. Design and advancement of the durable urban train infrastructures are of utmost importance for reliable mobility in the smart cities of the future. Given the importance of urban train lines, tunnels, and subway stations, these structures should be meticulously analyzed. In this research, two-dimensional modeling and analysis of the soil-structure mass of the Alan Dasht station of Mashhad Urban Train are studied. The two-dimensional modeling was conducted using Hashash's method and displacement interaction. After calculating the free-field resonance and side distortion of the soil mass, this resonance was entered into PLAXIS finite element program, and finally, stress and displacement contours together with the bending moment, shear force and axial force curves of the structure were obtained.

Keywords: Urban mobility · Urban train lines · Modeling · Soil mass-structure · Soil-structure interaction · PLAXIS · Computational mechanics · Simulation · Smart cities · Urban sustainable development · Urban rail transportation

1 Introduction

Today, due to the expansion of cities and urbanization, the need for efficient public transportation is continuously on the rise [1–13]. Among public transport vehicles, the urban train network is of great importance as a standard form of urban rail

A. R. Várkonyi-Kóczy (Ed.): INTER-ACADEMIA 2019, LNNS 101, pp. 361–381, 2020.
https://doi.org/10.1007/978-3-030-36841-8_35

transportation [14–17]. The construction of the metro network on the ground is less costly, but due to increasing surface congestion, heavy traffic, and scarcity of space, metro lines are mostly built underground [18–20]. On the one hand, due to the low depth of the underground tunnels, metro stations are usually built-in soft soils (i.e., loose surface soils [21–24]). Therefore, much research has been done to provide smart solutions to such problems. Thus, the optimal implementation of these underground spaces and ensuring their security during the long-term construction process is a factor that has been taken into account by designers of underground structures. Among these underground structures, transportation tunnels are of critical importance. Before 1995, tunnels were not designed to tolerate dynamic loads. However, the earthquakes of the 1990s wreaked havoc to some tunnels. For example, the 1999 earthquake in Chi-Chi, Taiwan, damaged the mountain tunnels [25]. Duzce earthquake in Turkey led to the partial destruction of Bolu twin tunnels [26], and Diakia metro station collapsed in Kobe, Japan in 1995 [27]. Sharma and Judd [28] investigated 192 cases of underground structural behavior in the face of 85 earthquakes worldwide, concluding that the vulnerability of an underground structure could be assessed in terms of the depth of overburden and earthquake parameters of the area. The extent of demolition and damage to tunnels declines with an increase in the depth of overburden [29]. In 2001, Hashash [30] noted that the tunnel structure should be considered in terms of plasticity and resistance to earthquake forces [31]. Moreover, to ensure sufficient plasticity in the tunnel cover during an earthquake, the force reduction factor should be equivalent to the tunnel plasticity [32]. In this research, a two-dimensional modeling of the Alan Dasht station of Mashhad Urban Train using Hashash's method along with the displacement interaction at two levels of seismic [33] is performed in PLAXIS software [34], and the results of two-dimensional analysis are presented as displacement and stress contours together with the curves of bending moment, shear force and axial force.

2 Data

Line 2 of the Mashhad urban train stretches from the end of the northern Tabarsi Blvd. to the south of Fazl bin Shazan Square with a total length of 12.985 km. The general slope of the ground is from south to north. Alan Dasht station is located between Alan Dasht Square and 9th Koohsangi Street (Edalat) in Mashhad, constituting the 10th station out of 12 stations of this line. The axis of the Alan Dasht station is planned to be positioned at a distance of 10 + 512.27 on the route of Line 2 of Mashhad urban train. Further, two platforms with a width of about 4 m and a length of 100 m on both sides of the railway represent another physical feature of the Alan Dasht station. The station is positioned in a northeast-southwest direction. Concerning the elevation, the station is categorized as the underground type and rails are located at a depth of about 20 m from the street and 976.89 m from the sea level. The slope of the project line of the rail is

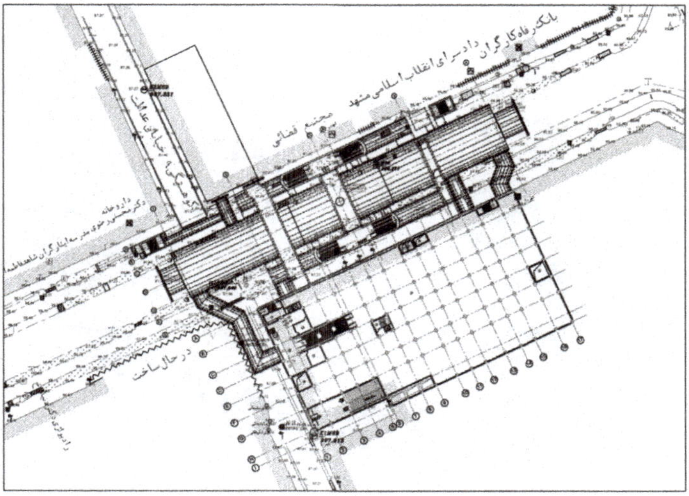

Fig. 1. The plan at the site of the Alan Dasht station

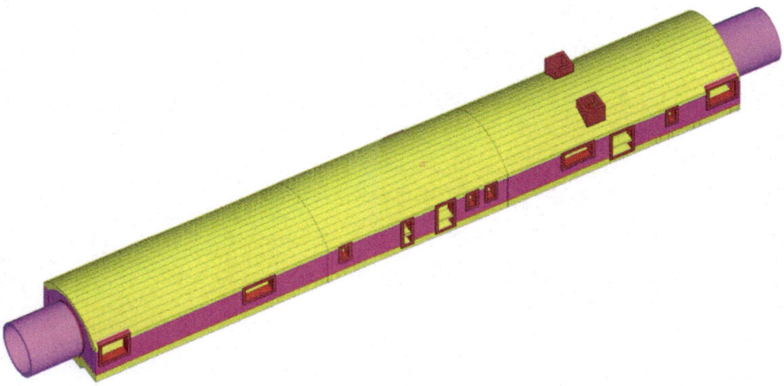

Fig. 2. Top view of Al-Dasht station

about 0.002 in the station, and in the longitudinal profile, the downward direction of the rail slope is towards the northeast. Figures 1 and 2 show the plan and the top view of the Alan Dasht station.

2.1 Geometry and Model Specifications

2 of the Mashhad urban train stretches from the end of the Northern Tabarsi Blvd. to the South of Mashhad (Table 1).

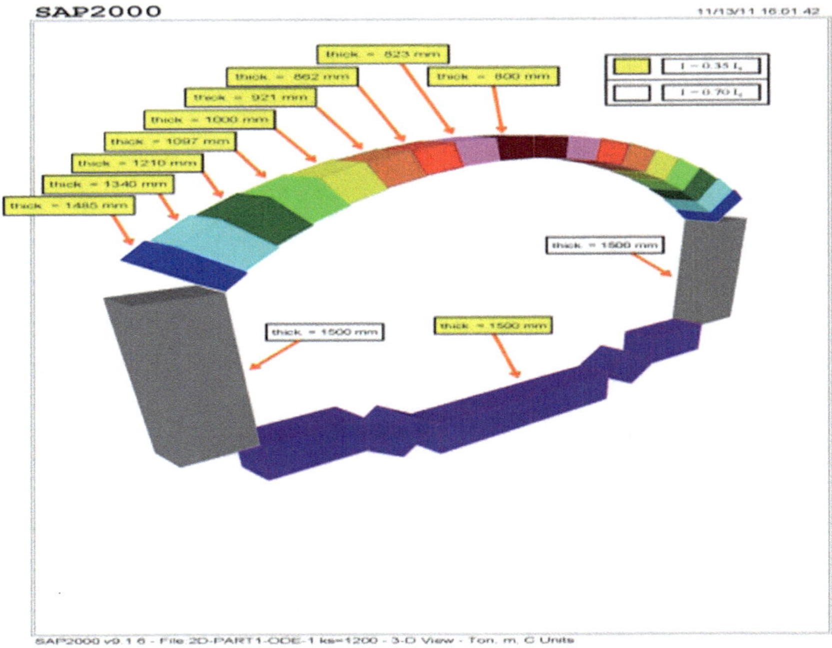

Fig. 3. Two-dimensional model view (MDE earthquake)

Table 1. Specifications of materials

Materials	Bar	Concrete
Resistance	F_y = 4000 Mpa	F_c = 250 Mpa

3 Soil-Structure Modeling

3.1 Hashash Method

The Hashash method [32] is based on imposing an absolute displacement on the structure buried in the soil. According to this method, the resonance of soil mass is determined based on soil seismic parameters [33, 35, 36]. Further, the resonance ratio imposed on the buried structure in the soil relative to the soil mass resonance is determined based on the ratio of structural stiffness to soil hardness, and then this displacement is applied to the structure based on the specific loading pattern. This method complies by the "principals of designing the stations of Line 2 of Mashhad urban train". The geodynamic parameters required in this method to estimate the load on the station are as follows:

(1) Propagation velocity of shear wave: The value of C_S of soil is determined from the geotechnical report at the site of each station by averaging of figures.

(2) **Dynamic Shear Modulus of Soil Mass:** The dynamic shear modulus of the soil mass [37] should also be determined from geotechnical studies. However, in the absence of this information, dynamic shear modulus can be calculated from Eq. (1):

$$G_m = \rho \times (C_s)^2 \tag{1}$$

The behavior of soil mass around the station is assumed to be linearly elastic, and given the dependence of the dynamic shear modulus on the strain level and normal level of strains during the earthquake, the effective dynamic shear modulus of the soil is estimated as $G_e = 0.6\ G_m$.

(3) The maximum shear strain of the original soil mass: The maximum shear strain of the original soil mass, $\gamma_{\mathbf{free-field}}$, is obtained from Eq. (2).

$$\gamma_{\text{free-field}} = \frac{\text{PGV}}{\text{C}_\text{s}} \tag{2}$$

where PGV is the maximum propagation velocity of soil particles at the depth of the station.

If this parameter is calculated in seismic studies, it can be directly used to compute $\gamma_{\mathbf{free-field}}$. Otherwise, PGV value at the depth of the station can be roughly determined using the following method:

(a) PGA calculation in the station depth during the desired return period at the mean depth of the station based on the acceleration distribution graphs in the soil profile.

Table 2. Variation of "PGV to PGA" ratio with distance from fault and soil type (after Power et al. 1996)

Moment magnitude (M_W)	Source-to-site distance (km)		
	0–20	20–50	50–100
Rock[a]			
6.5	66	76	86
7.5	97	109	97
8.5	127	140	152
Stiff soil[a]			
6.5	94	102	109
7.5	140	127	155
8.5	180	188	193
Soft soil[a]			
6.5	140	132	142
7.5	208	165	201
8.5	269	244	251

(b) PGV calculation from PGA based on station distance from the fault, type of soil and expected magnitude of the earthquake and its effect on the useful life of the structure according to Table 2.

In this table, the sediment types represent the following shear wave velocity ranges: rock $\geq$ 750 m/s, stiff soil 200–750 m/s, soft soil < 200 m/s. The relationship between peak ground velocity and peak ground acceleration is less certain in soft soils.

In the seismic risk analysis of the project, the baseline design earthquake (target earthquake) is defined as 6.8 on the Richter scale in the range of 6.3 to 7.3. However, considering the Shandiz fault as a fault with the seismic potential of up to 7.1 on the Richter scale at a distance of up to 20 km from the project, 7.1 Richter was determined as the basis for determining the ratio of the wave propagation velocity to the maximum seismic acceleration for the whole baseline project. Thus, we have:

PGV = 122 PGA (for Stiff Soil & SSD = 0–20 km & M_w = 7.1)

The distortion applied to the station is calculated by Eq. (3).

$$\gamma_{\text{structure}} = \mathrm{R} \times \gamma_{\text{free-field}} \tag{3}$$

Where, the coefficient R, which is the ratio of structural distortion to soil environment distortion, is a function of structural stiffness and soil hardness ratio, which is achieved from the displacement interaction method.

4 Displacement Interaction Method

In this method, the distortion applied to the underground structure is considered in terms of the structure-soil mass interaction using analytical or semi-analytical methods. In this method, considering the stiffness of the alternative structure relative to soil mass removed from the environment, a good approximation of soil-structure interaction and the distortion applied to the structure during an earthquake is obtained.

For interactive analysis, modeling should be performed in soil-structure interaction software such as PLAXIS [38, 39]. The steps and assumptions of modeling are as follows:

- Two-Dimensional modeling is performed in-plane strain mode.
- The soil environment is assumed to be elastic.
- G_e = 0.6 Gm.
- All tensions, including horizontal and vertical tensions, are assumed equal to zero.
- The width of the model is at least 10 times the width of the structure and its depth is at least one-fourth of the length of the shear wave.
- The modeling of the final coating structure is carried out using a bending beam element with elastic behavior.
- The connection of structural elements to the soil through interface elements is considered as elasto-plastic with environmental resistance in the static model.

To perform this analysis, it is necessary to construct two computational models which are described below.

4.1 Original Soil Model

In this step, the original soil model, irrespective of the structure, is subject to a compulsory displacement of $\Delta_{free-field}$, similar to the following (Figs. 3 and 4).

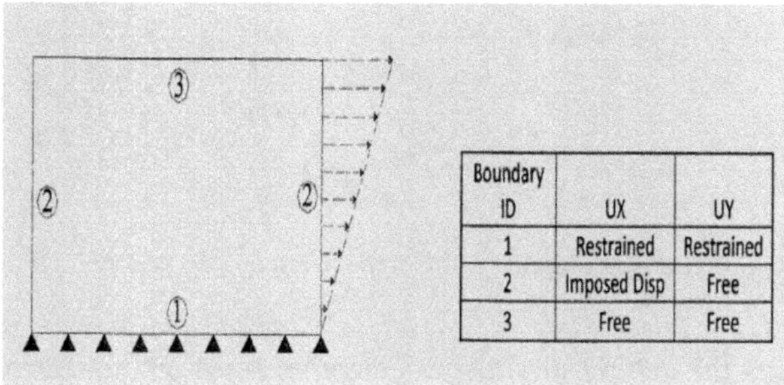

Boundary ID	UX	UY
1	Restrained	Restrained
2	Imposed Disp	Free
3	Free	Free

Fig. 4. Boundary conditions of the problem

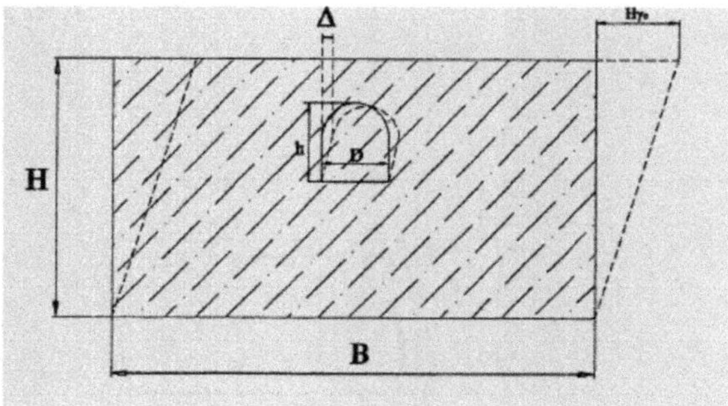

Fig. 5. Compulsory displacement of the structure

The relative displacement of the highest and lowest points of the construction site in the original soil is measured (Eq. 4) (Fig. 5).

$$\gamma = \frac{\Delta}{h}, \alpha = \frac{\gamma_{free-field}}{\gamma} \quad (4)$$

4.2 Soil-Structure Model

In this model, soil mass and structure are modeled simultaneously and subjected to a compulsory triangular displacement of size $\alpha\Delta_{free-field}$ in the vertical boundary of the soil mass, as shown in Fig. 6.

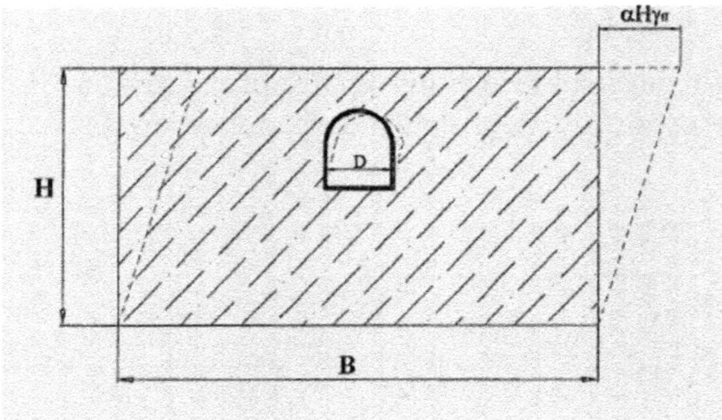

Fig. 6. Compulsory triangular displacement relative to soil mass and structure

By calculating the compulsory displacement imposed on the structure and applying this deformation relative to the depth of the station to one of the loading patterns discussed below, the internal forces for designing the elements of the station structure can be determined. For stations with overheads of >30 m, the loading pattern in Fig. 7a and, for stations with overheads of <30 m, the loading pattern in Fig. 7b should be used.

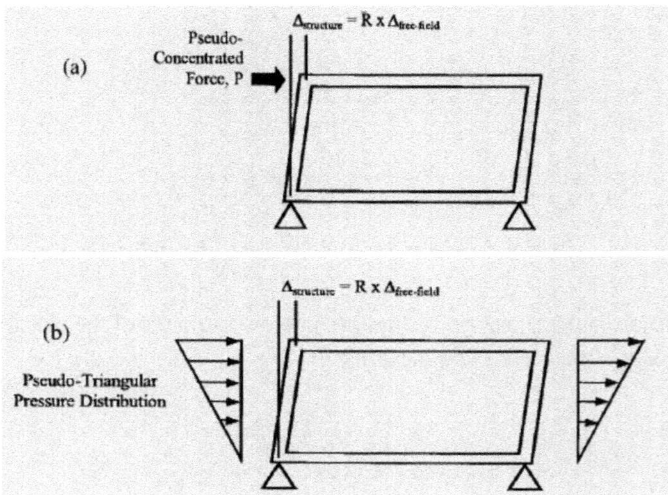

Fig. 7. Types of loading patterns [32]

In the seismic analysis of the structure, the structural analysis is performed with the unit length (for the lateral fixed part of the wall) or with a limited length on both moving sides of the sidewall, and the results are manually merged with the results of static analyses. The reinforcement of the structure is conducted on-site. A 3D modeling of the structure does not correspond to Hashash's assumptions in practice, and rigid performance of the sidewalls of the station will preclude any distortion in the station. The following assumptions will also be considered in the design of structural elements:

- In all analyses, the effective hardness of concrete cross-sections (cracked sections) is used.
- The behavior coefficient of the structure is R = 1.

5 Seismic Loading

The PLAXIS 8.2 [38] software package was used to perform displacement interaction analysis. This software is capable of two-dimensional modeling of drilling steps and installation of a maintenance system relative to the effects of soil-structural interaction. In this program, the numerical solution is the finite element, which allows analyzing the continuous environment with linear and elasto–plastic behavior in a plane strain model.

As described earlier, according to the basics of Line 2 of Mashhad Urban Train, two-dimensional modeling was performed in a plane strain mode with an elastic soil environment and $G_e = 0.6\ G_m$. For this modeling, three-node elements were used for the soil environment. All tensions in the model, including horizontal and vertical tensions, were considered to be zero. Also, all modeling was conducted in a single phase. The necessary calculations for the two-dimensional modeling of the Alan Dasht station are described in (5) to (10).

$$C_S = 695\,\frac{m}{s}, \rho = 2.0 \tag{5}$$

$$G_m = \rho(C_S)^2 = 2 \times 695^2 = 966050\,\text{kpa}$$

$$G_e = 0.6\,G_m = 579630\,\text{kpa} \tag{6}$$

In the ODE mode:

$$PGA = 0.17, PGV = 122\,PGA = 0.207\,\frac{m}{s} \tag{7}$$

$$\gamma_{free-field} = \frac{PGV}{C_S} = \frac{0.207}{695} = 2.978 \times 10^{-4} \tag{8}$$

In the MDE mode:

$$PGA = 0.37, PGV = 122\,PGA = 0.45\frac{m}{s} \tag{9}$$

$$\gamma_{free-field} = \frac{PGV}{C_S} = \frac{0.45}{695} = 6.455 \times 10^{-4} \tag{10}$$

The dimensions of the model should be chosen in a way to model the environment with good approximation. For this purpose, the width of the environment was assumed to be 10 times the width of the structural cross-section. The depth of the model was considered to be at least one-fourth of the length of the shear wave. The station structure and retaining structure were modeled using the bending beam element with elastic behavior. Also, the casing was modeled in the cracked form. The boundary conditions of the model were applied as shown below.

6 Boundary Conditions

In Fig. 8 and Table 3, the boundary conditions of the modeled sample are shown.

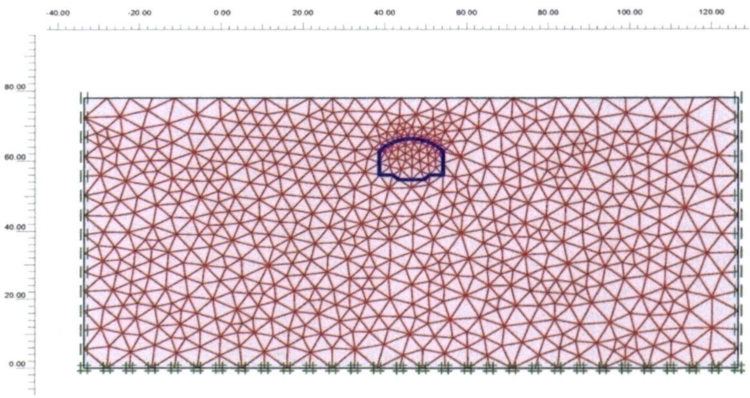

Fig. 8. How to apply boundary conditions on a two-dimensional model

Table 3. Model boundary conditions

Boundary ID	U_x	U_y
1	Restrained	Restrained
2	imposed displacement	Free
3	Free	Free

7 Results of Analysis

In this section, the results of the analysis of two-dimensional modeling of soil-structure are discussed. According to the resonances obtained in the ODE model ($\gamma_{free-field} = 6.455 \times 10^{-4}$) and MDE ($\gamma_{free-field} = 2.978 \times 10^{-4}$) in Sect. 5, the soil mass was subject to these resonances, Figs. 9 and 10 reveal the application of side displacement to the soil mass in PLAXIS software under ODE and MDE seismic conditions. In Figs. 11 and 12, the deformation of the meshed soil mass is observed under both ODE and MDE seismic conditions. For more meticulous analysis, stress and deformation graphs should be extracted. In Figs. 13 and 14, the displacement contours of free-field soil masses are observed under ODE and MDE seismic conditions. As can be seen, the maximum displacements are observed in the upper angles, with values reaching 54.18 mm in the ODE and 108.36 mm in the MDE seismic conditions. Also, according

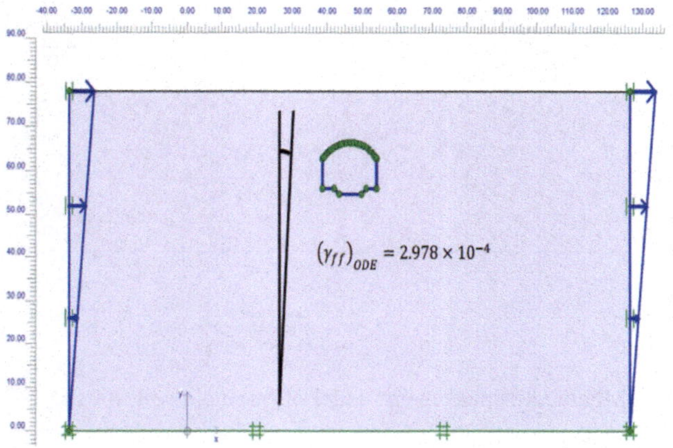

Fig. 9. Two-dimensional model of applying the free-field soil mass in ODE mode

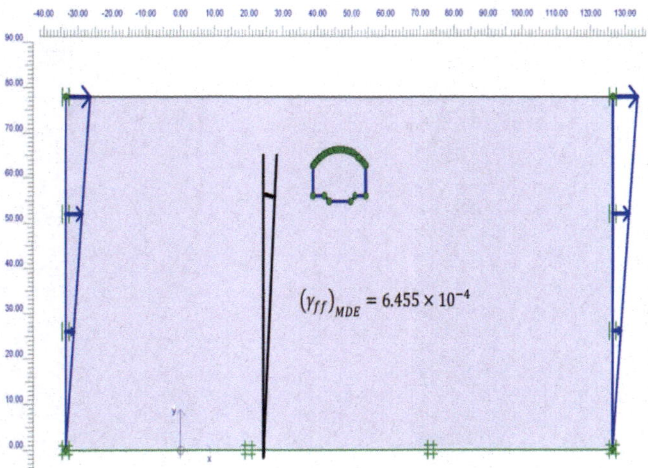

Fig. 10. Two-dimensional model of applying the free-field soil mass in MDE mode

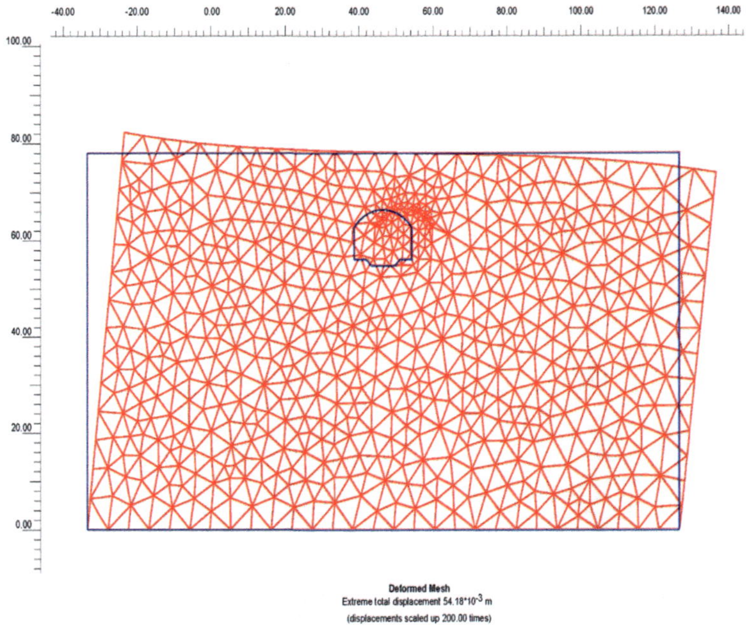

Fig. 11. Modified displacement model of free field soil mass in ODE mode

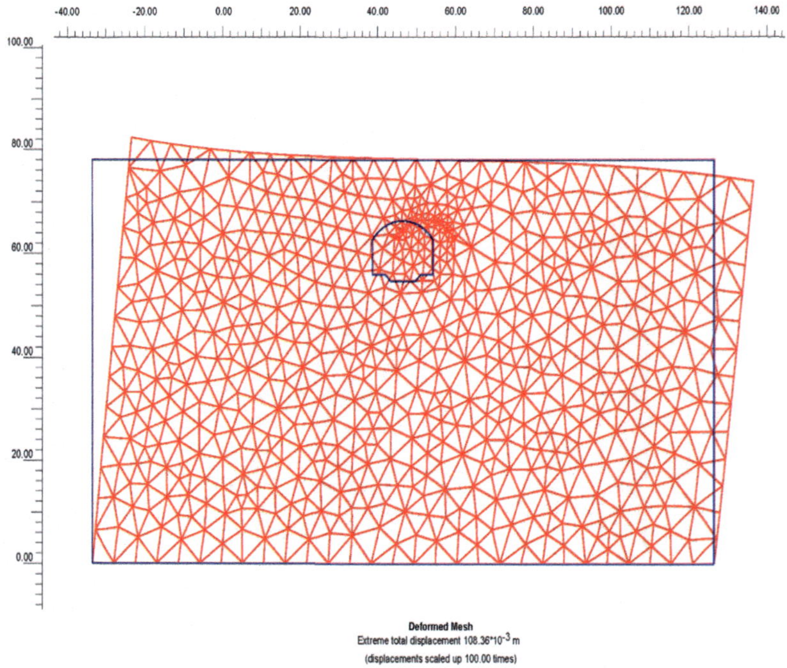

Fig. 12. Modified displacement model of free field soil mass in MDE mode

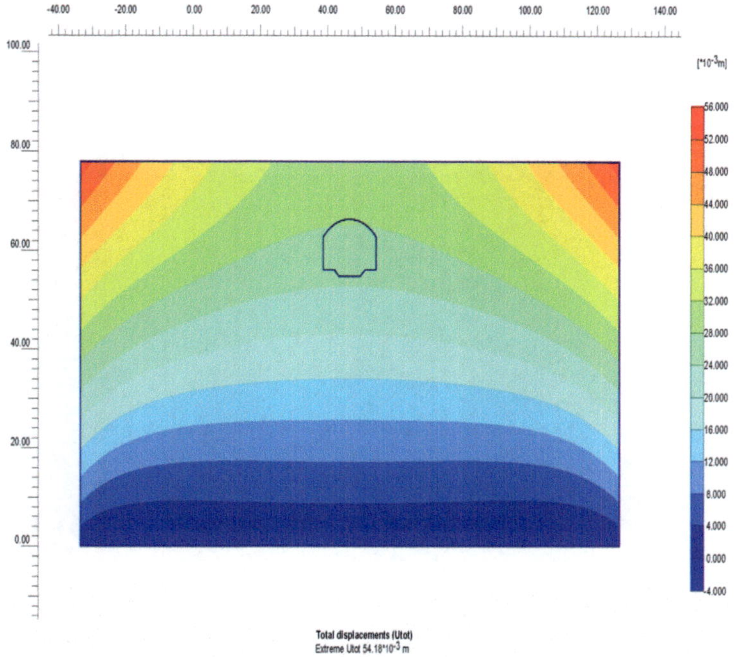

Fig. 13. Displacement contour of the free-field soil mass in ODE mode

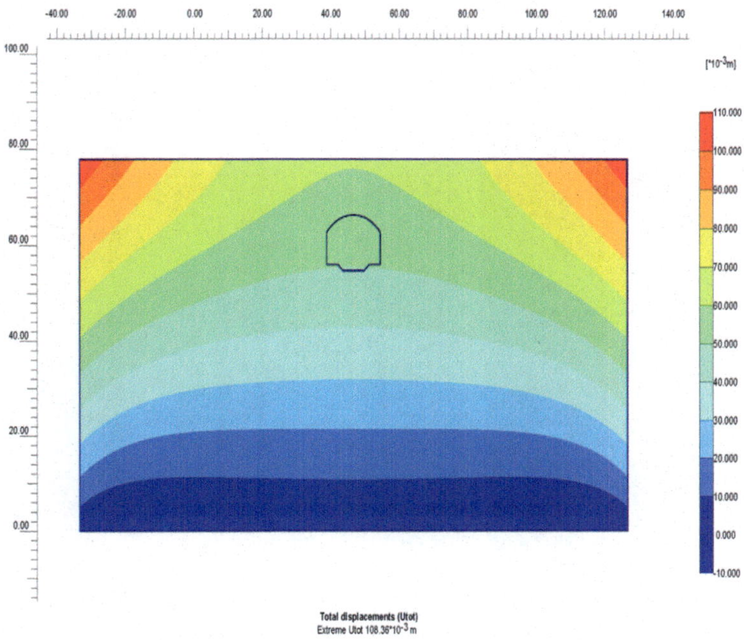

Fig. 14. Displacement contour of the free-field soil mass in MDE mode

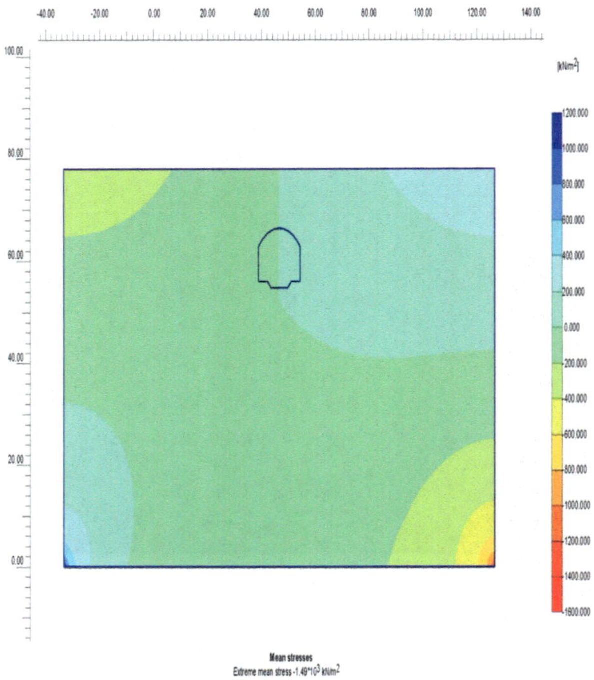

Fig. 15. Medium stress contour in free-field soil mass in ODE mode

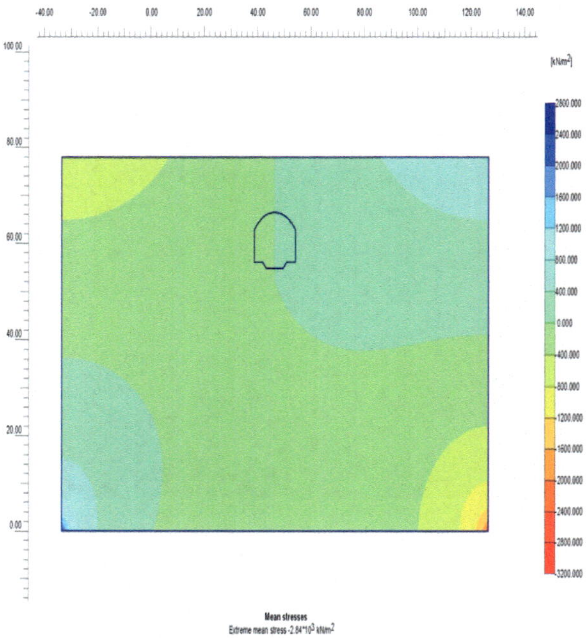

Fig. 16. Medium stress contour in free-field soil mass in MDE mode

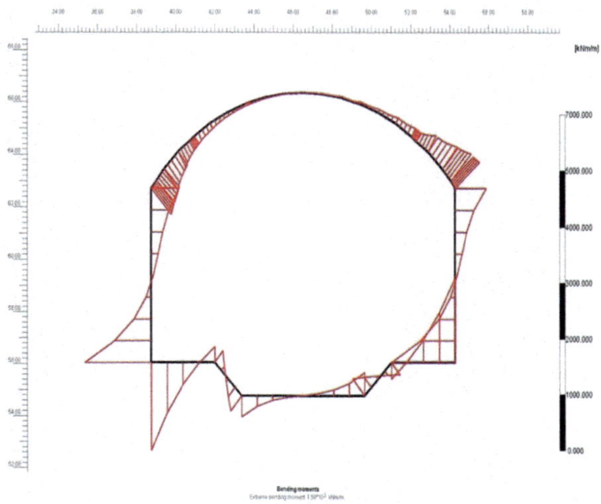

Fig. 17. Diagram of the main structure moments (without coefficient) under the impact of side distortion of soil-structure in ODE mode

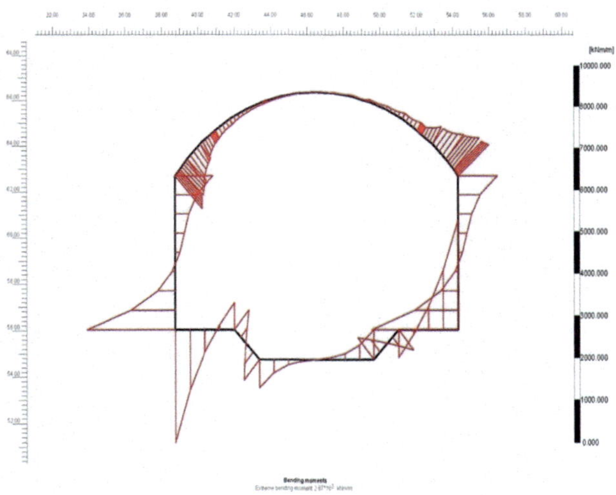

Fig. 18. Diagram of the main structure moments (without coefficient) under the impact of side distortion of soil-structure in MDE mode

to the mean stress contours derived from the analysis, as shown in Figs. 15 and 16, the maximum average stress in soil mass is 1490 KN/m^2 under ODE and maximum stress in soil mass is about 2840 KN/m^2 under MDE seismic conditions. After analyzing the soil-structure model underside distortion, the bending moment, shear force, and axial force curves were obtained. In Figs. 17 and 18, which exhibit the bending moment curves under both ODE and MDE seismic conditions, the maximum bending moment

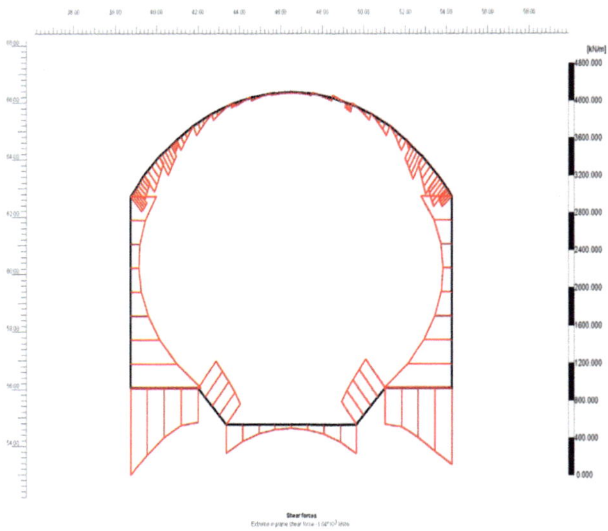

Fig. 19. Shear diagram of the main structure (without coefficient) under the impact of side distortion of soil-structure in ODE mode

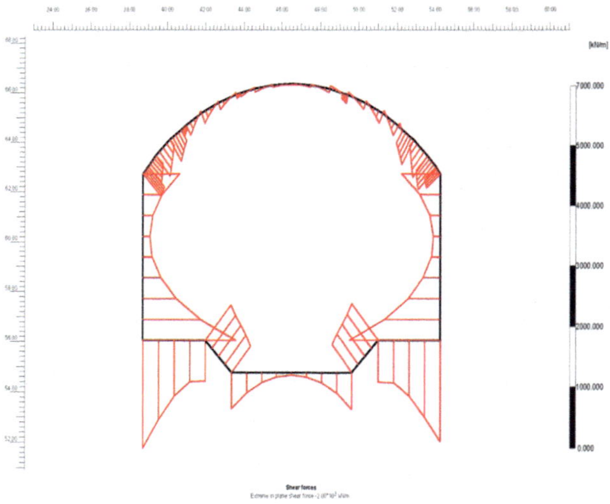

Fig. 20. Shear diagram of the main structure (without coefficient) under the impact of side distortion of soil-structure in MDE mode

was 1580 KN.m and 2670 KN.m in the ODE and MDE modes. Figures 19 and 20 illustrate the sheer force curves and Figs. 20 and 21 depict axial force curves of the structure under ODE and MDE conditions. In these curves, the maximum shear force and the maximum axial force are 2000 KN and 2330 KN, respectively (Fig. 22).

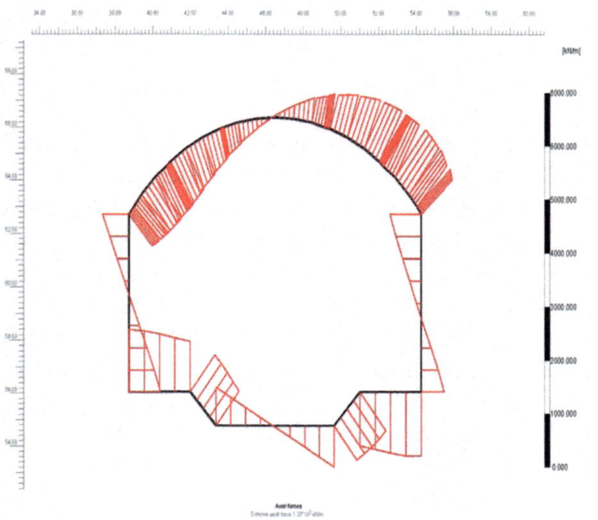

Fig. 21. Axial force of the main structure (without coefficient) under the impact of side distortion of soil-structure in MDE mode

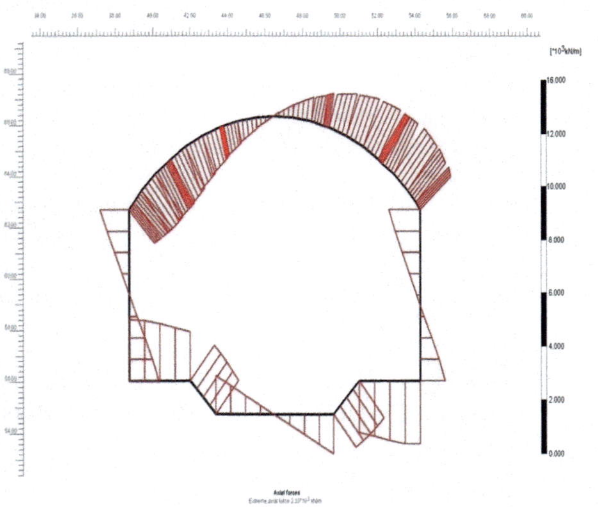

Fig. 22. The axial force of the main structure (without coefficient) under the impact of side distortion of soil-structure in MDE mode

8 Conclusion

In this research, two-dimensional analysis of the soil-structure of Alan Dasht station of the Mashhad Urban Train was conducted. The modeling was performed using PLAXIS finite element program, Hashash and displacement interaction methods for the modeling of soil mass- structure. In these methods, the free-field resonance and side distortion of the soil mass under ODE and MDE seismic conditions were obtained and then applied to a soil mass in accordance with the triangular load pattern. Then, under free-space resonance, the displacement and stress contours in the soil mass were achieved and underside distortion, bending moment, shear force and axial force of the structure were achieved. As indicated by the results, the maximum soil mass displacement (108.36 mm) and maximum soil mass stress (2840 KN/m^2) were recorded under MDE earthquake. Moreover, the bending moment, shear force and axial force curves were calculated from the soil mass analysis under side distortion, with maximum values of 2670 KN.m, 2000 KN and 2330 KN under MDE seismic conditions, respectively. For the future works, the application of artificial intelligence methods such as machine learning and deep learning methods can highly improve the quality of modeling in prediction of the optimal configurations [40–61]. In particular, the hybrid and ensemble machine learning methods are reported to provide higher performance [62–70].

Acknowledgments. This publication has been supported by the Project: "Support of research and development activities of the J. Selye University in the field of Digital Slovakia and creative industry" of the Research & Innovation Operational Programme (ITMS code: NFP313010T504) co-funded by the European Regional Development Fund.

References

1. Aguirre, E., et al.: Design and implementation of context aware applications with wireless sensor network support in urban train transportation environments. IEEE Sens. J. **17**(1), 169–178 (2017)
2. Bu, B., et al.: Research on method of cooperation among trains for energy saving in urban rail transportation. Tiedao Xuebao/J. China Railw. Soc. **40**(8), 43–51 (2018)
3. Fernández-Rodríguez, A., et al.: Energy efficiency and integration of urban electrical transport systems: EVS and metro-trains of two real European lines. Energies **12**(3), 366 (2019)
4. He, W., et al.: Effect of wind barrier's height on train-bridge system aerodynamic characteristic of cable-stayed bridge for urban railway transportation. Zhongnan Daxue Xuebao (Ziran Kexue Ban)/J. Central South Univ. (Science and Technology) **48**(8), 2238–2244 (2017)
5. Kim, K.: Exploring the difference between ridership patterns of subway and taxi: case study in Seoul. J. Transp. Geogr. **66**, 213–223 (2018)
6. Tang, T., et al.: VISOS: a visual interactive system for spatial-temporal exploring station importance based on subway data. IEEE Access **6**, 42131–42141 (2018)
7. Wang, J., et al.: IS2Fun: identification of subway station functions using massive urban data. IEEE Access **5**, 27103–27113 (2017)
8. Aram, F., et al.: Design and validation of a computational program for analysing mental maps: aram mental map analyzer. Sustainability (Switzerland) **11**(14), 3790 (2019)
9. Dehghani, M., et al.: Prediction of hydropower generation using Grey wolf optimization adaptive neuro-fuzzy inference system. Energies **12**(2), 289 (2019)
10. Mosavi, A., et al.: State of the art of machine learning models in energy systems, a systematic review. Energies **12**(7), 1301 (2019)

11. Nosratabadi, S., et al.: Sustainable business models: a review. Sustainability (Switzerland) **11**(6), 1663 (2019)
12. Taherei Ghazvinei, P., et al.: Sugarcane growth prediction based on meteorological parameters using extreme learning machine and artificial neural network. Eng. Appl. Comput. Fluid Mech. **12**(1), 738–749 (2018)
13. Torabi, M., et al.: A Hybrid clustering and classification technique for forecasting short-term energy consumption. Environ. Prog. Sustain. Energy **38**(1), 66–76 (2019)
14. Nosratabadi, S., et al.: Sustainable business models: a review. Sustain. **11**(6), 1663 (2019)
15. Peng, Y.T., Li, Z.C., Choi, K.: Transit-oriented development in an urban rail transportation corridor. Transp. Res. Part B Methodol. **103**, 269–290 (2017)
16. Wu, C., Pei, Y., Gao, J.: Analysis on transportation supply-demand adjustment ability of urban rail transit network. Wuhan Ligong Daxue Xuebao (Jiaotong Kexue Yu Gongcheng Ban)/J. Wuhan Univ. Technol. (Transportation Science and Engineering) **41**(1), 22–26 and 31 (2017)
17. Zhang, C., Xia, H., Song, Y.: Rail transportation lead urban form change: a case study of Beijing. Urban Rail Transit. **3**(1), 15–22 (2017)
18. Lee, Y., et al.: Generation characteristics of nanoparticles emitted from subways in operation. Aerosol Air Qual. Res. **18**(9), 2230–2239 (2018)
19. Liu, X., et al.: Evaluation of the utility efficiency of subway stations based on spatial information from public social media. Habitat Int. **79**, 10–17 (2018)
20. Zhao, X., et al.: Clustering analysis of ridership patterns at subway stations: a case in Nanjing, China. J. Urban Planning Dev. 145(2) (2019)
21. Liu, X., et al.: Evaluation of effects of static pile driving on existing metro tunnel structure. J. Perform. Constr. Facilities **33**(4) (2019)
22. Mostafaei, M., Rezaei Far, A.H., Rastegarnia, A.: Assessment of the impact of case parameters affecting abrasion and brittleness factors in alluviums of line 2 of theTabriz subway, Iran. Bull. Eng. Geology Environ. **78**(5), 3851–3861 (2019)
23. Zhou, S., et al.: An approach integrating dimensional analysis and field data for predicting the load on tunneling machine. KSCE J. Civil Eng. **23**(7), 3180–3187 (2019)
24. Mosleh, A., Nosratabadi, S., Bahrami, P.: Recognizing the business models types in tourism agencies: utilizing the cluster analysis. Int. Bus. Res. **8**(2), 173 (2015)
25. Wang, W., et al.: Assessment of damage in mountain tunnels due to the Taiwan Chi-Chi earthquake. Tunn. Undergr. Space Technol. **16**(3), 133–150 (2001)
26. Kontogianni, V.A., Stiros, S.C.: Earthquakes and seismic faulting: effects on tunnels. Turkish J. Earth Sci. **12**(1), 153–156 (2003)
27. Tajiri, M.: Damage done by the great earthquake disaster of the Hanshin. Awaji district to the Kobe municipal subway system and restoration works of the damage. Japanese Railway Eng. **36**(2), 19–23 (1997)
28. Sharma, S., Judd, W.R.: Underground opening damage from earthquakes. Eng. Geol. **30**(3–4), 263–276 (1991)
29. Hashash, Y.M., et al.: Seismic design and analysis of underground structures. Tunn. Undergr. Space Technol. **16**(4), 247–293 (2001)
30. Hashash, Y.M., Park, D.: Non-linear one-dimensional seismic ground motion propagation in the Mississippi embayment. Eng. Geol. **62**(1–3), 185–206 (2001)
31. Pakbaz, M.C., Yareevand, A.: 2-D analysis of circular tunnel against earthquake loading. Tunn. Undergr. Space Technol. **20**(5), 411–417 (2005)
32. Hashash, Y.M., et al.: Ovaling deformations of circular tunnels under seismic loading, an update on seismic design and analysis of underground structures. Tunn. Undergr. Space Technol. **20**(5), 435–441 (2005)
33. Wang, J.-N., Munfakh, G.: Seismic Design of Tunnels. Vol. 57. WIT Press (2001)

34. Plaxis, B.: Finite element code for soil and rock analysis. Users Manual, Version, 7 (2000)
35. Karballaeezadeh, N., et al.: Prediction of remaining service life of pavement using an optimized support vector machine (case study of Semnan-Firuzkuh road). Eng. Appl. Comput. Fluid Mech. **13**(1), 188–198 (2019)
36. Mohammadzadeh, S., et al.: Prediction of compression index of fine-grained soils using a gene expression programming model. Infrastructures **4**(2), 26 (2019)
37. Haase, D., et al.: Global urbanization. The Urban Planet: Knowledge Towards Sustainable Cities, **19** (2018)
38. Brinkgreve, R. et al.: PLAXIS 2D 2010. User manual, Plaxis bv (2010)
39. Manual, P.: Finite Element Code for Soil and Rock Analysis. Published and distributed by AA Balkema Publishers, Nederland's Comput. Geotech. **32**(5): p. 326–339 (2007)
40. Asadi, E., et al.: Groundwater Quality Assessment for Drinking and Agricultural Purposes in Tabriz Aquifer, Iran (2019)
41. Asghar, M.Z., Subhan, F., Imran, M., Kundi, F.M., Shamshirband, S., Mosavi, A., Csiba, P., Várkonyi-Kóczy, A.R.: Performance evaluation of supervised machine learning techniques for efficient detection of emotions from online content. Pre-prints 2019, 2019080019 https://doi.org/10.20944/preprints201908.0019.v1
42. Bemani, A., Baghban, A., Shamshirband, S., Mosavi, A., Csiba, P., Várkonyi-Kóczy, Applying, A.R.: ANN, ANFIS, and LSSVM Models for Estimation of Acid Sol-vent Solubility in Supercritical CO2. Preprints 2019, 2019060055. https://doi.org/10.20944/preprints201906.0055.v2
43. Choubin, B., et al.: Snow avalanche hazard prediction using machine learning methods. J. Hydrol., 577 (2019)
44. Choubin, B., et al.: An ensemble prediction of flood susceptibility using multivariate discriminant analysis, classification and regression trees, and support vector machines. Sci. Total Environ. **651**, 2087–2096 (2019)
45. Dineva, A., et al.: Multi-Label classification for fault diagnosis of rotating electrical machines (2019)
46. Ghalandari, M., et al.: Investigation of submerged structures' flexibility on sloshing frequency using a boundary element method and finite element analysis. Eng. Appl. Comput. Fluid Mech. **13**(1), 519–528 (2019)
47. Ghalandari, M., et al.: Flutter speed estimation using presented differential quadrature method formulation. Eng. Appl. Comput. Fluid Mech. **13**(1), 804–810 (2019)
48. Mosavi, A., Edalatifar, M.: A Hybrid Neuro-Fuzzy Algorithm for Prediction of Reference Evapotranspiration, in Lecture Notes in Networks and Systems, pp. 235–243, Springer (2019)
49. Mosavi, A., Ozturk, P., Chau, K.W.: Flood prediction using machine learning models: literature review. Water (Switzerland) **10**(11) (2018)
50. Mosavi, A., Rabczuk, T.: Learning and intelligent optimization for material design innovation, D.E. Kvasov, et al., Editors, pp. 358–363. Springer (2017)
51. Mosavi, A., Rabczuk, T., Varkonyi-Koczy, A.R.: Reviewing the novel machine learning tools for materials design, D. Luca, L. Sirghi, and C. Costin, Editors, pp. 50–58. Springer (2018)
52. Mosavi, A., et al.: Prediction of multi-inputs bubble column reactor using a novel hybrid model of computational fluid dynamics and machine learning. Eng. Appl. Comput. Fluid Mech. **13**(1), 482–492 (2019)
53. Mosavi, A., Varkonyi-Koczy, A.R.: Integration of machine learning and optimization for robot learning, R. Jablonski and R. Szewczyk, Editors, pp. 349–355. Springer (2017)
54. Qasem, S.N., et al.: Estimating daily dew point temperature using machine learning algorithms. Water (Switzerland) **11**(3) (2019)

55. Rezakazemi, M., Mosavi, A., Shirazian, S.: ANFIS pattern for molecular membranes separation optimization. J. Mol. Liq. **274**, 470–476 (2019)
56. Riahi-Madvar, H., et al.: Comparative analysis of soft computing techniques RBF, MLP, and ANFIS with MLR and MNLR for predicting grade-control scour hole geometry. Eng. Appl. Comput. Fluid Mech. **13**(1), 529–550 (2019)
57. Shabani, S., Samadianfard, S., Taghi Sattari, M., Shamshirband, S., Mosavi, A., Kmet, T., Várkonyi-Kóczy, A.R.: Modeling Daily Pan Evaporation in Humid Climates Using Gaussian Process Regression. Preprints 2019, 2019070351. https://doi.org/10.20944/preprints201907.0351.v1
58. Shamshirband, S., Hadipoor, M., Baghban, A., Mosavi, A., Bukor J., Varkonyi-Koczy, A. R.: Developing an ANFIS-PSO Model to predict mercury emissions in Combustion Flue Gases. Preprints 2019, 2019070165. https://doi.org/10.20944/preprints201907.0165.v1
59. Shamshirband, S., et al.: Ensemble models with uncertainty analysis for multi-day ahead forecasting of chlorophyll a concentration in coastal waters. Eng. Appl. Comput. Fluid Mech. **13**(1), 91–101 (2019)
60. Shamshirband, S., Mosavi, A., Rabczuk, T.: Particle swarm optimization model to predict scour depth around bridge pier. arXiv preprint arXiv:1906.08863 (2019)
61. Torabi, M., et al.: A Hybrid Machine Learning Approach for Daily Prediction of Solar Radiation, in Lecture Notes in Networks and Systems, pp. 266–274. Springer (2019)
62. Ardabili, S., Mosavi, A., Mahmoudi, Mesri Gundoshmian, T., Nosratabadi, S., Var-konyi-Koczy, A.: Modelling temperature variation of mushroom growing hall using artificial neural networks, Preprints 2019
63. Mesri Gundoshmian, T., Ardabili, S., Mosavi, A., Varkonyi-Koczy, A., Prediction of combine harvester performance using hybrid machine learning modeling and re-sponse surface methodology, Preprints 2019
64. Ardabili, S., Mosavi, A., Varkonyi-Koczy, A.: Systematic review of deep learning and machine learning models in biofuels research, Preprints 2019
65. Ardabili, S., Mosavi, A., Varkonyi-Koczy, A.: Advances in machine learning modeling reviewing hybrid and ensemble methods, Preprints 2019
66. Ardabili, S., Mosavi, A., Varkonyi-Koczy, A.: Building Energy information: demand and consumption prediction with Machine Learning models for sustainable and smart cities, Preprints 2019
67. Ardabili, S., Mosavi, A., Dehghani, M., Varkonyi-Koczy, A.: Deep learning and machine learning in hydrological processes climate change and earth systems a systematic review, Preprints 2019
68. Mohammadzadeh, D., Karballaeezadeh, N., Mohemmi, M., Mosavi, A., Varkonyi-Koczy A.: Urban Train Soil-Structure Interaction Modeling and Analysis, Preprints 2019
69. Mosavi, A., Ardabili, S., Varkonyi-Koczy, A.: List of deep learning models, Preprints 2019
70. Nosratabadi, S., Mosavi, A., Keivani, R., Ardabili, S., Aram, F.: State of the art survey of deep learning and machine learning models for smart cities and urban sustainability, Preprints 2019

Author Index

A. R. Várkonyi-Kóczy (Ed.): INTER-ACADEMIA 2019, LNNS 101, pp. 383–384, 2020.
https://doi.org/10.1007/978-3-030-36841-8

Printed by Printforce, the Netherlands